Student Study Guide/ Solutions Manual

to accompany

Foundations of General, Organic, and Biochemistry

Katherine J. Denniston
Towson University

Joseph J. Topping
Towson University

Timothy M. Dwyer
Villa Julie College

 **Higher Education**

Boston   Burr Ridge, IL   Dubuque, IA   New York   San Francisco   St. Louis
Bangkok   Bogotá   Caracas   Kuala Lumpur   Lisbon   London   Madrid   Mexico City
Milan   Montreal   New Delhi   Santiago   Seoul   Singapore   Sydney   Taipei   Toronto

The McGraw·Hill Companies

Student Study Guide/Solutions Manual to accompany
FOUNDATIONS OF GENERAL, ORGANIC, AND BIOCHEMISTRY
KATHERINE J. DENNISTON, JOSEPH J. TOPPING

Published by McGraw-Hill Higher Education, an imprint of The McGraw-Hill Companies, Inc., 1221 Avenue of the Americas, New York, NY 10020. Copyright © 2008 by The McGraw-Hill Companies, Inc. All rights reserved.

1 2 3 4 5 6 7 8 9 0 BKM/BKM 0 9 8 7

ISBN: 978-0-07-321743-7
MHID: 0-07-321743-3

www.mhhe.com

Contents

Contents

1   *Chemistry: Methods and Measurement*

Learning Goals

1. Discuss the approach to science, the scientific method, and distinguish among the terms *hypothesis*, *theory*, and *scientific law*.
2. Describe the properties of the solid, liquid, and gaseous states.
3. Provide specific examples of physical and chemical properties and physical and chemical change.
4. Distinguish between intensive and extensive properties.
5. Classify matter as element, compound, or mixture.
6. Distinguish between data and results.
7. Learn the major units of measure in the English and metric systems, and be able to convert from one system to another.
8. Report data and results using scientific notation and the proper number of significant figures.
9. Use appropriate units in problem solving.
10. Use density, mass, and volume in problem solving, and calculate the specific gravity of a substance from its density.

Introduction

The subject matter of chemistry deals with all material substances as well as all the changes that these materials undergo. The basic tools of chemistry will enable you to further your understanding of how chemistry works.

1.1 The Discovery Process

Chemistry is the study of matter and the changes that matter undergoes. **Matter** is anything that has mass and occupies space. The changes that matter undergoes always involve either gain or loss of energy. **Energy** is nonmaterial, and is the ability to do work (to accomplish some change). Thus, a study of chemistry involves matter, energy, and their interrelationship.

The Scientific Method

Learning Goal 1:	Discuss the approach to science, the scientific method, and distinguish among the terms *hypothesis*, *theory*, and *scientific law*.

The **scientific method** consists of six interrelated processes:

1. **Observation.** The description of the properties of a substance is a result of observation. The measurement of the temperature of a liquid or the size or mass of a solid results from observation.
2. **Formulation of a question.** The observation is the basis of the question.
3. **Pattern recognition.** Discerning a cause-and-effect relationship may give rise to a scientific **law.** A scientific law is a description of the orderly behavior observed in nature.
4. **Developing theories.** Observation of a phenomenon calls for some explanation. The process of explaining observed behavior begins with a **hypothesis**—an educated guess. If this hypothesis survives extensive testing, it may attain the status of a **theory.** A theory is a hypothesis supported by testing (experimentation) that explains scientific facts and is capable of predicting new facts.
5. **Experimentation.** The heart of the scientific method is the verification of theories. This verification process results from conducting carefully designed experiments intended to reinforce or refute the model system, the theory, or the hypothesis.
6. **Summarizing information.** Many phenomena have common causes and explanations. A scientific law summarizes and clarifies large amounts of information.

1.2 Matter and Properties

Matter and Physical Properties

Learning Goal 2:	Describe the properties of the solid, liquid, and gaseous states.

Properties (characteristics) of matter may be classified as either physical or chemical. **Physical properties** enable us to identify different kinds of matter without changing the identity (chemical composition) of the sample. Examples of physical properties include color, odor, taste, melting and boiling temperatures, and compressibility.

Three states of matter exist: the gaseous state, the liquid state, and the solid state. These states are distinguishable by differences in physical properties. They include gases, liquids, and solids.

1. **The gaseous state.** Gases have a very low mass per unit volume, density, because the individual particles that comprise the gas are separated by large distances. Consequently, gases can be compressed (pushed into a smaller volume) or expanded to a larger volume.
2. **The liquid state.** Particles that make up a liquid are much closer together than in the vapor state. Thus, liquids expand and contract only slightly, and the density of liquids is much greater than that of gases.
3. **The solid state.** Solids are characterized by particles that are very close together. Attractive forces between the particles are strong enough to provide a rigid shape or structure to the bulk material. The proximity of particles prevents significant expansion or compression. All common materials, except water, follow this order of decreasing particle separation.

Water is the most common example of a substance that may exist in all three states over a reasonable temperature range. The conversion of ice to liquid water or liquid water to the gaseous state is an example of a **physical change.** A physical change does not alter the composition or identity of the substance undergoing change.

Matter and Chemical Properties

Chemical properties result in a change in composition and can only be observed through chemical reactions. The process of photosynthesis is a common example of chemical change.

Learning Goal 3:	Provide specific examples of physical and chemical properties and physical and chemical change.

Example 1
Which of the following are physical changes?

1. An iron nail rusts.
2. A block of ice melts.
3. Water on the floor evaporates.
4. Digestion of foods in the small intestine.
5. Gasoline undergoes combustion.

Answer
Items 2 and 3

The terms rust, digestion, and combustion all indicate chemical change. The terms melt, boil, freeze, and evaporate all indicate a physical change.

Intensive and Extensive Properties

An **intensive property** is independent of the quantity of the substance. For example, the density of a drop of water is exactly the same as the density of a liter of water. Mass and volume are extensive properties. An **extensive property** depends on the quantity of a substance.

Classification of Matter

| **Learning Goal 5:** | **Classify matter as element, compound, or mixture.** |

All matter can be classified as either a pure substance or a mixture (see Figure 1.5 in the textbook). A **pure substance** is a form of matter that has identical composition and physical and chemical properties throughout. A **mixture** is a combination of two or more pure substances in which the combined substances retain their identity.

A mixture may be either homogeneous or heterogeneous matter. A **homogeneous mixture** has uniform composition. Its particles are well mixed, or thoroughly intermingled. A **heterogeneous mixture** has a nonuniform composition.

Pure substances may also be subcategorized as elements or compounds. An **element** is a pure substance that cannot be converted into a simpler form of matter by any chemical reaction. A **compound** is a substance resulting from the combination of two or more elements in a definite, reproducible fashion.

Example 2

Which of the following are mixtures?
1. soup   3. wine
2. tap water   4. blood

Answer

All, since they contain more than one pure substance.

1.3 Measurement in Chemistry

Data, Results, and Units

| **Learning Goal 6:** | **Distinguish between data and results.** |

A scientific experiment produces **data.** Each piece of data is the result of a single measurement. Examples include the mass of a sample and the time required for a chemical reaction to occur. Mass, length, volume, time, temperature, and energy are the most common types of data obtained from chemical experiments.

Results are the outcome of an experiment. Data and results may be identical, but more often several pieces of data are combined to produce a result.

A **unit** defines the basic quantity of mass, volume, time, and so on. A number not followed by the correct unit conveys no useful information.

English and Metric Units

Learning Goal 7:	Learn the major units of measure in the English and metric systems, and be able to convert from one system to another.

The English system of measurement has as its most commonly used unit of **weight** the standard pound (lb), its fundamental unit of length, the standard yard (yd), and its basic unit of volume, the standard gallon (gal). The English system is not used in scientific work primarily because of the difficulty involved in converting from one unit to another.

The metric system is a decimal-based system; it is inherently simpler and less ambiguous. In the metric system there are three basic units. **Mass** is represented as the gram, length as the meter, and volume as the liter. Any subunit or multiple unit contains one of these units preceded by a prefix indicating the power of ten by which the base unit is to be multiplied to form the subunit or multiple unit. The most common metric prefixes are shown in Table 1.1 of the textbook and should be memorized.

Unit Conversion: English and Metric Systems

A conversion factor or series of conversion factors relates two units when converting from one unit to another. The use of these conversion factors is referred to as the factor-label method.

This method is used either to convert from one unit to another within the same system, or convert units from one system to another.

Conversion of Units Within the Same System

The factor-label method is a self-indicating system; only if the factor is set up properly will the correct units result. Several commonly used English system conversion factors are included in Table 1.2 of the textbook. Most of these should already be familiar to you.

Conversion of units within the metric system may be accomplished using the factor-label method as well. Unit prefixes, which dictate the conversion factor, make conversion of units easy (refer to the textbook, Table 1.1).

Example 3
Convert 8.62 gallons into pints.

Example 3 continued

Work

$$8.62 \text{ gallons} \times \frac{4 \text{ quarts}}{1 \text{ gallon}} \times \frac{2 \text{ pints}}{1 \text{ quart}} = 68.96 \text{ pints}$$

Notice how the unit of gallon(s) cancels each other out, and the unit of quart(s) also cancels each other out, leaving "pints" as the final unit.

Study hint: You may be familiar with other ways to solve this problem, most notably by using ratios. Please spend some time becoming comfortable using this factor-label method, however. As problems become more complex, you may be more likely to make errors using methods other than the factor-label method. When a problem is set up using the factor-label method, all of the units cancel properly, leaving the unit that you want, and if the conversion factors are correct, you can be assured that your answer is correct.

Answer
69.0 pints (Three significant figures are needed.)

Example 4
Convert 16.4 kilometers into centimeters.

Work

$$16.4 \text{ kilometers} \times \frac{1000 \text{ meters}}{1 \text{ kilometer}} \times \frac{100 \text{ centimeters}}{1 \text{ meter}} = 1.64 \times 10^6 \text{ cm}$$

Answer
1.64×10^6 centimeters (Three significant figures are needed.)

Example 5
How many kilometers are in 1.568×10^6 centimeters?

Work

$$1.568 \times 10^6 \text{ cm} \times \frac{1 \text{ m}}{100 \text{ cm}} \times \frac{1 \text{ km}}{1000 \text{ m}} = 15.68 \text{ km}$$

$$1.568 \times 10^6 \text{ cm} \times \frac{1 \text{ m}}{1 \times 10^2 \text{ cm}} \times \frac{1 \text{ km}}{1 \times 10^3 \text{ m}} = 1.568 \times 10^1 \text{ km}$$

Answer
1.568×10^1 km, or 15.68 km (Four significant figures are needed.)

Conversion of Units from One System to Another

The conversion of a quantity expressed in units of one system to an equivalent quantity in the other system (English to metric or metric to English) requires a bridging conversion unit. For example:

Quantity	English		Metric
Mass	1 pound	=	454 grams
	2.2 pounds	=	1 kilogram
Length	1 inch	=	2.54 centimeters
	1 yard	=	0.91 meters
Volume	1 quart	=	0.946 liters
	1 gallon	=	3.78 liters

The conversion may be represented as a three-step process:

1. Conversion from units stated in the problem to a bridging unit.
2. Conversion to the other system using the bridge.
3. Conversion within the desired system to units required by the problem.

Example 6
A patient's weight is 129 pounds. What is the patient's weight in kilograms? (Use 454 grams = 1 pound and 1 kilogram = 1000 grams in your setup.)

Work

$$129 \text{ pounds} \times \frac{454 \text{ grams}}{1 \text{ pound}} \times \frac{1 \text{ kilogram}}{1000 \text{ grams}} = 58.666 \text{ kg}$$

Answer
58.6 kilograms (Three significant figures are needed. See Section 1.4.)

1.4 Significant Figures and Scientific Notation

Learning Goal 8: Report data and results using scientific notation and the proper number of significant figures.

Modern pocket calculators often generate more digits than are merited, given the data used in the calculation. It is important that the result of these calculations be reflective of the level of certainty of the measurement. Proper use of significant figures, scientific notation, and rounding-off are essential.

Significant Figures

Significant figures are all digits in a number representing data or results that are known with certainty, *plus the first uncertain digit.*

The number of significant figures associated with a measurement is determined by the measuring device. Conversely, the number of significant figures reported is an indication of the sophistication of the measurement itself.

Recognition of Significant Figures

Only significant digits should be reported as data or results. The six rules enumerated below describe the assignment of significant figures.

Rule 1: All nonzero digits are significant.

Rule 2: The number of significant digits is independent of the position of the decimal point.

Rule 3: Zeros located between nonzero digits are significant.

Rule 4: Zeros at the end of a number (often referred to as trailing zeros) are significant if the number contains a decimal point.

Rule 5: Trailing zeros are insignificant if the number does not contain a decimal point.

Rule 6: Zeros to the left of the first nonzero integer are not significant; they serve only to locate the position of the decimal point.

Scientific Notation

Very large numbers may be represented with the proper number of significant figures using **scientific notation.** Scientific notation, also referred to as exponential notation, involves the representation of a number as a power of ten; the usual convention shows the decimal point in standard position—to the right of the leading digit.

To convert a number greater than 1 to scientific notation, the original decimal point is moved x places to the left, and the resulting number is multiplied by 10^x. The exponent (x) is a positive number equal to the number of places the original decimal point was moved.

To convert a number less than 1 to scientific notation, the original decimal point is moved x places to the right, and the resulting number is multiplied by 10^{-x}. The exponent $(-x)$ is a negative number equal to the number of places the original decimal point was moved.

Example 7

Convert $30,000,000 \times 10^{-4}$ centimeters into correct scientific notation, and use the proper number of significant digits.

Example 7 continued

Work

$30,000,000 \times 10^{-4} cm = 3 \times 10^{7} \times 10^{-4} cm = 3 \times 10^{3} cm$

(Notice that only one significant digit, the "3", is present.)

Answer

3×10^{3} cm

Significant Figures in Calculation of Results

In the process of addition or subtraction, the position of the decimal point in the quantities being combined determines the number of significant figures in the answer.

In multiplication and division, this is not the case: the decimal point position is irrelevant. What is important is the number of significant figures. Remember: *The answer can be no more precise than the least precise number from which the answer is derived.*

Example 8

Do the following multiplication and division problems using scientific notation. Also, round off the answer to the proper number of significant digits.

$$\frac{68.0 \times 10^{-5}}{4.61 \times 10^{-6}} \times \frac{16 \times 10^{2}}{3.6} = ?$$

Work

Start by writing each part in scientific notation form:

$$\frac{6.80 \times 10^{1} \times 10^{-5}}{4.61 \times 10^{-6}} \times \frac{1.6 \times 10^{1} \times 10^{2}}{3.6} = ?$$

$$\frac{6.80 \times 10^{-4}}{4.61 \times 10^{-6}} \times \frac{1.6 \times 10^{3}}{3.6} = ?$$

Then multiply as below. First, combine the exponent parts.

$$\frac{(6.80)(1.6) \times 10^{-4} \times 10^{3}}{(4.61)(3.6) \times 10^{-6}} = \frac{(6.80)(1.6) \times 10^{-1}}{(4.61)(3.6) \times 10^{-6}} = \frac{10.88 \times 10^{-1}}{16.596 \times 10^{-6}} = ?$$

Example 8 continued

Then, divide and round off to two significant figures, since 3.6 contains only two significant figures.

$$0.6555795 \times \frac{10^{-1}}{10^{-6}} = 6.6 \times 10^{-1} \times \frac{10^{-1}}{10^{-6}} = 6.6 \times 10^{-1} \times 10^{5} = 6.6 \times 10^{4}$$

Answer
6.6×10^{4}

When a number is raised to a power, $n^{x} = y$, the number of significant figures in the answer (y) is identical to the number contained in the original term (n).

Defined or counted numbers do not determine the number of significant figures. The quantity being converted, not the conversion factor, determines the number of significant figures.

Rounding Off Numbers

A generally accepted rule for rounding off states that if the first digit dropped is 5 or greater, we raise the last significant digit to the next higher number. If the first digit dropped is 4 or less, the last significant digit remains unchanged.

Example 9
Round off 16.00468 grams to two significant figures.

Answer
The first zero past the 6 will not increase the value. Thus, the answer is 16 grams.

1.5 Experimental Quantities

Learning Goal 9:   Use appropriate units in problem solving.

Mass

Mass describes the quantity of matter in an object. The terms weight and mass, in common usage, are often considered synonymous. In fact, they are not. Weight is the manifestation of the force of gravity on an object.

weight = mass x acceleration due to gravity

The common conversion units for mass are as follows:

1 gram (g) = 1 x 10^{-3} kilogram (kg) = 1/454 pound (lb)

10

The atomic mass unit (amu) is a convenient way to represent the mass of very tiny quantities of matter, such as individual atoms and molecules.

$$1 \text{ atomic mass unit (amu)} = 1.661 \times 10^{-24} \text{ g}$$

Length

The standard metric unit of length is the meter. Large distances are measured conveniently in kilometers, and smaller distances are measured in millimeters or centimeters. Common conversions for length are as follows:

$$1 \text{ meter (m)} = 1 \times 10^{2} \text{ centimeters (cm)} = 3.94 \times 10^{1} \text{ inches (in.)}$$

In the same way that atomic mass units represent mass on an atomic scale, either the nanometer (nm) or the angstrom (Å) is a convenient measure of distance.

$$1 \text{ nm} = 10^{-7} \text{ cm} = 10^{-9} \text{ m}$$

$$1 \text{ Å} = 10^{-8} \text{ cm} = 10^{-10} \text{ m}$$

Volume

The standard metric unit of volume is the liter. A liter is the volume occupied by 1000 grams of water at 4° Celsius. A volume of 1 liter also corresponds to:

$$1 \text{ liter (1)} = 1000 \text{ milliliters (ml)} = 1.06 \text{ quarts (qt)}$$

Time

The standard metric unit of time is the second. The need for accurate measurement of time by chemists is necessary in many applications.

Temperature

Temperature is the degree of "hotness" of an object. Many substances, such as mercury, expand as their temperature increases, and this expansion provides us with a way to measure temperature and temperature changes. The height of the mercury in a thermometer is proportional to the temperature. A mercury thermometer may be calibrated, or scaled, in different units, just like a ruler. Three common temperature scales are Fahrenheit (°F), Celsius (°C), and Kelvin (K). Two convenient reference temperatures used to calibrate a thermometer are the freezing and boiling temperatures of water. Conversion from one temperature scale to another may be accomplished as shown below.

Fahrenheit to Celsius: $\quad °C = \dfrac{(°F - 32)}{1.8}$

Celsius to Fahrenheit: $°F = 1.8°C + 32$

Celsius to Kelvin: $K = °C + 273$

Example 10
What value, in °F, corresponds to 310.0 Kelvin?

Work
$K = °C + 273$
    or
$°C = K - 273$
$°C = 310.0 - 273$
$°C = 37.0$ (The rules of significant figures require that the result be expressed to the nearest tenth of a degree, since 310.0 is to the nearest tenth of a degree.)

$°F = 1.8°C + 32$
$°F = 1.8(37.0) + 32$
$°F = 98.6$

Answer
310.0 Kelvin = 98.6°F

(Note that "273" and "32" are exact numbers used in the conversion equations, and are not used to determine the number of significant figures.)

Density and Specific Gravity

Learning Goal 10: Use density, mass, and volume in problem solving, and calculate the specific gravity of a substance from its density.

Density, the ratio of mass to volume, $d = \dfrac{mass}{Volume} = \dfrac{m}{V} = \dfrac{gram}{milliliter}$

is independent of the amount of material and is characteristic of a substance; each substance has a unique density.

Values of density are often related to a standard, well-known reference, the density of pure water at 4°C. This "referenced" density is the **specific gravity.** Note that specific gravity has no units.

$$specific\ gravity = \frac{density\ of\ object\ (g/ml)}{density\ of\ water\ (g/ml)}$$

Example 11

A solid block that has a mass of 1267.4 grams was found to have the following measurements: length = 9.86 cm, width = 46.6 mm, and height = 0.224 m. What is its density?

Work

1. Convert each measurement into centimeters:

    length = 9.86 cm
    width 46.6 mm = 4.66 cm
    height 0.224 m = 22.4 cm

2. Then use the equation for the volume of a solid:

    V_{solid} = length x width x height
    V_{solid} = (9.86 cm) (4.66 cm) (22.4 cm) = 1.0292262 x 10^3 cm^3

    (Use only three significant figures.)

    V_{solid} = 1.03 x 10^3 cm^3

3. Then use the density equation:

    $$d = \frac{m}{V} = \frac{1267.4 \text{ g}}{1.03 \times 10^3 \text{ mL}}$$

    $$d = \frac{1.2674 \times 10^3 \text{ g}}{1.03 \times 10^3 \text{ mL}}$$

    $d = 1.23$ g/mL

Answer

The density equals 1.23 grams/milliliter.

Example 12

If the density of carbon tetrachloride is 1.59 grams/mL, what is the mass (in grams) of 2.00 liters of carbon tetrachloride?

Work

$$2.00 \text{ liters x } \frac{1000 \text{ mL}}{1 \text{ liter}} \text{ x } \frac{1.59 \text{ grams}}{1 \text{ mL}} = 3.18 \times 10^3 \text{ grams}$$

Example 12 continued

Answer

3.18×10^3 grams (Three significant figures are needed.)

Example 13

Calculate the mass in grams of 25 mL of mercury. The density of mercury is 13.5 g/mL.

Work

Use mass = $(d)(V)$, or use the factor-label method as below:

$$25 \text{ mL of mercury} \times \frac{13.5 \text{ g of mercury}}{1 \text{ mL of mercury}} = 337.5 \text{ g}$$

(Convert to scientific notation and round off to two significant figures.)

Answer

3.4×10^2 g

Example 14

Calculate the volume in mL of a liquid (ethyl alcohol) that has a density of 0.789 g/mL and a mass of 5.555 g.

Work

Use $V = \dfrac{m}{d}$, or use the factor label method as below:

$$5.555 \text{ g of liquid} \times \frac{1 \text{ mL of liquid}}{0.789 \text{ g of liquid}} = 7.0405576 \text{ mL}$$

(Round off to three significant figures.)

Answer

7.04 mL

Glossary of Key Terms in Chapter One

accuracy (1.4) the nearness of an experimental value to the true value.

chemical property (1.2) property of a substance that relates to the substance's participation in a chemical reaction.

chemical reaction (1.2) a process in which atoms are rearranged to produce new combinations.

chemistry (1.1) the study of matter and the changes that matter undergoes.

compound (1.2) a substance that is characterized by constant composition and can be chemically broken down into elements.

data (1.3) a group of facts resulting from an experiment.

density (1.5) mass per unit volume of a substance.

element (1.2) a substance that cannot be decomposed into simpler substances by chemical or physical means.

energy (1.1) the capacity to do work.

error (1.4) the difference between the true value and the experimental value for data and results.

extensive property (1.2) a physical property that depends on the quantity of the substance.

gaseous state (1.2) a physical state of matter characterized by the lack of fixed shape or volume and ease of compressibility.

heterogeneous mixture (1.2) a mixture of two or more substances and is characterized by nonuniform composition.

homogeneous mixture (1.2) a mixture of two or more substances and is characterized by uniform composition.

hypothesis (1.1) attempt to explain observations in a commonsense way.

intensive property (1.2) a property of a substance that is independent on the quantity of the substance.

law (1.1) a summary of a large quantity of information.

liquid state (1.2) a physical state of matter characterized by the fixed volume and absence of a fixed shape.

mass (1.5) a quantity of matter.

matter (1.1) the material component of the universe.

mixture (1.2) a material composed of two or more substances.

physical change (1.2) a change in the form of a substance but not in its chemical composition. No chemical bonds are broken in a physical change.

physical property (1.2) a characteristic of a substance that can be observed without the substance undergoing change (examples include color, density, and melting and boiling points).

precision (1.4) the degree of agreement between replicate measurements of the same quantity.

properties (1.2) physical or chemical characteristics of matter.

pure substance (1.2) a substance with constant composition.

result (1.3) the outcome of a designed experiment, often calculated from individual bits of data.

scientific method (1.1) the process of studying our surrounding that is based on experimentation.

scientific notation (1.4) a system used to represent numbers as powers of ten.

significant figures (1.4) all digits in a number known with certainty and the first uncertain digit.

solid state (1.2) a physical state of matter characterized by its rigidity and fixed volume and shape.

specific gravity (1.5) the ratio of the density of a substance to the density of water at a specified temperature.

temperature (1.5) a measure of the relative "hotness" or "coldness" of an object.

theory (1.1) a hypothesis supported by extensive testing that explains and predicts facts.

uncertainty (1.4) the degree of doubt in a single measurement.

unit (1.3) a determinate quantity (of length, time, etc.) adopted as a standard of measurement.

weight (1.5) the force exerted on an object by gravity.

Self Test for Chapter One

1. Calculate the density (in normal unit form) of a liquid with a volume of 348 mL and a mass of 0.3546 kg.

2. $\dfrac{68.0 \times 10^{-3}}{4.61 \times 10^{-1}} = ?$

3. $\dfrac{10.6 \times 10^{-4}}{0.0641 \times 10^{-2}} \times 16.4 \times 10^{-7} \times \dfrac{0.111 \times 10^{4}}{10.0 \times 10^{2}} = ?$

4. 50.0 miles/hour = ? cm/second

5. If one atom of uranium weighs 238.0 atomic mass units, and one atomic mass unit is equal to 1.66×10^{-24} gram, what is the mass, in grams, of one atom of uranium?

6. The speed of light is 186,000 miles per second. What is its speed in cm/second?
   USE: 5280 feet = 1 mile; 12 inches = 1 foot; 2.54 cm = 1 inch

7. $-4.6°F = ? °C$

8. If the density of carbon tetrachloride is 1.59 g/mL, what is the mass of 2.65 liters of carbon tetrachloride in grams?

9. $-40.0°C = ? °F$

10. What is the specific gravity of an object that weighs 13.35 g and has a volume of 25.00 mL? The density of water under the same conditions is 0.980 g/mL.

11. The ability to do work describes what term?

12. A scientific experiment produces what information?

13. Each piece of data is the individual result produced by what process?

14. Which system of measurement is a decimal-based system?

15. What factors are used in the factor-label method?

16. Which temperature system does not use a degree sign?

17. $98.6°F = ? K$

18. Round off 0.00369865 to two significant figures.

19. Which of the following are physical properties of an object?
    a. combustibility
    b. color
    c. melting point
    d. gas phase
    e. chlorine combines with hydrogen to form HCl
    f. density
    g. volume
    h. $S + O_2 \rightarrow SO_2$

20. Which of the following are mixtures?
    a. NaCl
    b. salt in water
    c. a soft drink
    d. tap water
    e. blood
    f. cake mix

21. What are the three states of matter?

22. Which state of matter will expand to fill any container?

23. Which state of matter has no definite shape or volume?

24. List six examples of physical properties of matter.

25. Salt and pepper is an example of what type of mixture?

Vocabulary Quiz for Chapter One

1. The study of matter and the changes that matter undergoes is referred to as _____.

2. _____ is a group of facts resulting from an experiment.

3. A suggested explanation of observed behavior of our surroundings is a(n)_____.

4. The _____ state is a physical state of matter characterized by the lack of fixed shape or volume and ease of compressibility.

5. Stored energy is called _____.

6. The outcome of a scientific experiment, often calculated from data, is the _____ of that experiment.

7. An _____ property depends on the quantity of matter.

8. The ratio of the density of a substance to the density of water at the same temperature is called the _____.

9. A(n) _____ is a hypothesis developed to explain observed behavior of matter that has been verified by using the scientific method.

10. _____ is a force exerted on an object by gravity.

2

The Structure of the Atom and the Periodic Table

Learning Goals

1 Describe the important properties of protons, neutrons, and electrons.
2 Calculate the number of protons, neutrons, and electrons in any atom.
3 Distinguish among atoms, ions, and isotopes and calculate atomic masses from isotopic abundance.
4 Trace the history of the development of atomic theory, beginning with Dalton.
5 Recognize the important subdivisions of the periodic table: periods, groups (families), metals, and nonmetals.
6 Use the periodic table to obtain information about an element.
7 Describe the relationship between the electronic structure of an element and its position in the periodic table.
8 Write electron configurations for atoms of the most commonly occurring elements.
9 Use the octet rule to predict the charge of common cations and anions.
10 Use the periodic table and its predictive power to estimate the relative sizes of atoms and ions, as well as relative magnitudes of ionization energy and electron affinity.

Introduction

In this chapter we will learn some of the properties of atomic particles and the experiments that form the basis of our current understanding of atomic structure. The atomic structure of each element is unique, but there exist structural similarities among these elements.

The **periodic law** states that properties of elements are periodic functions of their atomic numbers. The periodic table results from this relationship. The periodic table is an organized "map" of the elements that relates their structure to their chemical and physical properties. The chemical and physical properties of elements follow directly from the electronic structure of the atoms that make up these elements. Familiarity with the periodic table allows prediction of the structure and properties of the various elements, and serves as the basis for understanding chemical bonding.

2.1 Composition of the Atom

Electrons, Protons, and Neutrons

The basic structural unit of an element is the **atom,** which is the smallest unit of an element that retains the chemical properties of that element. An atom is composed of three primary particles: the **electron,** the **proton,** and the **neutron**. These particles are located in one of two distinct regions:

1. The **nucleus** is a small, dense, positively charged region in the center of the atom. The nucleus is composed of positively charged protons and uncharged neutrons.
2. Surrounding the nucleus is a diffuse region of negative charge populated by electrons, the source of the negative charge. Electrons are tiny in comparison to the protons and neutrons. The properties of these particles are summarized in Table 2.1 of the textbook.

The number of protons determines the identity of the atom. When the number of protons is equal to the number of electrons, the atom is neutral, because the charges are balanced and effectively cancel.

The **atomic number (Z)** is equal to the number of protons in the atom, and the **mass number (A)** is equal to the sum of the protons and neutrons (the mass of the electrons is so small as to be insignificant).

Isotopes

Isotopes are atoms of the same element which have different masses due to different numbers of neutrons (different atomic mass).

Inspection of the periodic table reveals that the atomic mass of many of the elements is not an integral number. For example the atomic mass of chlorine is actually 35.45 amu, not 35.00 amu. The existence of isotopes accounts for this difference. A natural sample of chlorine is principally composed of two isotopes, $^{35}_{17}Cl$ and $^{37}_{17}Cl$, in approximately a 3:1 ratio, and the tabulated atomic mass is the weighted average of the two isotopes.

Certain isotopes of elements emit particles and energy (**radioactivity,** Chapter 9) that may be useful in tracing the behavior of biochemical systems. These isotopes otherwise behave identically to any other isotope of the same element.

Example 1

Calculate the number of protons, neutrons, and electrons found in carbon-14 ($^{14}_{6}C$).

Example 1 continued

Work
The atomic number of carbon (from the periodic table) is 6. Thus, there are 6 protons and 6 electrons in one atom of carbon. The 14 in carbon-14 means that this specific isotope of carbon has a mass number of 14. The mass number of an atom is equal to the sum of the protons and neutrons:

mass number = number of protons + number of neutrons

number of neutrons = mass number – number of protons

Thus, for carbon-14:

number of neutrons = 14 – 6 = 8

Answer
Carbon-14 contains 6 protons, 6 electrons, and 8 neutrons.

Learning Goal 3:	**Distinguish among atoms, ions, and isotopes and calculate atomic masses from isotopic abundance.**

Example 2
Calculate the number of protons, neutrons, and electrons in the three isotopes of hydrogen:

$$^1_1H \qquad\qquad ^2_1H \qquad\qquad ^3_1H$$

normal hydrogen   deuterium   tritium

Work
Since all are hydrogen atoms, they each contain only 1 proton and 1 electron, because the atomic number of hydrogen is 1.
number of neutrons = mass number – number of protons
The mass number of normal hydrogen is 1. Thus, it contains no neutrons.
The mass number of deuterium is 2. Thus, it contains 1 neutron in its nucleus:
number of neutrons in deuterium = 2 – 1 = 1
The mass number of tritium is 3. Thus, it contains 2 neutrons in its nucleus:
number of neutrons in tritium = 3 – 1 = 2

Answer
The three isotopes each have 1 proton and 1 electron. Normal hydrogen has no neutrons. Deuterium has 1 neutron; tritium has 2 neutrons.

Ions

Ions are charged particles that result from a gain of one or more electrons by the parent atom (forming negative ions, or **anions**) or a loss of one or more electrons from the parent atom (forming positive ions, or **cations**).

For simplification, the atomic and mass numbers are often omitted. For example, the hydrogen cation would be written as H^+ and the anion as H^-.

Example 3

How many protons and electrons are present in each of the following atoms and ions?

$$H \quad + \quad 1e^- \quad \rightarrow \quad H^-$$

Work

The H with no charge means that it is an atom; atoms have no charge, since they have the same number of protons and electrons. The H^- is an ion that has a negative one charge. This means it has one more electron than it has protons. You can see in the reaction that the hydrogen atom (given as H) has gained one electron to form the ion H^-.

ATOM	ION

$$H \quad + \quad 1e^- \quad \rightarrow \quad H^-$$

$$
\begin{array}{ll}
+1 & = \text{proton} \\
\underline{-1} & = \text{electron} \\
0 & \text{charge}
\end{array}
\qquad
\begin{array}{l}
1 \text{ proton} = +1 \\
\underline{2 \text{ electrons} = -2} \\
\text{charge} = -1
\end{array}
$$

Answer

The hydrogen atom contains 1 proton and 1 electron. The –1 hydrogen ion contains 1 proton and 2 electrons.

Remember that if an ion is positive, it was formed by the loss of the required number of electrons from the given atom.

$Mg \rightarrow 2e^- + Mg^{2+}$ • Two electrons were lost by the magnesium atom.
$K \rightarrow 1e^- + K^+$ • One electron was lost by the potassium atom.

A negative ion, on the other hand, is formed by a gain of one or more electrons by the given atom.

$$S + 2e^- \rightarrow S^{2-}$$
$$Cl + 1e^- \rightarrow Cl^-$$

2.2 Development of the Atomic Theory

Learning Goal 4:	Trace the history of the development of atomic theory, beginning with Dalton.

Dalton's Theory

The first experimentally based theory of atomic structure was proposed by John Dalton in the early 1800s. Dalton postulated that:

1. All matter consists of tiny particles called atoms.
2. Atoms cannot be created, divided, destroyed, or converted to any other type of atom.
3. Atoms of a particular element have identical properties.
4. Atoms of different elements have different properties.
5. Atoms combine in simple, whole-number ratios.
6. Chemical change involves joining, separating, or rearranging atoms.

Postulates 1, 4, 5, and 6 are presently regarded as true. The discovery of the processes of nuclear fusion, fission, and radioactivity (Chapter 9) have disproved the postulate that atoms cannot be created or destroyed. Postulate 3, that the atoms of a particular element are identical, was disproved by the later discovery of isotopes.

Evidence for Subatomic Particles: Electrons, Protons, and Neutrons

Although Dalton pictured atoms as indivisible, various experiments, particularly those of William Crookes and Eugene Goldstein, indicated that the atom is composed of charged (+ and –) particles.

J. J. Thomson demonstrated the electrical and magnetic properties of cathode rays (Figures 2.4 and 2.5 in the textbook). Crookes observed rays, which he called cathode rays, emanating from the cathode (– charge) of an evacuated (vacuum) tube. Further experiments showed that the ability to produce cathode rays is a characteristic of all materials. In 1897, Thomson announced that cathode rays were streams of negative particles of energy. These particles are electrons.

Similar experiments, conducted by Goldstein, led to the discovery of protons, particles equal in charge to the electron, but opposite in sign.

The neutron has a mass virtually equal to that of the proton, and zero charge. The neutron was first postulated in the early 1920s, but it was not until 1932 that James Chadwick demonstrated its existence.

Evidence for the Nucleus

In the early 1900s, it was believed that protons and electrons were uniformly distributed throughout the atom. However, an experiment by Hans Geiger led Ernest Rutherford (in 1911) to propose that the majority of the mass and positive charge of the atom was actually located in a small, dense region, the nucleus, and that the small, negatively charged electrons spread across a much larger, diffuse area outside of the nucleus.

2.3 The Periodic Law and the Periodic Table

Numbering Groups in the Periodic Table

Two systems currently exist to number the groups in the periodic table. The older system uses Roman numerals followed by the letter A for the main-group elements or the letter B for the transition elements. The newer system, recommended by I.U.P.A.C., simply numbers the groups from 1 to 18. Both numbering systems are shown on the periodic table in your text.

Periods and Groups

Learning Goal 5:	Recognize the important subdivisions of the periodic table: periods, groups (families), metals, and nonmetals.

A horizontal row of elements in the periodic table is referred to as a **period.** The periodic table consists of seven periods, six of which contain 2, 8, 8, 18, 18, and 32 elements. The seventh period is still incomplete but potentially holds 32 elements. Note that the lanthanide series is a part of period six and the actinide series is a part of period seven.

The columns of elements in the periodic table are called **groups** or families. The elements of a particular family share many similarities in physical and chemical properties that are related to similarities in electronic structure. The various groups are labeled with Roman numerals, and each is subtitled with the letter A or B.

Group A elements are called **representative elements,** and Group B elements are **transition elements.** Certain families have common names as well as a Roman numeral-letter designation. Group IA elements are also known as the **alkali metals;** Group IIA, as the **alkaline earth metals;** Group VIIA, as the **halogens;** and Group VIIIA, as the **noble gases.**

Metals and Nonmetals

A bold zig-zag line runs from top to bottom of the table beginning to the left of boron (B) and ending between polonium (Po) and astatine (At). This line acts as the boundary between **metals,** to the left, and **nonmetals,** to the right. Elements straddling the boundary, such as Ge and As, have properties intermediate between metals and nonmetals and are often termed **metalloids.**

Atomic Number and Atomic Mass

The atomic number (Z, number of protons in the nucleus; the nuclear charge) and the atomic mass of each element are available from the periodic table. More detailed periodic tables may provide information such as electron arrangement, relative sizes of atoms, and most probable ion charges.

Learning Goal 6:	Use the periodic table to obtain information about an element.

Example 4
For each of the following symbols, provide the name of the element, its atomic number (Z), and the family to which it belongs.

1. Li   4. Si   7. I
2. Ra   5. N   8. Cu
3. Al   6. S

Answer
1. Li: lithium; Z = 3; family = IA   5. N: nitrogen; Z = 7; family = VA
2. Ra: radium; Z = 88; family = IIA   6. S: sulfur; Z = 16; family = VIA
3. Al: aluminum; Z = 13, family = IIIA   7. I: iodine; Z = 53; family = VIIA
4. Si: silicon; Z = 14; family = IVA   8. Cu: copper; Z = 29; family = IB

2.4 Electron Arrangement and the Periodic Table

Learning Goal 7:	Describe the relationship between the electronic structure of an element and its position in the periodic table.

The most important factor in chemical bonding is the arrangement of the electrons in the atoms that are combining. The periodic table provides us with a great deal of information about the electron arrangement, or electronic structure, of atoms.

Valence Electrons

Outermost electrons in an atom are **valence electrons.** For representative elements, the number of valence electrons in an atom corresponds to the number of the family in which the atom is found. Metals tend to have fewer valence electrons, and nonmetals tend to have more valence electrons.

The energy levels are symbolized by n, with the lowest energy level assigned a value of $n = 1$. Each energy level may contain up to a fixed maximum number of electrons. Two general rules of electron configuration are based upon the periodic law.

1. The number of valence electrons in a neutral atom equals the group number for all representative (A group) elements.
2. The energy level ($n = 1$, 2, etc.) in which the valence electrons are located corresponds to the period in which the element may be found.

Helium is an exception to rule 1, above. It cannot have eight valence electrons, since all of its electrons are in the $n = 1$ level, which has a maximum capacity of only two electrons.

The Quantum Mechanical Atom

Erwin Schrödinger described electrons in atoms in probability terms, and developed equations that emphasize the wavelike character of electrons. His theory, often described as quantum mechanics, incorporates Bohr's principle energy levels, which are made up of one or more sublevels. Each sublevel contains one or more atomic orbitals.

Energy Levels and Sublevels

The principal energy levels are designated $n = 1, 2, 3$, and so forth. The number of possible sublevels in a principal energy level is also equal to n. When $n = 1$, there can be only one sublevel; $n = 2$ allows two sublevels, and so forth. The total electron capacity of a principal level is $2(n)^2$.

The sublevels, or subshells, increase in energy: $s < p < d < f$

Both the principal energy level and type of sublevel are specified when describing the location of an electron. For example: $1s$, $2s$, $2p$. The first principal energy level ($n = 1$) has one possible subshell, $1s$. The second principal energy level ($n = 2$) has two possible subshells: $2s$ and $2p$. The third principal energy level ($n = 3$) has three possible subshells: $3s$, $3p$, and $3d$. The fourth principal energy level ($n = 4$) has four possible subshells: $4s$, $4p$, $4d$, and $4f$.

An **orbital** is a specific region of a subshell containing a maximum of two electrons. The s sublevel contains only one orbital, the p sublevel contains three orbitals, the d sublevel contains five orbitals, and the f sublevel contains seven orbitals. Each orbital may be empty, contain one electron, or be filled, containing two electrons. The s, p, d, and f sublevels have maximum capacities of 2, 6, 10, and 14 electrons.

Each type of orbital has its own characteristic shape:

- The s orbital is spherically symmetrical; a model appears as a Ping-Pong ball with its center corresponding to the intersection of imaginary x, y, and z coordinates.
- There exist three kinds of p orbitals, each identical in shape (often modeled as a dumbbell). They differ only in their orientation in space, along the hypothetical x-axis, (p_x), y-axis, (p_y), and z-axis, (p_z).
- The d and f orbitals are more complex; the textbook focuses exclusively on s and p orbitals in subsequent discussion.

Each atomic orbital has a maximum capacity of two electrons. The electrons are perceived to *spin* on an imaginary axis, and the two electrons in the same orbital must have opposite spins, clockwise and counterclockwise. Two electrons in one orbital that possess opposite spins are referred to as *paired* electrons.

Electron Configuration and the Aufbau Principle

Learning Goal 8:	Write electron configurations for atoms of the most commonly occurring elements.

The arrangement of electrons in atomic orbitals is referred to as the atom's **electron configuration.** We may represent the electron configuration of atoms of various elements using the aufbau, or building-up, principle. According to this principle, electrons fill the lowest-energy orbital that is available first. By knowing the order of filling of atomic orbitals, lowest to highest energy, you may write the electron configuration for any element.

Example 5
Name the two elements that have electrons only in the first energy level.

Answer
hydrogen and helium

Example 6
List the sublevels in order of lowest energy to highest energy.

Answer
$s < p < d < f$

Example 7
Write the symbols of all the possible sublevels found in the fourth principal energy level ($n = 4$).

Answer
$4s$, $4p$, $4d$, and $4f$

Example 8
Complete the following statement about an orbital:
Each orbital may be empty, _____, or be _____.

Answer
Each orbital may be empty, contain one electron, or be filled, containing two electrons.

Example 9
What is the maximum number of electrons that may be found in each sublevel?

Answer
$s = 2e^-$; $p = 6e^-$; $d = 10e^-$; $f = 14e^-$

Example 10
Give the electronic configuration for each of the following elements:

1.	H	5.	Na	9.	S
2.	He	6.	Al	10.	Ar
3.	O	7.	P		
4.	F	8.	B		

Answer

1. H: $1s^1$
2. He: $1s^2$
3. O: $1s^2 2s^2 2p^4$
4. F: $1s^2 2s^2 2p^5$
5. Na: $1s^2 2s^2 2p^6 3s^1$

6. Al: $1s^2 2s^2 2p^6 3s^2 3p^1$
7. P: $1s^2 2s^2 2p^6 3s^2 3p^3$
8. B: $1s^2 2s^2 2p^1$
9. S: $1s^2 2s^2 2p^6 3s^2 3p^4$
10. Ar: $1s^2 2s^2 2p^6 3s^2 3p^6$

Shorthand Electron Configurations

Electron configurations may be abbreviated by substituting the symbol for a noble gas (in brackets, [ ]) for the part of the electron configuration represented by that noble gas. For example, the electron configuration of strontium, Sr, would be quite long: $1s^2 2s^2 2p^6 3s^2 3p^6 4s^2 3d^{10} 4p^6 5s^2$. However, the noble gas that is located before Sr on the periodic table is Kr, which has the following electron configuration: $1s^2 2s^2 2p^6 3s^2 3p^6 4s^2 3d^{10} 4p^6$. The electron configuration of Sr, then, can be abbreviated as follows:

Sr   [Kr] $5s^2$

2.5 The Octet Rule

Elements in the last family, the noble gases, have either two valence electrons (helium) or eight valence electrons (neon, argon, krypton, xenon, and radon). Their most important property is their extreme stability. A full n = 1 energy level (as in helium) or an outer octet of electrons is responsible for this unique stability.

Atoms of elements in other groups are more reactive than the inert gases because they are, in the process of chemical reaction, trying to achieve a more stable "noble gas" configuration by gaining or losing electrons. This is the basis of the **octet rule.** In chemical reactions they will gain, lose, or share the minimum number of electrons necessary to achieve a more stable energy state.

Ion Formation and the Octet Rule

Learning Goal 9:	Use the octet rule to predict the charge of common cations and anions.

     Metallic elements tend to form positively charged ions called **cations.** Positive ions are formed when an atom loses one or more electrons. These ions are more stable than their corresponding neutral atoms. The ion is isoelectronic (that is, it has a similar number of electrons) with its nearest noble gas neighbor and has an octet of electrons in its outermost energy level. Nonmetallic elements tend to gain electrons to become isoelectronic with the nearest noble gas element, forming negative ions referred to as **anions.**

Example 11

Show how a sodium atom will form a sodium ion.

Answer

1.   Na $\rightarrow$ Na$^+$ + 1e$^-$
      Atom   Ion

2.   Na (atom); it has 11e$^-$
3.   Na$^+$ (ion); it has 10e$^-$

Example 12

Show how an oxygen atom will form an oxide ion.

Answer

1.   O   + 2 e$^-$ $\rightarrow$ O^{2-}
      oxygen atom   oxide ion

2.   O (atom); it has 8e$^-$
3.   O^{2-} (ion); it has 10e$^-$

     The transition metals tend to form positive ions by losing electrons, just like the representative metals. However, the transition elements are characterized as "variable valence" elements; depending upon the type of substance they react with, they may form more than one stable ion. For example, iron has two stable ionic forms: Fe^{2+} and Fe^{3+}.

2.6 Trends in the Periodic Table

Learning Goal 10:	Use the periodic table and its predictive power to estimate the relative sizes of atoms and ions, as well as the relative magnitudes of ionization energy and electron affinity.

Atomic Size

The size of the atom will be determined principally by two factors:
1. The energy level (n-level) in which the outermost electron(s) is located increases as we go down a group (recall that the outermost n-level correlates with period number).
2. As the magnitude of the positive charge of the nucleus increases, its "pull" on all of the electrons increases, and the electrons are drawn closer to the nucleus.

Consequently, atomic size increases down a group and decreases across a period.

Ion Size

Three generalizations can be made about the size of ions:
1. Positive ions (cations) are smaller than the parent atom.
2. Negative ions (anions) are larger than the parent atom.
3. Ions with multiple positive charge (such as Cu^{2+}) are even smaller than their corresponding monopositive ion (Cu^+); ions with multiple negative charge (such as O^{2-}) are larger than their corresponding less-negative ion.

Ionization Energy

The energy required to remove an electron from an isolated atom in the gas phase is the **ionization energy.** The magnitude of the ionization energy correlates with the strength of the attractive force between the nucleus and the outermost electron.

1. As we go down a group, the ionization energy decreases, since the atom's size is increasing. The outermost electron is progressively farther from the nuclear charge and hence easier to remove.
2. As we go across a period, atomic size decreases, as the outermost electrons are closer to the nucleus, more tightly held, and more difficult to remove. Therefore, the ionization energy must increase.

A correlation does indeed exist between trends in atomic size and ionization energy. Atomic size decreases from bottom to top of a group and left to right in a period. Ionization energies increase in the same periodic way. Note also that ionization energies are highest for the noble gases; this accounts for the extreme stability and nonreactivity of the noble gases.

Electron Affinity

The energy released when a single electron is added to a neutral atom in the gaseous state is known as the **electron affinity.** Electron affinity is a measure of the ease of forming negative ions. A large value of electron affinity (energy released) indicates that the atom becomes more stable as it becomes a negative ion (through the process of gaining an electron).

Periodic trends for electron affinity are as follows:
1. Electron affinities generally decrease as we go down a group.
2. Electron affinities generally increase as we go across a period.

Be aware that exceptions to these general trends do exist.

Glossary of Key Terms in Chapter Two

alkali metal (2.3) occupy group IA (1) of the periodic table.

alkaline earth metal (2.3) occupy group IIA (2) of the periodic table.

anion (2.1) a negatively charged atom or group of atoms.

atom (2.1) the smallest unit of an element that retains properties of that element.

atomic mass (2.1) the mass of the atom expressed in atomic mass units.

atomic number (2.1) the number of protons in the nucleus of an atom. It is a characteristic identifier of an element.

atomic orbital (2.4) a specific region of space where an electron may be found.

cathode rays (2.2) a stream of electrons that are given off by the cathode (negative electrode) in a cathode ray tube.

cation (2.1) a positively charged atom or group of atoms.

electron (2.1) a negatively charged particle outside of the nucleus of an atom.

electron affinity (2.6) the energy released when an electron is added to an isolated atom.

electron configuration (2.4) the arrangement of electrons in atomic orbitals.

energy levels (2.4) atomic region where electrons may be found.

group (2.3) any one of 18 vertical columns of elements. (Note that a group is often referred to as a *family*.)

halogen (2.3) occupy group VIIA (17) of the periodic table.

ion (2.1) an electrically charged particle that is formed by the gain or loss of an electron.

ionization energy (2.6) the energy needed to remove an electron from an atom in the gas phase.

isoelectronic (2.5) atoms, ions, and molecules containing the same number of electrons.

isotope (2.1) atom of the same element that differs in mass because it contains different numbers of neutrons.

mass number (2.1) the sum of the number of protons and neutrons in an atom.

metal (2.3) element located on the left side of the periodic table (left of the "staircase" boundary).

metalloid (2.3) an element along the "staircase" boundary between metals and nonmetals; they exhibit both metallic and nonmetallic properties.

neutron (2.1) an uncharged particle, with the same mass as the proton, found in the nucleus of an atom.

noble gas (2.3) elements in group VIIIA (18) of the periodic table.

nonmetal (2.3) element located on the right side of the periodic table (right of the "staircase" boundary).

nucleus (2.1) the small, dense center of positive charge in the atom.

octet rule (2.5) a rule predicting that atoms form the most stable molecules or ions when they are surrounded by eight electrons in their highest occupied energy level.

period (2.3) any one of seven horizontal rows of elements in the periodic table.

periodic law (2.3) a law stating that properties of elements are periodic functions of their atomic numbers. (Note that Mendeleev's original statement was based on atomic masses.)

proton (2.1) a positively charged particle in the nucleus of an atom.

quantization (2.4) a characteristic that energy can occur only in discrete units (quanta).

representative element (2.3) member of the group of the periodic table designated as "A".

sublevel (2.4) set of equal-energy orbitals within a principal energy level.

transition element (2.3) occurs between groups IIIB (3) and IIB (12) in the long period of the periodic table.

valence electron (2.4) electron in the outermost shell (principal quantum level) of an atom.

Self Test for Chapter Two

1. How many protons and electrons are found in a magnesium atom?

2. How many protons and electrons are found in a chloride ion, Cl⁻ ?

3. Give the number of protons and electrons found in a lithium atom and a lithium ion, Li⁺. Write the reaction showing how a lithium atom can form a lithium ion.

4. What is the dense positive center of an atom?

5. Which particle of an atom has the least mass?

6. What number is equal to the number of protons in an atom?

7. The mass number minus the number of protons equals the number of what particle?

8. If an atom loses an electron, what charge does the resulting ion have?

9. Symbolically represent the three isotopes of the element hydrogen.

10. What do we call the outermost electrons of an element, which are involved in chemical bonding?

11. How many valence electrons are in sulfur?

12. Name the only noble gas with two valence electrons.

13. Name the group which contains elements frequently isoelectronic with stable cations and anions.

14. In chemical reactions, elements will gain, lose, or share the minimum number of electrons necessary to achieve what kind of energy state?

15. What do we call an ion that has the same electronic arrangement as its nearest noble gas neighbor?

16. What are the Group A elements often called?

17. What are the Group B elements often called?

18. List the values of the principal (main) energy levels of an atom.

19. How many sublevels are present in the second principal energy level?

20. What is the name of a specific region of space of a sublevel where a maximum of two electrons may be found?

21. What orbital has a spherical shape?

22. What is the relationship of the spins of two electrons in the same orbital?

23. What term refers to the arrangement of electrons in atomic orbitals?

24. Write the electron configuration of carbon.

25. Which elements do not readily bond to other elements?

26. All of the ions in this group always form a + 1 ion.

27. What do we call the ability of an atom to attract electrons to itself?

Vocabulary Quiz for Chapter Two

1. The sum of the number of protons and neutrons in an atom is the _____.

2. A(n)_____ is a positively charged atom (or group of atoms).

3. A(n)_____ is an integer used to describe the orbitals of an atom.

4. A(n)_____ is a negatively charged particle outside of the nucleus of an atom.

5. A(n)_____ is a substance that cannot be decomposed into simpler substances by chemical or physical means.

6. According the Bohr theory, a region of the atom where electrons may be found is called a(n)_____.

7. A neutral atom that gains one electron has a _____ charge and is called a(n) _____.

8. Atoms of the same element that differ in mass because they contain different numbers of neutrons are called _____.

9. A(n) _____ is an uncharged particle in the nucleus of an atom that has the same mass as the proton.

10. A positively charged particle in the nucleus of an atom is called a (n)_____.

11. The _____ occupy group IA of the periodic table.

12. The _____ is an organized "map" of the elements.

13. A horizontal row of elements in the periodic table is referred to as a _____.

14. Columns in the period table are also called _____.

15. The outermost electrons in an atom are _____ electrons.

16. Neon, argon, krypton, xenon, and radon are termed _____.

17. Energy released when an electron is added to an atom is _____.

18. Energy required to remove an electron from an atom is _____.

3

Structure and Properties of Ionic and Covalent Compounds

Learning Goals

1. Classify compounds as having ionic, covalent, or polar covalent bonds.
2. Write the formula of a compound when provided with the name of the compound.
3. Name common inorganic compounds using standard conventions and recognize the common names of frequently used substances.
4. Predict differences in physical state, melting and boiling points, solid-state structure, and solution chemistry that result from differences in bonding.
5. Draw Lewis structures for covalent compounds.
6. Describe the relationship between bond order, bond energy, and bond length.
7. Predict the geometry of molecules using the octet rule and Lewis structures.
8. Understand the role that molecular geometry plays in determining the solubility and melting and boiling points of compounds.
9. Use the principles of VSEPR theory and molecular geometry to predict relative melting points, boiling points, and solubilities of compounds.

Introduction

This chapter describes the role of valence electrons in bond formation between atoms. Systems of naming the resultant compounds are discussed, as well as the procedure for writing formulas based on the names of the compounds. The chemical and physical properties of these compounds are related to structure and bonding.

3.1 Chemical Bonding

When two atoms are joined together to make a chemical compound, the force of attraction between the two species is referred to as a **chemical bond.** Interactions involving valence electrons are responsible for the chemical bond.

Lewis Symbols

The **Lewis symbol** is a convenient way of representing atoms singly or in combination. Its principal advantage is that only valence electrons are shown. This results in simpler structures and greater clarity. The chemical symbol of the atom is written; this symbol represents the nucleus and all of the lower-energy nonvalence electrons, which do not directly participate in bonding. The valence electrons are indicated by dots (•) or crosses (x) arranged around the atomic symbol.

Principal Types of Chemical Bonds: Ionic and Covalent

Learning Goal 1:	Classify compounds as having ionic, covalent, or polar covalent bonds.

Ionic bonding is characterized by an electron transfer process occurring prior to bond formation. In **covalent bonding,** electrons are shared between atoms in the bonding process.

The essential features of ionic bonding are as follows:

- Elements with low ionization energy and low electron affinity tend to form positive ions.

- Elements with high ionization energy and high electron affinity tend to form negative ions.

- Ion formation takes place by an electron transfer process.

- The resulting positive and negative ions are held together by the electrostatic force between ions of opposite charge in an ionic bond.

- Reactions between metals and nonmetals (elements far to the left and right, respectively, on the periodic table) tend to result in ionic bonds.

When electrons are shared rather than transferred, the shared electron pair is referred to as a covalent bond. Compounds characterized by covalent bonding are called covalent compounds. Covalent bonds tend to form among atoms with similar tendencies to gain or lose electrons. The most obvious examples are the diatomic molecules H_2 as well as N_2, O_2, F_2, Cl_2, I_2, and Br_2. Bonding in these molecules is totally covalent because there is no net tendency for electron transfer between identical atoms.

Two atoms do not have to be identical in order to form a covalent bond. Compounds such as hydrogen fluoride, water, methane, and ammonia are common examples.

Polar Covalent Bonding and Electronegativity

Polar covalent bonding, like covalent bonding, is based on the concept of electron sharing; however, the sharing is unequal and based on the electronegativity difference between joined atoms. **Electronegativity** is a measure of the tendency of an atom in a molecule to attract shared electrons.

Electronegativity is represented by a scale derived from the measurement of energies of chemical bonds. The most electronegative element is fluorine, F, with a value of 4.0, and the least electronegative elements are Cs and Fr, each of which has a value of 0.7. The periodic trends for electronegativity, which increases from left to right and decreases from top to bottom, are similar to both ionization energy and electron affinity.

3.2 Naming Compounds and Writing Formulas of Compounds

Learning Goal 2:	Write the formula of a compound when provided with the name of the compound.

Learning Goal 3:	Name common inorganic compounds using standard conventions and recognize the common names of frequently used substances.

Proper use of **nomenclature,** the assignment of a correct and unambiguous name to every chemical compound, is fundamental to the study of chemistry. The student must be able to write the name corresponding to a compound and the formula, when provided only with the name.

Ionic Compounds

The "shorthand" symbol for a compound is its **formula.** Examples are: $NaCl$, $MgBr_2$, and $NaIO_3$. The formula identifies the number and type of the various atoms that compose the compounds. The number of like atoms is denoted by a subscript. The presence of one atom is implied when no subscript is present.

Remember that positive ions are formed from elements that
1. are located to the left on the periodic table.
2. are referred to as metals.
3. have low ionization energies and low electron affinities, and hence easily lose electrons.

Elements that form negative ions, on the other hand,
1. are located to the right on the periodic table (but exclude the noble gases).
2. are referred to as nonmetals.
3. have high ionization energies and high electron affinities, and hence easily gain electrons.

Metals and nonmetals react to produce ionic compounds as a result of electron transfer. Although we refer to ionic compounds as **ion pairs,** in the solid state these ion pairs do not actually exist as individual units. Positive and negative ions arrange themselves in a regular, three-dimensional, repeating array known as a **crystal lattice.**

The names given to ionic compounds are based upon their formulas, with the name of the cation appearing first, followed by the anion name. The positive ion is simply the name of the element, while the negative ion is named as the stem of the element's name joined to the suffix *-ide*.

Example 1

Give the correct name for each of the following ionic compounds:

1. NaCl   3. $AlCl_3$ 5. Ba_3N_2
2. Li_2S 4. CaO   6. MgO

Answer

1. sodium chloride   3. aluminum chloride   5. barium nitride
2. lithium sulfide   4. calcium oxide   6. magnesium oxide

If the cation and anion exist in only one common charged form, there is no ambiguity between formula and name. Sodium chloride must be NaCl, and lithium sulfide must be Li_2S, so that the sum of positive and negative charges is zero. With many elements, such as the transition metals, several ions of different charge may exist. Fe^{2+}, Fe^{3+} and Cu^+, Cu^{2+} are a few common examples. Clearly, an ambiguity exists if we use the name iron for both Fe^{2+} and Fe^{3+} or copper for Cu^+ and Cu^{2+}. Two systems have been developed that avoid this problem: the stock system and the common nomenclature system.

In the stock system for naming an ion (the systematic name), a Roman numeral indicates the magnitude of charge of the cation. In the older common nomenclature, the suffix *-ous* indicates the lower of the ionic charges, and the suffix *-ic* indicates the higher ionic charge. Systematic names are preferred; they are easier and less ambiguous.

Example 2

Give the correct names for each of the following ionic compounds that contain transition metal ions in their structures.

                    1. $FeCl_3$ 2. $FeCl_2$

Work

When a transition metal is found in an ionic compound, we must first find the charge on the transition metal before we can name the compound.

1. $FeCl_3$
    a. First use the formula: we find that one Fe ion and three Cl ions are present.
       Fe + 3 Cl
    b. From the periodic table, we find that the Cl is in family VIIA. These elements always form a −1 ion.
    c. $Fe^{?+}$ + 3 Cl^-
       We now know that there is a total charge of −3 in the compound. Thus, the iron ion must be Fe^{3+} so it can balance the −3 charge.

Note: The total + charges must always equal the total − charges in any ionic compound!

Formula = $FeCl_3$ Ion-pair form = Fe^{3+} + 3 Cl^- Correct name: iron(III) chloride

Example 2 continued

2. $FeCl_2$
   a. $FeCl_2$ means $Fe^{?+}$ + 2 Cl^-
      Since there are a total of –2, then the $Fe^{?+}$ must be an Fe^{2+} to balance the charges:
      +2 balances –2
   b. Correct name: iron(II) chloride

Answer
1. iron(III) chloride   2. iron(II) chloride

    Ions consisting of only a single atom are said to be **monatomic.** In contrast, **polyatomic ions,** such as the hydroxide ion, OH^-, are composed of two or more atoms bonded together. The polyatomic ion has an overall positive or negative charge. Some common polyatomic ions are listed in Table 3.3 of the textbook. Your instructor may suggest that several of the most common polyatomic ions be committed to memory.
    It is equally important to write the correct formula when given the compound name. It is essential to be able to predict the charge of monatomic ions and the charge and formula of polyatomic ions. Remember, the relative number of positive and negative ions in the unit must result in a unit (compound) charge of zero.

Example 3
Name the following ionic compounds. Use Roman numerals when needed.

1. $Ca_3(PO_4)_2$ 3. $CsOH$ 5. Na_2SO_4
2. $Mg(C_2H_3O_2)_2$ 4. $FeSO_4$ 6. $Fe_2(CO_3)_3$

Answer
1. calcium phosphate
2. magnesium acetate
3. cesium hydroxide
4. $FeSO_4 = Fe^{2+} + SO_4^{2-}$; iron(II) sulfate
5. sodium sulfate
6. $Fe_2(CO_3)_3 = 2\ Fe + 3\ CO_3^{2-} = 2\ Fe^{3+} + 3\ CO_3^{2-}$; iron(III) carbonate

Example 4
Write the formulas of the following ionic compounds:
1. sodium chloride   3. iron(III) hydroxide
2. magnesium phosphate   4. ammonium carbonate

Answer
1. sodium chloride
   a. Na^+ + Cl^- IA family = +1 charge; VIIA = –1 charge
   b. formula = $NaCl$

Example 4 continued

Answer
2. magnesium phosphate
   a. $Mg^{2+} + PO_4^{3-}$ IIA = +2 charge; phosphate ion = –3
   b. need 3 Mg^{2+} to balance 2 PO_4^{3-}
      +6 balances –6
   c. formula = $Mg_3(PO_4)_2$

3. iron(III) hydroxide
   a. $Fe^{3+} + OH^-$ (Charges are not balanced.)
      need 3 OH^- to balance the +3 charge
      $Fe^{3+} + 3\ OH^-$
      +3 balances –3
   b. formula = $Fe(OH)_3$

4. ammonium carbonate
   a. $NH_4^+ + CO_3^{2-}$ (Charges are not balanced.)
   b. formula = $(NH_4)_2\ CO_3$

Covalent Compounds

Covalent compounds are formed by the reaction of nonmetals. Covalently bonded compounds are **molecules.** The existence of this compound unit is a major distinctive feature of covalently bonded substances. The convention used for naming covalent compounds is as follows:

1. The names of the elements are written in the order in which they appear in the formula.

2. A prefix indicating the number of each kind of atom found in the unit is placed before the name of the element.

3. If only one atom of a particular kind is present in the molecule, the prefix *mono-* is usually omitted.

4. The stem of the name of the last element is used with the suffix *-ide*.

Common names are often used. For example, H_2O is water, NH_3 is ammonia, C_2H_5OH (ethanol) is alcohol, and $C_6H_{12}O_6$ is glucose. It is useful to be able to correlate both systematic and common names with the corresponding molecular formula and vice versa.

Writing formulas of covalent compounds can only be done from memory when common names are used. You must remember that water is H_2O, ammonia is NH_3, and so forth. This is the major disadvantage of common names; however, they cannot be avoided.

Example 5

Name the following covalent compounds:

1. CO_2 3. SO_2 5. N_2O_4
2. CO 4. SO_3 6. N_2O_3

Work

A prefix must be used to indicate the number of atoms of each nonmetal in the covalent compound.

Answer

1. carbon dioxide
2. carbon monoxide (The prefix mono- is usually not used. However, carbon monoxide has become a common name for CO.)
3. sulfur dioxide
4. sulfur trioxide
5. dinitrogen tetraoxide
6. dinitrogen trioxide

3.3 Properties of Ionic and Covalent Compounds

Learning Goal 4:   **Predict differences in physical state, melting and boiling points, solid-state structure, and solution chemistry that result from differences in bonding.**

The differences in ionic and covalent bonding account for the different properties of ionic and covalent compounds. Covalently bonded molecules are discrete units, and they have less tendency to form an extended structure in the solid state. Ionic compounds do not have definable units, but form a **crystal lattice** composed of hundreds of billions of positive and negative ions in an extended three-dimensional network.

Major differences in the properties of these compounds are summarized below.

Physical State

All ionic compounds are solids at room temperature, while covalent compounds may be solids (sugar, silicon dioxide), liquids (water, ethyl alcohol), or gases (carbon monoxide, carbon dioxide).

Melting and Boiling Points

The **melting point** is the temperature at which a solid is converted to a liquid, and the **boiling point** is the temperature at which a liquid is converted to a gas. Considerable energy is needed to break apart an ionic crystal lattice and convert an ionic solid to a liquid or a gas. Therefore, the melting and boiling temperatures for ionic compounds are generally higher than those of covalent compounds, which interact less strongly in the solid state.

Structure of Compounds in the Solid State

Ionic solids are crystalline, characterized by a regular structure, whereas covalent solids may be either crystalline or amorphous (no regular structure).

Solutions of Ionic and Covalent Compounds

Many ionic solids dissolve in solvents, such as water. If soluble, an ionic solid will dissociate in solution to form positive and negative ions (ionization). Because these ions are capable of conducting electric current, these compounds are **electrolytes,** and the solution is termed an **electrolytic solution.** Covalent solids in solution are neutral and are **nonelectrolytes.** The solution is not an electrical conductor.

3.4 Drawing Lewis Structures of Molecules

Learning Goal 5:   Draw Lewis structures for covalent compounds.

A Strategy for Drawing Lewis Structures of Molecules

    Rule 1:   Use chemical symbols for the various elements to write the skeletal structure of the compound.

    Rule 2:   Determine the number of valence electrons associated with each atom; combine them to determine the total number of valence electrons in the compound.

    Rule 3:   Connect the central atom to each of the surrounding atoms using electron pairs.

    Rule 4:   If the octet rule is not satisfied for the central atom, move one or more electron pairs from the surrounding atoms to create double or triple bonds until all atoms have an octet.

    Rule 5:   After you are satisfied with the Lewis structure that you have constructed, perform a final electron count verifying that the total number of electrons and the number around each atom are correct.

Example 6
Draw the Lewis dot and line structures of carbon dioxide. The carbon is in the middle of two oxygen atoms. The carbon atom is the central atom.

Example 6 continued

Work
1. O C O (arrangement as in nature)
2. O = 6 valence e⁻
    C = 4 valence e⁻
    O = 6 valence e⁼
       16 total valence e⁻ in CO_2

3. Dot structure: :Ö::C::Ö:
    (All atoms must have 8 valence e⁻ in their final valence shell.)

4. Line structure: **O=C=O**

    This structure is also acceptable: ⏐Ō=C=Ō⏐

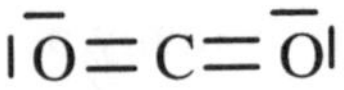

Example 7
Draw the Lewis dot and line structures of SO_3. Sulfur is the central atom in the compound.

Work
          O
1. O S O

2. 3 oxygen atoms =   18 valence e⁻
    1 sulfur atom =   6 valence e⁼
                    24 total valence e⁻ in SO_3
3. Dot structure (Any one of the following is correct.)

    :Ö: :Ö: :O:
    :Ö:S::Ö: or   :Ö::S:Ö: or   :Ö:S:Ö:

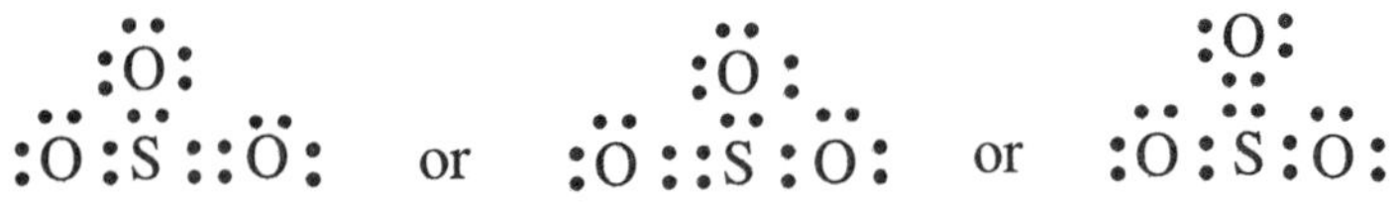

4. Line structure (Any one of the following is correct.)

      O   O   O
      ‖   ‖   ‖
    O−S=O   or   O−S=O   or   O−S=O

41

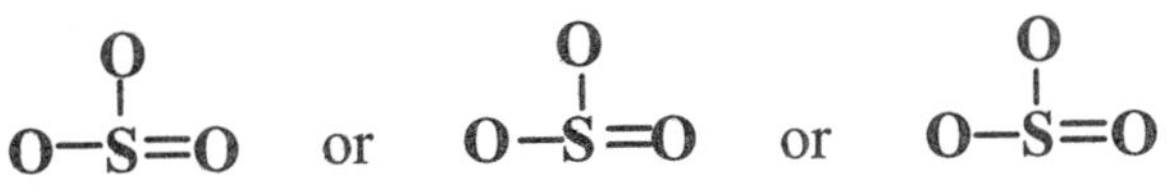

Multiple Bonds and Bond Energies

Learning Goal 6:	Describe the relationship between bond order, bond energy, and bond length.

The order of stability often parallels the bond order (the number of bonds between adjacent atoms). Stability is also related to the bond energy, where the **bond energy** is defined as the amount of energy in kilocalories required to break a bond holding two atoms together. The magnitude of the bond energy decreases in the order triple bond > double bond > single bond. The bond length decreases in the order single bond > double bond > triple bond.

Lewis Structures and Exceptions to the Octet Rule

Molecules with fewer than four and as many as five or six electron pairs around the central atom also exist. They are exceptions to the octet rule. In general, if one atom in a molecule (or polyatomic ion) must be an exception to the octet rule, it will be the central atom.

Example 8

Draw the Lewis structure for BF_3. Boron is the central atom in this molecule.

Work

Boron has 3 valence electrons, and each fluorine has 7 valence electrons, for a total of: $3 + (3 \times 7) = 3 + 21 = 24$ valence electrons, which are not enough for each atom to have a full octet. One atom—the central atom, boron—must be an exception to the octet rule.

Note: This is why it's important to count valence electrons and then to double-check your total. Drawing a full octet around each atom may *look* correct, but for BF_3, that would be incorrect.

Answer

Lewis Structures and Molecular Geometry; VSEPR Theory

Learning Goal 7:	Predict the geometry of molecules using the octet rule and Lewis structures.

The shape of a molecule contributes to its properties and reactivity. We may predict the shapes of various molecules by inspecting their Lewis structures for the orientation of their electron pairs. The covalent bond, in which bonding electrons are localized between the atoms, is directional: the bond has a specific orientation in space among the bonded atoms. Electrostatic forces in ionic bonds, in contrast, are nondirectional: they have no specific orientation in space. As a result, the specific orientation of electron pairs in covalent molecules imparts a characteristic shape to the molecules.

Electron pairs around the central atom of the molecule arrange themselves to minimize repulsion; the electron pairs are as far as possible from each other. This is termed the valence shell electron pair repulsion theory (**VSEPR Theory**). Two electron pairs around the central atom lead to a **linear** arrangement of the attached atoms; three indicate a **trigonal planar** arrangement; and four result in a **tetrahedral** geometry. Molecules with five and six electron pairs also exist. They may have structures that are trigonal bipyramidal or octahedral.

Compounds containing the same central atom will have structures with similar geometries. This is an approximation; it is not always true, but useful in writing reasonable, geometrically accurate structures for a large number of compounds. Consider the following:

BeH_2 (linear)
   BeI_2, $BeCl_2$, $Be(OH)_2$ are also linear.

BF_3 (trigonal planar)
   BCl_3, BBr_3, BI_3 are also trigonal planar.

CH_4 (tetrahedral)
   CF_4, CCl_4, CBr_4 are also tetrahedral.

Furthermore, the periodic similarity of group members is useful in predictions involving bonding. Consider the following:

H_2O (angular)
   H_2S, H_2Se are also angular because O, S, and Se are in the same group, VI, 6 valence electrons.

Lewis Structures and Polarity

Learning Goal 8:	Understand the role that molecular geometry plays in determining the solubility and melting and boiling points of compounds.

A molecule is **polar** if its centers of positive and negative charges do not coincide. On the other hand, the positive and negative centers are coincident in nonpolar molecules. No dipole exists, and the dipole moment is equal to zero.

This property may also be described by considering the equality of electron sharing between the atoms being bonded. The degree of electron sharing can be described using the property of electronegativity (which was introduced in section 3.1). The atoms of H_2 are identical; their electronegativity is the same. The electrons remain, on average, at the center of the molecule, and the bond is nonpolar. The hydrogen molecule itself is nonpolar as well, with a dipole moment of zero. O_2, N_2, Cl_2, and F_2 are also nonpolar molecules with nonpolar bonds.

Bonds formed from atoms of unequal electronegativity are polar. The more electronegative end of the bond is designated δ^- (partial negative) and the less electronegative end δ^+ (partial positive). A polar covalent molecule is characterized by **polar covalent bonding.** This implies that the electrons are shared unequally.

A molecule containing all nonpolar bonds must itself be nonpolar. A molecule containing polar bonds may be either polar or nonpolar depending on the relative orientation of the bonds.

3.5 Properties Based on Electronic Structure and Molecular Geometry

Learning Goal 9:   **Use the principles of VSEPR theory and molecular geometry to predict relative melting points, boiling points, and solubilities of compounds.**

The effects of molecular polarity result from the strength of attractive forces among individual molecules of a compound. These attractions are **intermolecular forces.** The student should not confuse intermolecular forces with intramolecular forces. **Intramolecular forces** are the attractive forces within molecules. It is the intermolecular forces that determine such properties as solubility, melting point, and boiling point.

Solubility

Solubility is defined as the maximum amount of solute that dissolves in a given amount of solvent. (The solute is present in lesser quantity, and the solvent is present in the greater amount.) Polar molecules are most soluble in polar solvents, and nonpolar molecules are most soluble in nonpolar solvents. This is the rule of "like dissolves like."

Boiling Points of Liquids and Melting Points of Solids

Boiling is the conversion of a liquid to a vapor. Such a process requires energy, overcoming the intermolecular attractive forces in the liquid. The energy required is related to the magnitude of the boiling point and depends upon the strength of the intermolecular attractive forces in the liquid. These attractive forces are directly related to polarity. Molecular size is also an important consideration. The larger or heavier the molecule, and the more polar the molecule, the more difficult it becomes to convert to the gas phase.

The melting points of solids may be described on the basis of intermolecular forces as well. As a general rule, polar compounds have strong attractive (intermolecular) force, and their boiling and melting points tend to be higher than for nonpolar substances.

Glossary of Key Terms in Chapter Three

angular structure (3.4) planar molecule with bond angles less than 180°.
boiling point (3.3) the temperature at which the vapor pressure of a liquid is equal to the atmospheric pressure.
bond energy (3.4) the amount of energy necessary to break a chemical bond.
chemical bond (3.1) the attractive force holding two atomic nuclei together in a chemical compound.
covalent bonding (3.1) a pair of electrons shared between two atoms.
crystal lattice (3.1) a unit of a solid characterized by a regular arrangement of components.
dissociation (3.3) production of positive and negative ions when an ionic compound dissolves in water.
double bond (3.4) a bond in which two pairs of electrons are shared by two atoms.
electrolyte (3.3) a material that dissolves in water to produce a solution that conducts an electrical current.
electrolytic solution (3.3) a solvent composed of an electrolytic solute dissolved in water.
electronegativity (3.1) a measure of the tendency of an atom in a molecule to attract shared electrons.
formula (3.2) the representation of the fundamental compound unit using chemical symbols and numerical subscripts.
intermolecular force (3.5) attractive force that occurs between molecules.
intramolecular force (3.5) attractive force that occurs within molecules.
ionic bonding (3.1) an electrostatic attractive force between ions resulting from electron transfer.
ion pair (3.1) the simplest formula unit for an ionic compound.
Lewis symbol (3.1) representation of an atom or ion using the atomic symbol (for the nucleus and core electrons) and dots to represent valence electrons.
linear structure (3.4) the structure of a molecule in which the bond angle(s) about the central atom(s) is (are) 180°.
lone pair (3.4) an electron pair that is not involved in bonding.
melting point (3.3) the temperature at which a solid converts to a liquid.
molecule (3.2) a collection of two or more atoms held together by covalent bonds.
monatomic ion (3.2) an ion formed by electron gain or loss from a single atom.
nomenclature (3.2) a system for naming chemical compounds.
nonelectrolyte (3.3) a substance that, when dissolved in water, produces a solution that does not conduct an electrical current.
polar covalent bonding (3.4) a covalent bond in which the electrons are not equally shared.
polar covalent molecule (3.4) a molecule that has permanent electric dipole moment resulting from an unsymmetrical electron distribution; a dipolar molecule.
polyatomic ion (3.2) an ion containing a number of atoms.
single bond (3.4) a bond in which one pair of electrons is shared by two atoms.
solubility (3.5) the amount of a substance that will dissolve in a given volume of solvent at a specified temperature.
tetrahedral structure (3.4) a molecule consisting of four groups attached to a central atom which occupy the four corners of an imagined tetrahedron.
trigonal pyramidal molecule (3.4) a molecular structure resulting from three groups bonded to a central atom; all groups are equidistant from the central atom.
triple bond (3.4) a bond in which three pairs of electrons are shared by two atoms.
valence shell electron pair repulsion theory (3.4) a model that predicts molecular geometry using the premise that electron pairs will arrange themselves as far apart as possible, to minimize electron repulsion.

Self Test for Chapter Three

1. The transfer of electrons from one atom to another forms which type of chemical bond?

2. The sharing of electrons between two atoms forms which type of chemical bond?

3. What particles are found in ionic compounds?

4. How many sodium ions are in Na_2S?

5. On what side of the periodic table are elements found that usually form positive ions?

6. On what side of the periodic table are elements found that usually form negative ions?

7. The name of an ionic compound always contains the name of which ion first?

8. Provide the best name of Na_2O.

9. Provide the best name of Li_2S.

10. Provide the best name of $AlBr_3$.

11. What is the common name of the iron(II) ion?

12. What is the best name of Fe^{3+}?

13. What does the ending "ic" on the common names of ions mean?

14. Give the common name of $FeSO_4$.

15. Give the systematic name of CuO.

16. Write the formula of sodium sulfate.

17. How many bonding electrons are in H_2O?

18. How many nonbonding electrons are in CO_2?

19. Which end of HCl is slightly more negative?

20. Why is HF a polar molecule?

21. When H_2O is compared to C_2H_4, which has the higher boiling point?

22. When Cl_2 is compared to ICl, which has the higher boiling point?

23. Draw the dot and line structures of ammonia.

24. Draw the dot and line structures of water.

25. Which electrons of an atom are involved in chemical bond formation?

26. What type of bond is formed by the sharing of valence electrons?

27. Write the formula of the ions that result when sodium atoms react with chlorine atoms.

28. Write the formula of the ions that result when calcium atoms react with sulfur atoms.

29. What process results in ion formation in ionic reactions?

30. How many electrons are jointly shared in the hydrogen molecule?

31. How many electrons are shared in H_2O?

Vocabulary Quiz for Chapter Three

1. A(n) _____ force is a attraction between molecules.

2. A(n) _____ is a unit of solid characterized by regular arrangement of components.

3. A material that dissolves in water to produce a solution that conducts an electrical current is a(n)_____.

4. _____ are attractive forces that occur between molecules.

5. A(n) _____ is a diagram of an atom, ion, or molecule showing valence electron arrangement.

6. A(n) _____ is a electron pair that is not involved in bonding.

7. A covalent bond in which the electrons are not equally shared is referred to as a(n) _____ bond.

8. A molecule consisting of four groups attached to a central atom usually has a(n) _____ structure.

9. A bond in which three pairs of electrons are shared by two atoms is called a(n) _____.

4 Calculations and the Chemical Equation

Learning Goals

 1 Know the relationship between the mole and Avogadro's number and the usefulness of these quantities.
 2 Perform calculations using Avogadro's number and the mole.
 3 Write chemical formulas for common inorganic substances.
 4 Calculate the formula weight and molar mass of a compound.
 5 Know the major function served by the chemical equation, the basis for chemical calculations.
 6 Classify chemical reactions by type: combination, decomposition, or replacement.
 7 Recognize the various classes of chemical reactions: precipitation, reactions with oxygen, acid-base, and oxidation-reduction.
 8 Balance chemical equations given the identity of products and reactants.
 9 Calculate the number of moles or grams of product resulting from a given number of moles or grams of reactants or the number of moles or grams of reactant needed to produce a certain number of moles or grams of product.
10 Calculate theoretical and percent yield.

Introduction

You should be able to predict the quantity of a product produced from the reaction of a given amount of reactant. You should also be able to calculate how much of a reactant, or group of reactants, would be required to produce a desired amount of product.

These calculations are based on the **chemical equation.** The chemical equation provides all of the needed information: the combining ratio of elements or compounds (reactants) that are necessary to produce a particular product or products as well as the relationship between the amounts of reactants and products formed.

4.1 The Mole Concept and Atoms

The Mole and Avogadro's Number

Learning Goal 1:	Know the relationship between the mole and Avogadro's number and the usefulness of these quantities.

Atomic masses have been experimentally determined for each of the elements. Their unit of measurement is the **atomic mass unit,** abbreviated **amu.**

$$1 \text{ amu} = 1.661 \times 10^{-24} \text{ grams}$$

A more practical unit for defining a "collection" of atoms is the **mole.**

$$1 \text{ mole of atoms} = 6.022 \times 10^{23} \text{ atoms of an element}$$

This number is **Avogadro's number.**

The mole and the amu are related. The atomic mass of a given element corresponds to the average mass of a single atom in amu *and* the mass of a mole of atoms in grams. For example, the average mass of one copper atom is 63.54 amu, while the mass of one *mole* of copper atoms is 63.54 grams. One mole of atoms of *any element* contains the same number, Avogadro's number, of atoms.

Calculating Atoms, Moles, and Mass

Learning Goal 2:   Perform calculations using Avogadro's number and the mole.

Calculations based on the chemical equation (Section 4.6 of the textbook) utilize the relationship of the number of atoms, number of moles, and mass in grams. Conversion factors are used to proceed from the information *provided* in the problem to the information *requested* by the problem.

Before you begin any calculations, map out a pattern for the required conversion. You may be given the number of grams and need the number of atoms that corresponds to that mass. Begin by tracing a path to the answer:

$$\text{grams} \quad \xrightarrow{\text{Step 1}} \quad \text{moles} \quad \xrightarrow{\text{Step 2}} \quad \text{atoms}$$

Two transformations, or conversions are required:

Step 1:   grams to moles

Step 2:   moles to atoms

The conversion between the three principal measures of quantity of matter, the number of grams (mass), the number of moles, and the number of individual particles (atoms, ions, or molecules) is depicted in Figure 5.2 of the textbook.

Example 1
Calculate the mass of one mole of iron atoms.

Work
From the periodic table, we find that the average mass of one iron atom is 55.85 atomic mass units. Thus, one mole of iron atoms has a mass of 55.85 grams.

4.2 The Chemical Formula, Formula Weight, and Molar Mass

The Chemical Formula

Learning Goal 3:   Write chemical formulas for common inorganic substances.

Chemical compounds are represented by their **chemical formulas,** a combination of symbols of the various elements that make up the compounds. The chemical formula is based upon the **formula unit,** the smallest collection of atoms that provides the following information:

1. the identity of the atoms present in the compound and
2. the relative numbers of each type of atom.

The term *formula* may be used in reference to any ionic or covalent compound. In Chapter 3 compounds were classified as nonpolar covalent, polar covalent, or ionic, depending upon the type of bonding that holds the unit together.

A covalent compound is composed of discrete molecules, and the term *molecular formula* is most appropriate. Ionic compounds, on the other hand, are composed of ion pairs, not molecules, and *cannot* be described as having a molecular formula.

Formula Weight and Molar Mass

Learning Goal 4:   Calculate the formula weight and molar mass of a compound.

A mole of a compound is based upon the formula mass or **formula weight.** The formula weight is calculated by adding the masses of all the atoms of which the unit is composed. To calculate the formula weight, the formula unit must be known.

Example 2

Calculate the mass of one mole of carbon dioxide (the formula weight of carbon dioxide).

Work

From the periodic table, we find that one atom of carbon has a mass of 12.01 amu, and one atom of oxygen has a mass of 16.00 amu.

CO_2 = 1 carbon atom + 2 oxygen atoms
CO_2 = 1 C + 2 O
CO_2 = 1(12.01 amu) + 2(16.00 amu) = 44.01 amu

Answer

Thus, one molecule of CO_2 has a mass of 44.01 amu, and one mole of carbon dioxide molecules weighs 44.01 grams.

Example 3
How many grams of carbon dioxide are contained in 10.00 moles of carbon dioxide?

Work
Since 1 mole of CO_2 = 44.01 grams, we can use the factor-label method to convert grams into mole units.

$$10.00 \text{ mol } CO_2 \times \frac{44.01 \text{ g } CO_2}{1 \text{ mol } CO_2} = 4.401 \times 10^2 \text{ g } CO_2$$

Answer
10.00 moles of CO_2 = 4.401×10^2 grams of CO_2

Example 4
Calculate the formula weight of water.

Work
Water (H_2O) is composed of two hydrogen atoms plus one oxygen atom. From the periodic table, we find that O = 16.00 amu, and H = 1.008 amu.

H_2O = 2 hydrogen atoms + 1 oxygen atom
H_2O = 2 H + 1 O
H_2O = 2(1.008 amu) + 1(16.00 amu) = 18.01 amu

Answer
The formula weight of water equals 18.01 amu. Notice that amu are used for formula weights. A mole of water, however, is equal to 18.01 grams.

4.3 Chemical Equations and the Information They Convey

A Recipe for Chemical Change

Learning Goal 5: Know the major function served by the chemical equation, the basis for chemical calculations.

The chemical equation describes all of the substances that react to produce the product(s). The chemical equation also describes the physical state of the reactants and products as solid, liquid, or vapor. It tells us whether the reaction occurs, and identifies the solvent and experimental conditions. Most important, the relative number of moles of reactants and products appears in the equation. According to the law of conservation of mass, matter cannot be either gained or lost in the process of a chemical reaction. The law of conservation of mass states that we must have a balanced chemical equation.

51

Features of a Chemical Equation

1. The identity of products and reactants must be specified.
2. Reactants are written to the left of the reaction arrow and products to the right.
3. The physical state of reactants and products is shown in parentheses.
4. The symbol Δ over the reaction arrow means that heat energy is necessary for the reaction to occur.
5. The equation must be balanced.

The Experimental Basis of a Chemical Equation

The chemical equation represents a real chemical transformation. Evidence for the reaction may be based on observations such as the formation of a gas, a solid precipitate, the evolution of heat, or a color change in the solution.

Writing Chemical Reactions

Learning Goal 6:	Classify chemical reactions by type: combination, decomposition, or replacement.

Spontaneous chemical reactions occur for a variety of reasons, linked by the tendency to achieve the lowest (most stable) electronic energy state. Strong electrolytes will react to form weak (less dissociated) electrolytes, if possible. Reactions forming gaseous products are favored. Formation of insoluble solid products favors a chemical reaction.

Chemical reactions involve the combination of reactants to produce products, the decomposition of reactant(s) into products, or the replacement of one or more elements in a compound to yield products.

Combination reactions: Combination reactions involve the joining of two or more atoms or compounds, producing a product of different composition.

Decomposition reactions: Decomposition reactions produce two or more products from a single reactant.

Replacement reactions: Replacement reactions are subcategorized as either single-replacement or double-replacement. Single-replacement reactions occur between atoms and compounds; the atom replaces another in the compound, producing a new compound and the replaced atom. Double-replacement reactions, on the other hand, involve two compounds undergoing a "change of partners."

Types of Chemical Reactions

Learning Goal 7:	Recognize the various classes of chemical reactions: precipitation, reactions with oxygen, acid-base, and oxidation-reduction.

Precipitation reactions: Precipitation reactions involve the conversion of soluble reactants into one or more insoluble products. Information contained in solubility tables enables prediction of possible insoluble products.

Reactions with oxygen: Reactions involving combination of metals or nonmetals with oxygen are generally exothermic; they are often used specifically for their heat-generating properties. Examples include the reaction of oxygen with carbon/hydrogen-containing compounds (natural gas, gasoline, and oil).

Acid-base and oxidation-reduction reactions: Another approach to the classification of chemical reactions is based upon a consideration of charge transfer. Acid-base reactions involve the transfer of a hydrogen ion, H^+, from one reactant to another. Another important reaction type, oxidation-reduction, takes place because of the transfer of one or more electrons from one reactant to another.

4.4 Balancing Chemical Equations

Learning Goal 8:	Balance chemical equations given the identity of products and reactants.

A chemical equation tells us both the identity and state of the reactants and products. **Reactants,** or starting materials, are all substances that undergo change in a chemical reaction, while **products** are substances produced by a chemical reaction.

The chemical equation also shows the *molar quantity* of reactants needed to produce a certain *molar quantity* of products.

The number of moles of each product and reactant is indicated by placing a whole number *coefficient* before the formula of each substance in the chemical equation.

Many equations are balanced by trial and error. The following steps provide a method for correctly balancing a chemical equation:

Step 1: Count the number of atoms of each element on both product and reactant side.

Step 2: Determine which atoms are not balanced.

Step 3: Balance one atom at a time, using coefficients.

Step 4: After you believe that you have successfully balanced the equation, check, as in Step 1, to be certain that mass conservation has been achieved.

Example 5

Write the balanced reaction of an aqueous solution of hydrochloric acid (HCl) reacting with solid calcium to produce calcium chloride plus hydrogen gas.

Work

1. First write the correct symbols and formulas needed for the reactants and products:

Example 5 continued

$$HCl + Ca \rightarrow CaCl_2 + H_2$$

2. Next, count the number of atoms of each element on the right and left side of the reaction arrow:

$$HCl + Ca \rightarrow CaCl_2 + H_2$$

H	Cl	Ca	‖	H	Cl	Ca
1 atom	1 atom	1 atom	‖	2 atoms	2 atoms	1 atom

Notice that the numbers of atoms on each side are not equal; the equation is **not** balanced.

3. After the correct formulas of all reactants and products are given, you can now use coefficients alone to balance the equation.

4. $2\,HCl + Ca \rightarrow CaCl_2 + H_2$

H	Cl	Ca	‖	H	Cl	Ca
2	2	1	‖	2	2	1

The equation is now balanced.

Answer
$2\,HCl + Ca \rightarrow CaCl_2 + H_2$

Note that the equation becomes more complete when the states of each reactant and product are added:

$$2\,HCl(aq) + Ca(s) \rightarrow CaCl_2(aq) + H_2(g)$$

Example 6
Balance the following reaction: $FeCl_3(aq) + Cu(s) \rightarrow CuCl_2(aq) + Fe(s)$

Work
1. First, find the number of atoms of each element:

$$FeCl_3 + Cu \rightarrow CuCl_2 + Fe$$

Fe	Cl	Cu	‖	Fe	Cl	Cu
1	3	1	‖	1	2	1

The equation is **not** balanced.

54

Example 6 continued

2. Next, use only coefficients to balance the equation:

$$2\,FeCl_3 + 3\,Cu \rightarrow 3\,CuCl_2 + 2\,Fe$$

Fe	Cl	Cu	‖	Fe	Cl	Cu
2	6	3	‖	2	6	3

The equation is now balanced.

Answer

$$2\,FeCl_3(aq) + 3\,Cu(s) \rightarrow 3\,CuCl_2(aq) + 2\,Fe(s)$$

4.5 Calculations Using a Chemical Equation

> **Learning Goal 9:**   Calculate the number of moles or grams of product resulting from a given number of moles or grams of reactants or the number of moles or grams of reactant needed to produce a certain number of moles or grams of product.

General Principles

In doing calculations involving chemical reactions, we apply the following rules:

1. The basis for the calculations is a balanced equation.

2. The calculations are performed in terms of moles.

3. The conservation of mass must be obeyed.

The mole is the basis for calculations. However, masses are generally measured in grams (or kilograms). A facility for interconversion of moles and grams is therefore necessary in calculations involving chemical reactions.

Example 7

In the following balanced equation, how many moles of each element or compound are involved?

$$2\,C_8H_{18} + 25\,O_2 \rightarrow 16\,CO_2 + 18\,H_2O$$

Work

The coefficients in a balanced equation tell us the number of moles of each reactant and product.

Example 7 continued

Answer
Two moles of octane (C_8H_{18} is found in gasoline) react with 25 moles of oxygen (always O_2 in nature) to form 16 moles of carbon dioxide plus 18 moles of water.

Use of Conversion Factors

1. *Conversion between moles and grams.* Conversion from moles to grams and vice versa requires only the formula weight of the compound of interest.

2. *Conversion of moles of reactants and products.* Conversion factors, based on the chemical equation, permit us to perform a variety of calculations.

A general problem-solving strategy is summarized in Figure 4.2 (p. 112) of the textbook.

Example 8
The reaction of propane (C_3H_8 is found in bottled gas) with oxygen produces carbon dioxide and water. The balanced equation is given below:
$$C_3H_8 \ + \ 5\,O_2 \ \rightarrow \ 3\,CO_2 \ + \ 4\,H_2O$$
Calculate the number of grams of oxygen (O_2, as in nature) needed to completely react with 1 mole of propane.

Work
1. From the balanced equation, we see that 1 mole of propane reacts with 5 moles of O_2 to form 3 moles of CO_2 plus 4 moles of water.
2. One mole of O_2 equals 32.00 grams.
3. Since 5 moles of O_2 are needed to react with 1 mole of propane, then 160.0 grams of oxygen (O_2) are needed.

$$1 \text{ mol } C_3H_8 \text{ x } \frac{5 \text{ mol } O_2}{1 \text{ mol } C_3H_8} \text{ x } \frac{32.00 \text{ g } O_2}{1 \text{ mol } O_2} = 160.0 \text{ g } O_2$$

Answer
One mole of propane will react with 160.0 grams of oxygen (O_2).

Learning Goal 10:   Calculate theoretical and percent yield.

Theoretical and Percent Yield

The **theoretical yield** is the *maximum* amount of product that can be produced (in an ideal world). In the real world it is difficult to produce the amount calculated as the theoretical yield.

A percent yield, the ratio of the actual and theoretical yields multiplied by 100%, is often used to show the relationship between predicted and experimental quantities. Thus

$$\% \text{ yield} = \frac{\text{actual yield}}{\text{theoretical yield}} \times 100\%$$

The percent yield for most chemical reactions is less than 100%. This is due, in part, to experimental limitations in isolation, transfer, and recovery of the product. Many reactants do not completely convert to products. We will study these reactions in detail in Chapter 8.

Glossary of Key Terms in Chapter Four

acid-base reaction (4.3) reactions that involve the transfer of a hydrogen ion (H^+) from one reactant to another.

atomic mass unit (4.1) the unit of measure of atomic mass, one-twelfth of the mass of the ^{12}C atom, equivalent to 1.6606×10^{-24} gram.

Avogadro's number (4.1) 6.022×10^{23} particles of matter contained in one mole of a substance.

chemical equation (4.3) a record of chemical change, showing the conversion of reactants to products.

chemical formula (4.2) the representation of a compound or ion in which elemental symbols represent types of atoms and subscripts show the relative numbers of atoms.

combination reaction (4.3) a reaction in which two substances join to form another substance.

corrosion (4.3) the deterioration of metals caused by an oxidation-reduction process.

decomposition reaction (4.3) the breakdown of a substance into two or more substances.

double-replacement reaction (4.3) a chemical change in which cations and anions "exchange partners."

formula unit (4.2) the smallest collection of atoms from which the formula of a compound can be established.

formula weight (4.2) the mass of a formula unit of a compound relative to a standard (carbon-12).

hydrate (4.2) any substance that has water molecules incorporated in its structure.

law of conservation of mass (4.3) a law stating that, in chemical change, matter cannot be created or destroyed.

molar mass (4.1, 4.2) the mass in grams of one mole of a substance.

mole (4.1) the amount of substance containing Avogadro's number of particles.

oxidation (4.3) a loss of electrons; in organic compounds it may be recognized as a loss of hydrogen atoms or the gain of oxygen atoms.

oxidizing agent (4.3) a substance that oxidizes, or removes electrons from, another substance; the oxidizing agent is reduced in the process.

percent yield (4.5) the ratio of the actual and theoretical yields of a chemical reaction multiplied by 100%.

product (4.3) the chemical species that results from a chemical reaction and that appears on the right side of the chemical equation.

reactant (4.3) starting material for a chemical reaction that appears on the left side of the chemical equation.

reducing agent (4.3) a substance that reduces, or donates electrons to, another substance; the reducing agent is oxidized in the process.

reduction (4.3) the gain of electrons; in organic compounds it may be recognized by a gain of hydrogen atoms or a loss of oxygen atoms.

single-replacement reaction (4.3) also called substitution reaction, one in which one atom in a molecule is displaced by another.

theoretical yield (4.5) the maximum amount of product that can be produced from a given amount of reactant.

Self Test for Chapter Four

1. 6.02×10^{23} molecules of a covalent compound equal how many moles of that compound?

2. One mole of hydrogen molecules weighs how many grams? Give answer to the nearest tenth of a gram.

3. How many iron atoms are present in 1 mole of iron?

4. How many grams of sulfur are found in 0.150 mole of sulfur? Use S = 32.06 amu.

5. Write the formula of the smallest unit of nitrogen as it is normally found in nature.

6. The term *molecular formula* is used for what type of compound?

7. What units are used for the formula weights of atoms and individual molecules?

8. What is the formula weight of one molecule of H_2O (nearest hundredth)?

    Use H = 1.01; O = 16.00 amu.

9. The formula weight of a covalent compound may also be called what weight?

10. What term means the starting materials in a chemical reaction?

11. How many moles of hydrogen are needed to react with oxygen to form 2 moles of water?

    $$2 H_2 + O_2 \rightarrow 2 H_2O$$

12. How many moles of oxygen are needed to react with hydrogen to form 1 mole of water?

    $$2 H_2 + O_2 \rightarrow 2 H_2O$$

13. The number of moles of reactants and products in a balanced chemical equation is given by what kind of numbers?

14. What number will be found in front of HCl when the following equation is balanced?

    $$Ca + HCl \rightarrow CaCl_2 + H_2$$

15. What number will be found in front of H_2O when the following equation is balanced?

    $$Mg(OH)_2 + HCl \rightarrow MgCl_2 + H_2O$$

16. What number will be found in front of Cl_2 when the following equation is balanced?

    $$Na + Cl_2 \rightarrow NaCl$$

17. How many grams of sodium hydroxide will react with 73.00 grams of hydrochloric acid?

    Use NaOH = 40.00; HCl = 36.45 amu.

    $$NaOH + HCl \rightarrow NaCl + H_2O$$

18. Calculate the number of grams of oxygen that must react with 46.85 grams of C_3H_8 to produce carbon dioxide and water.

Use C = 12.01, H = 1.01, and O = 16.00 amu.

19. Iron reacts with oxygen to form iron (III) oxide (Fe_2O_3). How many grams of product will be formed from 5.00 grams of Fe?

Use Fe = 55.85, and 0 = 16.00 amu.

20. How many moles of sulfur are found in 1.81×10^{24} atoms of sulfur?

Use S = 32.06 amu.

21. What is the name given to the number of atoms in a mole of iron?

22. Write the formula of the smallest unit of oxygen as it is normally found in nature.

Vocabulary Quiz for Chapter Four

1. _____ is 6.022 x 10^{23} particles of matter contained in one mole of a substance.

2. A record of chemical change showing the conversion of reactants to products is called a(n) _____.

3. A(n) _____ is the representation of a compound in which elemental symbols represent types of atoms and subscripts show the relative numbers of atoms.

4. The smallest collection of atoms from which the formula of a compound can be established is called _____.

5. The mass of a formula unit of a compound relative to a standard (carbon-12) is called _____.

6. In chemical change, "matter cannot be created or destroyed" is a statement of the _____.

7. A(n) _____ is the amount of substance containing Avogadro's number of particles.

8. The _____ is the mass of a molecule relative to a standard (carbon-12).

9. _____ result from a chemical reaction and appear on the right side of a chemical equation.

10. _____ are starting materials for a chemical reaction, appearing on the left side of a chemical equation.

5 Energy, Rate, and Equilibrium

Learning Goals

1 Correlate the terms *endothermic* and *exothermic* with heat flow between a system and its surroundings.
2 State the meaning of the terms *enthalpy* and *entropy* and know their implications.
3 Describe experiments that yield thermochemical information and calculate fuel values based on experimental data.
4 Describe the concept of reaction rate and the role of kinetics in chemical change.
5 Describe the importance of *activation energy* and the *activated complex* in determining reaction rate.
6 Predict the way reactant structure, concentration, temperature, and catalysis affect the rate of a chemical reaction.
7 Recognize and describe equilibrium situations.
8 Use LeChatelier's principle to predict changes in equilibrium position.

Introduction

Three concepts play an important role in determining the extent and speed of a chemical reaction:

1. Thermodynamics deals with energy changes associated with chemical reactions.
2. Kinetics concerns itself with the rate or speed of a chemical reaction.
3. Equilibrium considerations determine the completeness of a reaction.

A reaction may be thermodynamically favored but very slow; conversely, a fast reaction may result in very little product being formed. Thus, each is considered independently.

In a reaction that does not go to completion, a system is at equilibrium when the rates of the forward and reverse reactions are equal.

The concepts of energy, rate, and equilibrium can be applied to changes in state as well as to chemical reactions. The behavior of the states of matter (gas, liquid, and solid), their properties, and their interconversion are considered in this chapter.

5.1 Energy, Work, and Heat

Energy: The Basics

Energy, the ability to do work, may be categorized as either **kinetic energy,** the energy of motion, or **potential energy,** the energy of position. Kinetic energy may be considered energy in process; potential energy is stored energy. All energy is either kinetic or potential.

Another useful way of classifying energy is by form. The principle forms of energy include light, heat, electrical, mechanical, and chemical energy. All of these forms of energy share the following set of characteristics:

- In chemical reactions, energy cannot be created or destroyed.

- Energy may be converted from one form to another.

- Conversion of energy from one form to another always occurs with less than 100% efficiency. Energy is not lost (remember, energy cannot be destroyed) but rather, is not useful. We buy gasoline to move our car from place to place; however, much of the energy stored in the gasoline is released as heat.

- All chemical reactions involve either a "gain" or a "loss" of energy.

Energy absorbed or liberated in chemical reactions is usually in the form of heat energy. Heat energy may be represented in units of *calories* or *joules*; their relationship is

$$1 \text{ calorie (cal)} = 4.18 \text{ joules (J)}$$

One calorie is defined as the amount of heat energy required to increase the temperature of 1 gram of water 1°C.

Heat energy measurement is a quantitative measure of heat content. It is an extensive property, dependent upon the quantity of material. Temperature, as we have mentioned, is an intensive property, independent of quantity.

Not all substances have the same capacity for holding heat; 1 gram of iron and 1 gram of water do *not* contain the same amount of heat energy even if they are at the same temperature. The **specific heat,** the quantity of heat required to raise the temperature of one gram of a substance one degree Celsius, is a convenient measure of a substance's capacity for holding heat.

Thermodynamics

Thermodynamics is the study of energy, work, and heat. Thermodynamics may be applied to chemical change or physical change. There exist three basic laws of thermodynamics; only the first two are of concern in Chapter 5.

The Chemical Reaction and Energy

Every chemical reaction involves a change in energy. The first law of thermodynamics, also known as the law of conservation of energy, states that energy cannot be created or destroyed in the course of the reaction. It may only be converted from one form to another or transferred from one component of the system to another.

Our current view of events at the molecular level is consistent with the first law. When substances react:

- molecules and atoms in a reaction mixture are in constant, random motion;
- these molecules and atoms frequently collide with each other;
- only some collisions, those with sufficient energy, will break bonds in molecules; and
- when reactant bonds are broken, new bonds may be formed and products result.

Exothermic and Endothermic Reactions

Learning Goal 1:	Correlate the terms *endothermic* and *exothermic* with heat flow between a system and its surroundings.

A reaction that absorbs energy (exhibits a net gain in energy) is termed **endothermic.** The products of the reaction possess more energy than the reactants. On the other hand, a reaction that releases energy (exhibits a net loss of energy) is termed **exothermic.** The products of the reaction possess less energy than the reactants.

Figure 5.1 in the textbook graphically represents energy changes in endothermic and exothermic reactions.

The energy released in exothermic reactions is often in the form of **heat energy.** It may also take the form of light or even electricity. A battery is a common example of a chemical reaction producing electrical energy. The energy absorbed during chemical change is stored as **chemical energy.**

Enthalpy

Learning Goal 2:	State the meaning of the terms *enthalpy* and *entropy* and know their implications.

Enthalpy is the term used to represent heat energy and is symbolized by H°. In chemical change, we are primarily interested in the heat gained or released, and that is the enthalpy change, ΔH°. For endothermic reactions, the enthalpy change is positive; $\Delta H^\circ = +$. For exothermic reactions, the enthalpy change is negative; $\Delta H^\circ = -$. The enthalpy change is most often reported in units of joules (or kilojoules) or calories (or kilocalories).

Spontaneous and Nonspontaneous Reactions

Spontaneous reactions are just that: they occur without any external energy input. Nonspontaneous reactions must be persuaded; they need an input of energy.

It seems that all exothermic reactions should be spontaneous. After all, an external supply of energy does not appear to be necessary; in fact, energy is a product of the reaction. It also seems that all endothermic reactions should be nonspontaneous: energy is a reactant that we must provide. However, these hypotheses are not supported by experimentation.

Experimental measurement has shown that most, but not all, exothermic reactions are spontaneous; likewise, most, but not all, endothermic reactions are not spontaneous. There must be some factor in addition to enthalpy that will help us to explain the less-obvious cases of nonspontaneous exothermic reactions and spontaneous endothermic reactions. This other factor is entropy.

Entropy

The second law of thermodynamics states that a system and its surroundings spontaneously tend toward increasing disorder (randomness). A measure of the randomness of a chemical system is referred to as **entropy.** A random, or disordered system is characterized by high entropy; a well-ordered system is said to have low entropy.

Often, reactions that are exothermic and whose products are more disordered (higher in entropy) occur spontaneously, while endothermic reactions that produce products of lower entropy are not spontaneous. If they occur at all, they will require external energy input.

Disorder or randomness increases as we proceed from the solid to liquid to the vapor state. Solids often have an ordered crystalline structure, and liquids have a loose structure, while gas particles are virtually random in their distribution.

5.2 Experimental Determination of Energy Change in Reactions

Learning Goal 3: Describe experiments that yield thermochemical information and calculate fuel values based on experimental data.

The measurement of the energy demand or energy release in a chemical reaction is **calorimetry.** This technique involves the measurement of the change in the temperature of a quantity of water that is in contact with the reaction of interest and also isolated from its surroundings. A device used in these types of measurements is a calorimeter, which measures heat changes in calories (or energy, since the calorie is a unit of heat energy).

For an exothermic reaction, heat is released by the reaction to the surrounding water, which is isolated from its surroundings. The **specific heat** of water (SH_w) is defined as the number of calories needed to raise the temperature of 1 gram of water 1 degree Celsius. This, coupled with the total number of grams of water (m_w) and the temperature increase (ΔT), allows one to calculate the heat released, Q.

$$Q = m_w \times \Delta T \times SH_w$$

Many chemical reactions that produce heat are combustion reactions. In our bodies, many food substances, principally carbohydrates and fats, are oxidized to produce energy. In these cases, the amount of energy per gram of food is referred to as its fuel value. A special type of calorimeter, a bomb calorimeter, is useful for the measurement of the **fuel value** (calories) of foods.

Example 1

When equal moles of hydrochloric acid (HCl) and sodium hydroxide (NaOH) are mixed in a calorimeter, the temperature of 50.00 grams of solution increased from 25.00°C to 38.00°C. If the specific heat of the solution is the same as the specific heat of water, which is 1.000 calories/(gram)(°C), calculate the quantity of energy in calories involved in this reaction.

Work

1. The change in temperature is $\Delta T = 38.00°C - 25.000°C = 13.00°C$
2. Since heat energy is released from the reaction to the solution, the reaction is exothermic.
3. The quantity of heat absorbed or released by a reaction (Q) is the product of the mass of water or solution in the calorimeter (m_w), the specific heat of the water or solution (SH_w), and the change in temperature (ΔT) of the water or solution as the reaction proceeds from the initial to final states. The above relationship is given by the following equation:

$$Q = (m_w)(SH_w)(\Delta T)$$

4. $Q = 50.00 \text{ g} \times \dfrac{1.000 \text{ calories}}{(1 \text{ gram} [1°C])} \times 13.00°C = 650.0 \text{ calories}$

Answer

650.0 calories are released during the reaction.

Example 2

When 1.000 gram of glucose ($C_6H_{12}O_6$) was burned in a bomb calorimeter, the temperature of 100.0 grams of water in the calorimeter increased from 25.00°C to 63.00°C. Calculate the fuel value of one gram of glucose. Remember that the SH_w for water is 1.000 calorie per gram of water per 1°C temperature change.

Work

1. Use the equation $Q = (m_w)(SH_w)(\Delta T)$
2. $\Delta T = 63.00°C - 25.00°C = 38.00°C$

3. $Q = 100.0 \text{ g} \times 38.00°C \times \dfrac{1.000 \text{ calories}}{(1 \text{ gram} (1°C))}$

    $Q = 3.800 \times 10^3 \text{ calories}$

4. In food-calorie books, we see the value of 1 gram of carbohydrate (glucose is a pure carbohydrate) to be 4 calories. These calorie values are really kilocalories. Thus, our value of 3800 calories, or 3.8 kilocalories, is close to those given in food-calorie handbooks.

Answer

One gram of glucose (or any carbohydrate) has a fuel value of 3.8×10^3 calories, or 3.8 kilocalories, or just 3.8 nutritional calories.

5.3 Kinetics

Chemical kinetics is the study of the rate (speed) of a chemical reaction.

Learning Goal 4:	Describe the concept of reaction rate and the role of kinetics in chemical change.

The Chemical Reaction

If the energy made available by the collision of reacting molecules exceeds the bond energies, the bonds will break, and the resulting atoms will recombine in a lower energy configuration. A collision meeting the above conditions and producing one or more product molecules is referred to as an effective collision. Only effective collisions lead to chemical reaction.

Activation Energy and the Activated Complex

Learning Goal 5:	Describe the importance of *activation energy* and the *activated complex* in determining reaction rate.

The minimum amount of energy required to produce a chemical reaction is the **activation energy** for the reaction. As implied above, a large component of the activation energy is the bond energy of the reacting molecules.

The chemical reaction may be represented in terms of the changes in potential energy that occur as a function of the time of the reaction. Several important characteristics of this relationship follow:

1. The reaction proceeds from reactants to products through an extremely unstable intermediate state that we term the **activated complex.**
2. Formation of the activated complex requires energy. The difference between the energy of reactants and activated complex is the activation energy.
3. For an exothermic reaction, the overall energy change must be a net release of energy. The net release of energy is the difference in energy between products and reactants.

Factors That Affect Reaction Rate

Learning Goal 6:	Predict the way reactant structure, concentration, temperature, and catalysis affect the rate of a chemical reaction.

Five major experimental conditions influence the rate of a chemical reaction.

1. Structure of the Reacting Species
   The size and shape of reactant molecules influence the rate of the reaction. Large molecules, containing bulky groups of atoms, may block the reactive part of the molecule from interacting with another reactive substance, causing the reaction to proceed slowly.

Only molecular collisions that have the correct collision orientation, as well as sufficient energy, lead to product formation. These collisions are termed *effective collisions*.

2. The Concentration of Reactants
   In a very general sense, **concentration** is a measure of the quantity of a substance of interest (perhaps a reactant or product) contained in a specified volume of a mixture. The rate of a chemical reaction is often a complex function of the concentration of one or more of the reacting substances. The rate will generally *increase* as concentration *increases* simply because a higher concentration means more reactant molecules in a given volume and therefore a greater number of collisions per unit time. If we assume that other variables are held constant, a larger number of collisions leads to a larger number of effective collisions.

3. The Temperature of Reactants
   The rate of a reaction *increases* as the temperature increases because the average kinetic energy of the reacting particles is directly proportional to the Kelvin temperature. Increasing the speed of particles increases the likelihood of collision, and the higher kinetic energy means that a higher percentage of these collisions will result in product formation (effective collisions).

4. The Physical State of Reactants
   The rate of a reaction depends on the physical state of the reactants: solid, liquid, or gas. For a reaction to occur, the reactants must collide frequently and have sufficient energy to react. In the solid state, the atoms, ions, or molecules are restricted in their motion. In the gaseous and liquid states, the particles have both free motion and proximity to each other. There are many more collisions per unit time. Consequently, there are more effective collisions as well. Hence reactions tend to be fastest in the liquid and gaseous states and slowest in the solid state.

5. The Presence of a Catalyst
   A **catalyst** is a substance that *increases* the reaction rate. If added to a reaction mixture, the catalytic substance undergoes no net change, nor does it alter the outcome of the reaction. However, the catalyst interacts with the reactants to create an alternative pathway for production of products. This alternative path has a lower activation energy. This makes it easier for the reaction to take place and thus increases the rate.

5.4 Equilibrium

Learning Goal 7:   Recognize and describe equilibrium situations.

Many chemical reactions do not proceed to "completion." After a period of time, determined by the kinetics of the reaction, the concentration of reactants no longer decreases and the concentration of products no longer increases. At this point, a mixture of products and reactants exists, and its composition would remain constant unless the experimental conditions were changed. This mixture is in a state of chemical equilibrium.

Rate and Reversibility of Reactions

Many reactions may proceed in either direction, left to right or right to left. The concentration of the various species is fixed at equilibrium because product is being consumed and formed at the same rate. In other words, the reaction continues indefinitely (dynamic) but the concentration of products and reactants is fixed (equilibrium). This is a dynamic equilibrium.

Chemical Equilibrium

Chemical change, as well as physical change, can attain equilibrium. For example:

$$H_2(g) + I_2(g) \rightleftharpoons 2HI(g)$$

At equilibrium, the rate of *disappearance* of H_2 and I_2 is equal to the rate of formation of H_2 and I_2. The rates of the forward and reverse reactions are equal, and the concentrations of H_2, I_2, and HI are fixed. The process is dynamic, with a continuous interconversion of products and reactants.

LeChatelier's Principle

Learning Goal 8:	Use LeChatelier's principle to predict changes in equilibrium position.

LeChatelier's principle states that if a stress is placed on an equilibrium system, the system will respond by altering the equilibrium in such a way as to minimize the stress.

Product introduced: equilibrium

← shifted

Reactant introduced: equilibrium

→ shifted

Effect of Concentration

Addition of extra product or reactant to a fixed reaction volume is just another way of saying that we have increased the concentration of product or reactant. Removal of material from a fixed volume decreases the concentration. Therefore, changing the concentration of one or more components of a reaction mixture is a way to alter the equilibrium composition of an equilibrium mixture.

Effect of Heat

The change in equilibrium composition caused by the addition or removal of heat from an equilibrium mixture can be explained by treating heat as a product or reactant. Adding heat to an exothermic reaction is similar to increasing the amount of product, shifting the equilibrium to the left. Removing heat from an exothermic reaction shifts the equilibrium to the right.

Heat is a reactant in an endothermic reaction; its removal shifts the equilibrium to the left. Adding heat favors product formation.

Effect of Pressure

Only gases are affected significantly by changes in pressure, because gases are free to expand and compress in accordance with Boyle's law. However, liquids and solids are not compressible, so their volumes are unaffected by pressure.

Therefore, pressure changes will alter equilibrium composition only when they involve a gas or variety of gases as products and/or reactants. If the reaction shifts to conserve volume:

- If the number of moles of gaseous product is <u>greater</u> than the number of moles of gaseous reactant, an increase in pressure shifts the equilibrium to the left.
- If the number of moles of gaseous product is <u>less</u> than the number of moles of gaseous reactant, an increase in pressure shifts the equilibrium to the right.
- If the number of moles of gaseous product and reactant is identical, pressure will have no effect on the equilibrium because there is no volume advantage.

Effect of a Catalyst

A catalyst has no effect on the equilibrium composition. A catalyst increases the rates of both forward and reverse reactions to the same extent. The equilibrium composition and equilibrium concentration do not change when a catalyst is used.

Glossary of Key Terms in Chapter Five

activated complex (5.3) the arrangement of atoms at the top of the potential energy curve as a reaction proceeds.
activation energy (5.3) the threshold energy that must be overcome to produce a chemical reaction.
calorimetry (5.2) the measurement of changes in heat energy during a chemical reaction.
catalyst (5.3) a substance that speeds up a reaction without undergoing change.
contentration (5.3) a measure of the quantity of a substance contained in a specified volume of solution.
dynamic equilibrium (5.4) the state that exits when the rate of change in the concentration of products and reactants is equal, resulting in no net concentration change.
endothermic reaction (5.1) a chemical or physical change in which energy is absorbed.
enthalpy (5.1) a term that represents heat energy.
entropy (5.1) a term that represents a level of randomness or disorder.
equilibrium reaction (5.4) a reaction that is reversible and the rates of the forward and reverse reactions are equal.
exothermic reaction (5.1) a chemical or physical change that releases energy.
fuel value (5.2) the amount of energy derived from a given mass of material.

kinetics (5.3) the study of rates of chemical reactions.

LeChatelier's principle (5.4) states that when a system at equilibrium is disturbed, the equilibrium shifts in the direction that minimizes the disturbance.

nutritional Calorie (5.2) equivalent to one kilocalorie (1000 calories); also know as a large Calorie.

rate of chemical reaction (5.3) the change in concentration of a reactant or product per unit time.

specific heat (5.2) the quantity of heat (calories) required to raise the temperature of 1 g of a substance one degree Celsius.

surroundings (5.1) the universe outside of the system.

system (5.1) the process under study.

thermodynamics (5.1) the branch of science that deals with the relationship between energies of systems, work, and heat.

Self Test for Chapter Five

1. When equal moles of HCl and NaOH were mixed in a calorimeter, the temperature of 100.0 grams of water surrounding the reaction increased from 22.5°C to 36.5°C. If the specific heat of water is 1.000 calories/(gram)(°C), calculate the quantity of heat energy in calories that was released.

2. A random, or disordered, system is characterized by what kind of entropy?

3. Are endothermic reactions producing products of lower entropy usually spontaneous or nonspontaneous?

4. Which of the three states of matter has the highest entropy?

5. Are the ΔH values often positive or negative for nonspontaneous reactions?

6. What device is used to measure heat changes in calories?

7. What is the amount of energy per gram of food substance?

8. What is the name of an extremely unstable intermediate state of a chemical reaction?

9. Many chemical reactions do not proceed to completion. After a period of time, the concentration of reactants no longer decreases, and the concentration of products ceases to increase. At this point, what state describes the chemical reaction?

10. What French chemist in the nineteenth century discovered that changes in equilibrium were dependent upon stress that might be applied to the system?

11. In what direction will the addition of heat shift an exothermic reaction?

12. In what direction will the addition of heat shift an endothermic reaction?

Vocabulary Quiz for Chapter Five

1. _____ is the study of energy, work, and heat.

2. Exothermic reactions (release, absorb) heat.

3. Endothermic reactions (release, absorb) heat.

4. _____ is a measure of the disorder of a system.

5. The amount of energy per gram of food is termed _____.
6. A _____ measures heat changes in chemical reactions.
7. _____ is the study of the rate of chemical reactions.
8. The minimum energy needed for a reaction is the _____.
9. A _____ increases the rate of a reaction.
10. _____ describes a chemical reaction's response to stress.

6

States of Matter
Gases, Liquids, and Solids

Learning Goals

1 Describe the major points of the kinetic molecular theory of gases.
2 Explain the relationship between the kinetic molecular theory and the physical properties of macroscopic quantities of gases.
3 Describe the behavior of gases expressed by the gas laws: Boyle's law, Charles's law, combined gas law, Avogadro's law, the ideal gas law, and Dalton's law.
4 Use gas law equations to calculate conditions and changes in conditions of gases.
5 Describe the properties of the liquid state in terms of the properties of the individual molecules that comprise the liquid.
6 Describe the processes of melting, boiling, evaporation, and condensation.
7 Describe the dipolar attractions known collectively as van der Waals forces.
8 Describe hydrogen bonding and its relationship to boiling and melting temperatures.
9 Relate the properties of the various classes of solids (ionic, covalent, molecular, and metallic) to the structure of these solids.

Introduction

The major differences between solid, liquid, and gaseous substances lie in the following properties:

1. The average distance of separation of particles in each state
2. The strength of the attractive forces between the particles
3. The degree of organization of particles

The behavior of the states of matter, their properties, and their interconversion are considered in this chapter.

6.1 The Gaseous State

Ideal Gas Concept

An ideal gas is a model that describes how molecules or atoms behave at the microscopic level. The model works well for most real gases that are *not* at high concentration, high pressure, or close to the temperature at which they would condense, that is, for gases that are not about to become a liquid! (Also, attractive forces between gas particles can cause deviation from the ideal gas law. This is addressed later in the chapter.)

The ideal gas model has been used to determine the interrelationship of temperature, volume, pressure, and quantity (mass or number of moles) in gaseous systems. These relationships are summarized in the ideal gas law and are described in this chapter.

Description of a Gas

Measuring the temperature and volume of a gas is as familiar as using a thermometer and looking at a one-liter container, for example. (More precise measurements of volume can be taken by using glassware that is carefully marked, such as a graduated cylinder.)

Pressure can be measured with a mercury barometer, in which a long tube filled with mercury (Hg) is inverted into a dish of mercury. The atmospheric pressure pushing on the surface of the mercury in the dish supports the column of mercury in the tube. So, the height of the mercury in the tube is proportional to the atmospheric pressure, and "mm Hg" can be used as a unit of pressure.

In addition to *mm Hg (millimeters of mercury)*, other common units of pressure are: *atmosphere (atm)* and *torr*. Other units of pressure include the *pascal (Pa)* or *kilopascal (kPa)*, *lb/in² (pounds per square inch, or psi)*, and *in Hg (inches of mercury)*. These units are related as follows:
1 atm = 760 mm Hg = 760 torr = 14.7 lb/in² = 29.9 in Hg = 1.01×10^5 Pa = 101 kPa

Properties of Gases and the Kinetic Molecular Theory

Learning Goal 1:	Describe the major points of the kinetic molecular theory of gases.

Learning Goal 2:	Explain the relationship between the kinetic molecular theory and the physical properties of macroscopic quantities of gases.

The fundamental model of particle behavior in the gas phase is the kinetic-molecular theory. This theory describes an ideal gas, in which gas particles exhibit no interactive or repulsive forces, and the volumes of the individual gas particles are assumed to be negligible. The theory may be summarized as follows:

1. A gas consists of particles that are far apart. The volume of the individual particles is assumed to be small in comparison to the average distance of separation of the particles.
2. The gas particles are in continuous, rapid, random motion. Particles change direction only as a result of collisions with other particles or with the wall of a container.
3. Upon collision, there is no net loss of energy; energy may only be transferred from one particle to another.

4. The speed of the particles is directly proportional to the absolute (Kelvin) temperature; as a result, the average kinetic energy is directly proportional to the absolute temperature.

Learning Goal 3:	Describe the behavior of gases expressed by the gas laws: Boyle's law, Charles's law, combined gas law, Avogadro's law, the ideal gas law, and Dalton's law.

Learning Goal 4:	Use gas law equations to calculate conditions and changes in conditions of gases.

Boyle's Law: Relationship of Pressure and Volume

Robert Boyle found that the volume of a gas varies inversely with the pressure exerted by the gas, if the number of moles and the temperature of the gas are held constant. This relationship is known as Boyle's law.

Mathematically, the product of pressure (P) and volume (V) is a constant:

$$PV = k_1$$

Boyle's law is often used to calculate the volume resulting from a pressure change or vice versa. We consider $P_iV_i = k_1$ the initial condition, and $P_fV_f = k_1$ the final condition. Since (PV), initial or final, is constant and is equal to k_1,

$$P_iV_i = P_fV_f$$

Note that the calculation can be done with volume units of mL or L. It is only important that the units be the same on both sides of the equation. This is also true of pressure; however, the most commonly used unit of pressure measurement is the atmosphere. The relationship between the various pressure units follows: 1 atmosphere (atm) = 76 cm Hg = 760 mm Hg = 760 torr = 29.9 inches Hg = 14.7 psi (pounds per square inch).

Example 1
Convert each of the following into atmospheres.

1. 77.0 mm of Hg
2. 30.2 inches of Hg
3. 16.8 psi (The abbreviation psi means pounds per square inch.)
4. 800.00 torr

Example 1 continued

Work

One atmosphere of pressure equals 760 mm of Hg pressure, which also equals 760 torr of pressure, which also equals 14.7 psi, or

$$1 \text{ atmosphere} = 760 \text{ mm Hg} = 760 \text{ torr} = 14.7 \frac{lb}{in^2}$$

1. $77.0 \text{ mm Hg} \times \dfrac{1 \text{ atm}}{760 \text{ mm Hg}} = 1.01 \times 10^{-1} \text{ atm}$

2. $30.2 \text{ in. Hg} \times \dfrac{2.54 \text{ cm}}{1 \text{ inch}} \times \dfrac{1 \text{ meter}}{100 \text{ cm}} \times \dfrac{1000 \text{ mm}}{1 \text{ meter}} \times \dfrac{1 \text{ atm}}{760 \text{ mm Hg}} = 1.01 \text{ atm}$

3. $16.8 \text{ psi} \times \dfrac{1 \text{ atm}}{14.7 \text{ psi}} = 1.14 \text{ atm}$

4. $800.0 \text{ torr} \times \dfrac{1 \text{ atm}}{760 \text{ torr}} = 1.053 \text{ atm}$

Answer
The pressure is 1.053 atm.

Example 2

A given mass of carbon dioxide at 25°C occupies a volume of 500.0 mL at 2.00 atmospheres of pressure. What pressure must be applied to compress the gas to a volume of 50.0 mL, assuming no temperature change?

Work

1. Use Boyle's Law: $P_i V_i = P_f V_f$
    P_i represents the starting pressure; P_f represents the final pressure.
    V_i represents the starting volume; V_f represents the final volume.

2. From the given information, we have the following:

    $P_i = 2.00$ atmospheres $V_i = 500.0$ mL (the initial volume)

    $P_f = ?$ (not known yet) $V_f = 50.0$ mL (the final volume)

Example 2 continued

3. Rearrange the equation as follows: $P_iV_i = P_fV_f$
$$P_fV_f = P_iV_i$$

$$P_f = \frac{P_iV_i}{V_f}$$

We can thus find the pressure needed to compress the 500.0 mL of gas to 50.0 mL.

4. $P_f = \dfrac{P_iV_i}{V_f}$

$$P_f = \frac{(2.00 \text{ atm})(500.0 \text{ mL})}{50.0 \text{ mL}}$$

$P_{final} = 20.0$ atm

Answer
20.0 atmospheres of pressure will be needed.

Charles's Law: Relationship of Temperature and Volume

Jacques Charles found that the volume of a gas varies directly with the absolute temperature (K) if pressure and number of moles of gas are constant. This relationship is Charles's Law.

Mathematically the ratio of volume (V) and temperature (T) is a constant:

$$V/T = k_2$$

In a way that is analogous to Boyle's Law, we may use this expression to solve some practical problems.

Example 3
A balloon filled with helium has a volume of 5.00 liters at 0.00°C. What would be the balloon's volume at 25.00°C if the pressure surrounding the balloon remained constant?

Work

1. Use Charles's Law: $\dfrac{V_i}{T_i} = \dfrac{V_f}{T_f}$

V_i represents starting volume; V_f means final volume.
T_i represents starting temperature in Kelvin units.
T_f represents final temperature in Kelvin units.

75

Example 3 continued

2. Change the °C values into Kelvin values. Use the following equation:

K = °C + 273

$T_i = 0.00°C$; in K, $T_i = 0.00°C + 273 = 273.00$ K

$T_f = 25.00°C$; in K, $T_f = 25.00°C + 273 = 298.00$ K

3. Rearrange the equation: $\dfrac{V_i}{T_i} = \dfrac{V_f}{T_f}$

$$(V_f)(T_i) = (V_i)(T_f)$$

$$V_f = \frac{(V_i)(T_f)}{T_i}$$

4. $V_{final} = \dfrac{(5.00 \text{ liters})(298 \text{ K})}{273 \text{ K}} = 5.46 \text{ L}$

Answer
5.46 liters

Combined Gas Law

Often a sample of gas (a fixed number of moles of gas) undergoes change involving volume, pressure, and temperature simultaneously. It would be useful to have one equation that describes such processes.

A **combined gas law** is such an equation. It can be derived from Boyle's law and Charles's law and takes the form:

$$\frac{P_i V_i}{T_i} = \frac{P_f V_f}{T_f}$$

Avogadro's Law: Relationship of the Number of Moles and Volume

The relationship between the volume and number of moles of a gas is known as Avogadro's law. This law states that equal volumes of a gas contain the same number of moles, if measured under the same conditions of temperature and pressure.

Mathematically, the ratio of volume (V) and number of moles (n) is a constant:

$$V/n = k_3$$

A relationship comparing initial and final conditions, similar to Boyle's and Charles's laws, may be derived:

$$\frac{V_i}{n_i} = \frac{V_f}{n_f}$$

Molar Volume of a Gas

The volume occupied by 1 mole of any gas is referred to as its molar volume. The basis for this relationship is Avogadro's law. At **standard temperature and pressure (STP),** the molar volume of any gas is 22.4 L. STP conditions are defined as follows:

$$T = 273 \text{ K (or } 0°\text{C)} \qquad P = 1 \text{ atm}$$

Thus, 1 mole of N_2 (28g), O_2 (32g), H_2 (2g), and He (4g) all occupy the same volume, 22.4 L at STP.

Gas Densities

It is also possible to compute the density of various gases at STP if one recalls that density is the mass/unit volume:

$$d = m/V$$

The Ideal Gas Law

Boyle's law (relating volume and pressure), Charles' law (relating volume and temperature) and Avogadro's law (relating volume to the number of moles) may be combined into a single expression relating all four terms. This expression is the **ideal gas law:**

$$PV = nRT$$

where R is the ideal gas constant.

$$R = 0.0821 \text{ L•atmK}^{-1}\text{mole}^{-1}$$

if the units P (atmospheres), V (liters), n (number of moles), and T (Kelvin) are used.

Example 4
Calculate the number of grams of helium in a 1.00-liter balloon at 27.0°C and under 1.00 atmosphere of pressure.

Work
1.   Use the ideal gas law:
      $PV = nRT$

Example 4 continued

P is the pressure in atmospheres.
V is the volume of the gas in liters.
R is the ideal gas constant, which is equal to $\dfrac{0.0821 \ (L)(atm)}{(K)(mole)}$

T is the temperature given in Kelvin units.
n is the number of moles of gas present.

2. Convert 27.0°C into Kelvin units:
$K = °C + 273$
$K = 27.0 + 273 = 300 \ K$

3. Rearrange $PV = nRT$ so we can find the number of moles (n in the equation) of gas present:

$$n = \frac{PV}{RT}$$

4. $n_{helium} = \dfrac{(1.00 \ atm)(1.00 \ L)}{\left(\dfrac{0.0821 \ (L)(atm)}{(K)(mole)}\right)(300 \ K)}$

$n_{helium} = 4.06 \times 10^{-2}$ moles of helium are present.

5. Convert moles of helium to grams of helium using its molar mass, 4.00 g/mole.
$(4.06 \times 10^{-2}$ moles$)(4.00$ g/mole$) = 0.162$ grams (or, 1.62×10^{-1} grams)

Answer
0.162 grams of helium are present.

Dalton's Law of Partial Pressures

A mixture of gases exerts a pressure that is the sum of the pressure that each gas would exert if it were present alone under similar conditions. This is known as Dalton's law of partial pressures.

$$P_t = P_1 + P_2 + P_3 + \bullet \bullet \bullet + P_n$$

where P_t = total pressure and P_1, P_2, P_3, and so on are the partial pressures of the component gases.

Ideal Gases Versus Real Gases

We have assumed so far, both in theory and calculations, that all gases behave as ideal gases. However, in reality, there is no such thing as an ideal gas. Interactive forces, even between the widely spaced particles of gas, are not totally absent in any sample of gas.

Attractive forces are particularly significant in gases composed of polar molecules. Calculations involving polar gases such as HF, NO, and SO_2, using ideal gas equations (which presume no such interactions), are approximate at best.

Nonpolar molecules exhibit a temporary dipole-dipole interaction known as a London force. The temporary dipole-dipole interaction is weak, and London forces are much weaker than permanent dipole interactions. As a result, the behavior of nonpolar molecules, such as N_2, O_2, and H_2, is explained rather well by the ideal-gas equations.

6.2 The Liquid State

Molecules in the liquid state are close to one another. Attractive forces are large enough to keep the molecules together, in contrast to gases, in which cohesive forces are so low that a gas expands to fill any volume. These attractive forces in a liquid are, however, not large enough to restrict movement, as in solids.

Learning Goal 5: Describe the properties of the liquid state in terms of the properties of the individual molecules that comprise the liquid.

Compressibility

Liquids are almost incompressible. In fact, the spacing between molecules is so small that even the application of many atmospheres of pressure does not significantly decrease the volume of a liquid.

Viscosity

The **viscosity** of a liquid is a measure of its resistance to flow. Viscosity is a function of the attractive forces present between molecules as well as the molecular geometry. Complex molecules, which do not "slide" smoothly past each other, as well as polar molecules, tend to have higher viscosity than less structurally complex, less polar liquids.

Surface Tension

The surface tension of a liquid is a measure of the attractive forces at its surface. Intermolecular attraction is stronger at the surface of a liquid. This increased surface force is responsible for the spherical shape of drops of liquid. Substances known as surfactants may be added to a liquid to decrease surface tension.

Vapor Pressure of a Liquid

Learning Goal 6: Describe the processes of melting, boiling, evaporation, and condensation.

According to the kinetic theory, the molecules of a liquid are in continuous motion, with their average kinetic energy proportional to the temperature. Although the average kinetic energy is too small to allow molecules to "escape" from liquid to vapor, a few high-energy molecules possess sufficient energy to escape from the bulk liquid.

At the same time, a fraction of these vaporized molecules lose energy (perhaps by collision with the walls of the container) and return to the liquid state. The process of conversion of liquid to vapor, at a temperature too low to boil, is **evaporation.** The reverse process, conversion of the gas to the liquid state, is **condensation.** After some period of time, the rates of evaporation and condensation become equal, and this constitutes a dynamic equilibrium between liquid and vapor states. The vapor pressure of the liquid is therefore defined as the pressure exerted by the vapor at equilibrium.

The boiling point of a liquid is defined as the temperature at which the vapor pressure of the liquid becomes equal to the prevailing atmospheric pressure. The "normal" atmospheric pressure is 760 torr, or one atmosphere, and the **normal boiling point** is the temperature at which the vapor pressure of the liquid is equal to one atmosphere.

Van der Waals Forces

Learning Goal 7:	Describe the dipolar attractions known collectively as van der Waals forces.

In 1930 Fritz London demonstrated that he could account for a weak attractive force between any two molecules, whether polar or nonpolar. He postulated that the electron distribution in molecules is not fixed; electrons are in continuous motion, relative to the nucleus. So, for a short period of time a nonpolar molecule could experience an instantaneous dipole, a short-lived polarity due to a temporary dislocation of the electron cloud. These temporary dipoles could interact with other temporary dipoles, just permanent dipoles interact in polar molecules. We now call these intermolecular forces **London forces.**

London forces and dipole-dipole interactions are collectively known as **van der Waals forces.** London forces exist among polar and nonpolar molecules because electrons are in constant motion in all molecules. Dipole-dipole attractions occur only among polar molecules. In the next section we will see a third type of intermolecular force, the **hydrogen bond.**

Hydrogen Bonding

Learning Goal 8:	Describe hydrogen bonding and its relationship to boiling and melting temperatures.

Molecules in which a hydrogen atom is bonded to a highly electronegative atom such as nitrogen, oxygen, or fluorine exhibit **hydrogen bonding.** This arrangement of atoms produces a very polar bond, often resulting in a polar molecule with strong intermolecular attractive forces.

Example 5

Illustrate the hydrogen bonding found between a water molecule and its closest three neighboring water molecules.

Work

1. The polar nature of water is caused by the large electronegativity difference between oxygen and hydrogen atoms.
2. One water molecule has the following polar structure:

$$\delta^+ \ H \overset{\overset{\displaystyle \delta^-}{\displaystyle O}}{\diagdown} H \ \delta^+$$

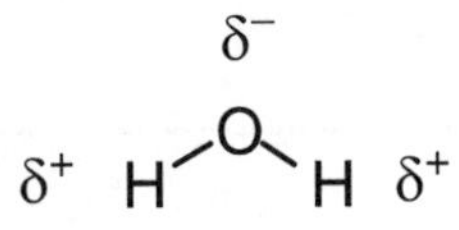

Notice that the oxygen end is slightly negative, and the two hydrogen ends are slightly positive.
3. The hydrogen bonding in liquid water is due to the attraction of the slightly negative oxygen atom to a slightly positive hydrogen in another water molecule. In similar fashion each slightly positive hydrogen atom attracts an oxygen atom in two other water molecules.
4. Thus, one water molecule has three hydrogen bonds with three other water molecules.

Answer

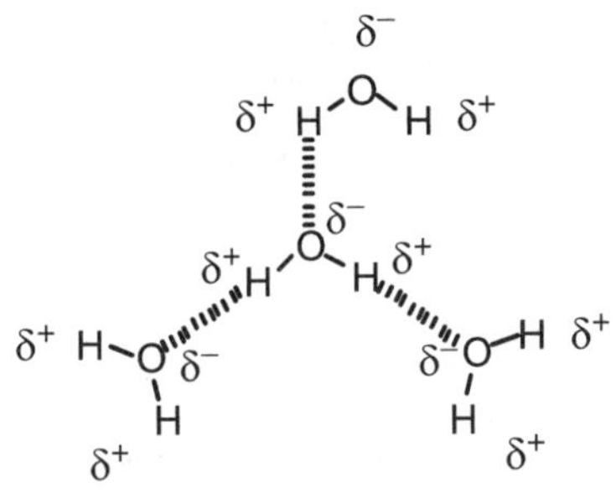

Example 6

Show the hydrogen bonding that will occur between one molecule of methyl alcohol (CH_3OH) and water molecules. The methyl alcohol has the following polar structure:

$$\delta^+ \ H \overset{\overset{\displaystyle \delta^-}{\displaystyle O}}{\diagdown} CH_3$$

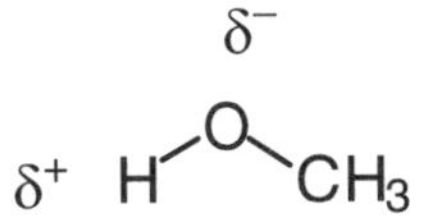

Example 6 continued

Answer

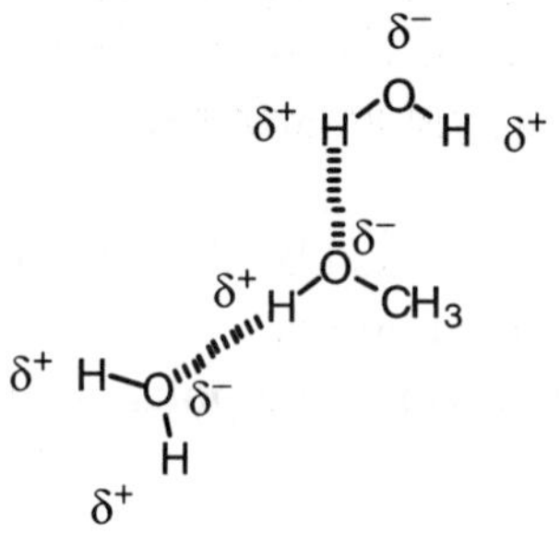

$$\delta^+ \; H\text{---}\overset{\delta^-}{O}\text{---}H \; \delta^+$$
$$\vdots$$
$$\delta^+ \; H\text{---}\overset{\delta^-}{O}\text{---}CH_3$$
$$\delta^+ \; H\text{---}\overset{\delta^-}{O}\underset{\underset{\delta^+}{H}}{}$$

Example 7

Show the hydrogen bonding that will occur between water and the amide structure that is found in proteins.

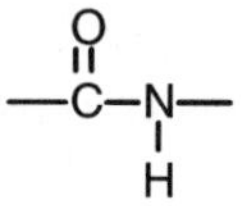

$$\underset{\underset{H}{|}}{-\overset{\overset{O}{\|}}{C}-N-}$$

Work

The amide structure has the following polar arrangement:

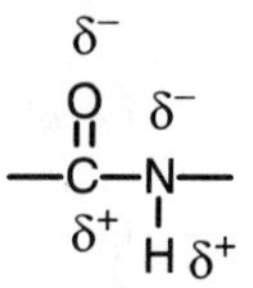

$$-\overset{\overset{\delta^-}{\overset{O}{\|}}}{C}-\overset{\delta^-}{N}-$$
$$\delta^+ \; H \; \delta^+$$

Answer

$$\delta^+ \; H\text{---}\overset{\delta^-}{O}\text{---}H \; \delta^+ \qquad \overset{\delta^-}{H\text{---}O\text{---}H} \; \delta^+$$
$$\overset{\delta^-}{O}=\underset{\delta^+ \; H \; \delta^+}{\overset{\delta^-}{C}-\overset{\delta^-}{N}-}$$
$$\delta^+ \; H\text{---}\overset{\delta^-}{O}\text{---}H \; \delta^+$$

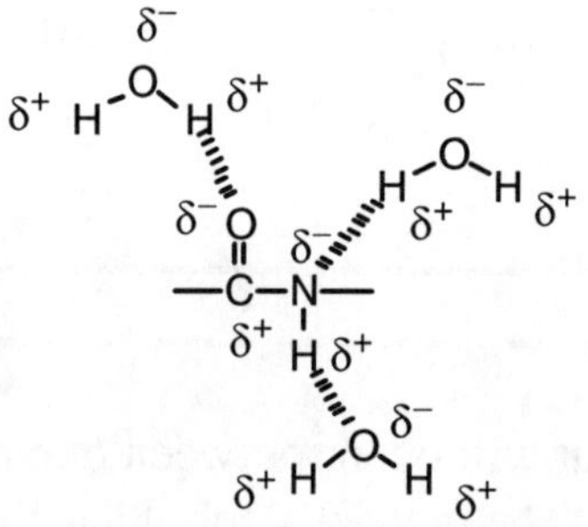

6.3 The Solid State

The close packing of the particles of a solid results from attractive forces that are strong enough to restrict motion. The particles are "locked" together in a defined and highly organized fashion. This results in a fixed shape and volume (recall that gases have no fixed shape or volume, and liquids have a fixed volume but no fixed shape).

Properties of Solids

Learning Goal 9:	Relate the properties of the various classes of solids (ionic, covalent, molecular, and metallic) to the structure of these solids.

Solids are incompressible, due to the small distance between particles. Solids may be **crystalline,** having a regular repeating structure, or **amorphous,** having no organized structure.

Types of Crystalline Solids

Crystalline solids may exist in one of four general groups:
1. **Ionic solids.** The units that make up an **ionic solid** are positive and negative ions. Electrostatic forces hold the crystal together.
2. **Covalent solids.** The units that make up a **covalent solid** are atoms held together by covalent bonds.
3. **Molecular solids.** The units composing **molecular solids** are molecules held together by intermolecular forces (dipole-dipole and London).
4. **Metallic solids. Metallic solids** are composed of metal atoms held together by metallic bonds. Metallic bonds are formed by the overlap of orbitals of metal atoms, resulting in regions of high electron density surrounding the positive metal centers. Electrons in these regions are extremely mobile, resulting in the high conductivity (ability to carry electrical current) exhibited by many metallic solids.

Glossary of Key Terms in Chapter Six

amorphous solid (6.3) solids that have no organized, regular structure.

Avogadro's law (6.1) states that the volume is directly proportional to the number of moles of gas particles if the pressure and temperature are constant.

barometer (6.1) a device for measuring pressure.

Boyle's law (6.1) states that the volume of a gas varies inversely with pressure exerted if the temperature and number of moles are constant.

Charles's law (6.1) states that the volume of a gas is directly proportional to the temperature of the gas if the pressure and number of moles of the gas are constant.

combined gas law (6.1) an equation that describes the behavior of a gas when volume, pressure, and temperature may change simultaneously.

condensation (6.2) the conversion of gas to a liquid.

covalent solid (6.3) a collection of atoms held together by covalent bonds.

crystalline solid (6.3) a solid having a regular repeating structure.

Dalton's law (6.1) the law of partial pressure, states that the total pressure exerted by a gas mixture is the sum of the partial pressures of the component gases.

dipole-dipole interactions (6.2) attractive forces between polar molecules.

evaporation (6.2) the conversion of a liquid to a gas below the boiling point of the liquid.

hydrogen bonding (6.2) the attractive force between a hydrogen atom covalently bonded to a small, highly electronegative atom and another atom containing an unshared pair of electrons.

ideal gas (6.1) a gas in which the particles do not interact and the volume of the individual gas particles is assumed to be negligible.

ideal gas law (6.1) states that for an ideal gas the product of pressure and volume is proportional to the product of the number of moles of the gas and its temperature. The proportionality constant for an ideal gas is symbolized as R.

ionic solid (6.3) solids composed of positive and negative ions in a regular three-dimensional crystalline arrangement.

kinetic-molecular theory (6.1) the fundamental model of particle behavior in the gas phase.

London forces (6.2) weak attractive forces between molecules that result from short-lived dipoles that occur because of continuous motion of electrons in the molecules.

melting point (6.3) the temperature at which a solid converts to a liquid.

metallic bond (6.3) bond that results from the orbital overlap of metal atoms.

metallic solid (6.3) solid composed of metal atoms held together by metallic bonds.

molar volume (6.1) the volume occupied by 1 mol of a substance.

molecular solid (6.3) solid in which the molecules are held together by dipole-dipole and London forces.

normal boiling point (6.2) the temperature at which a substance will boil at one atmosphere of pressure.

partial pressure (6.1) the pressure exerted by one component of a gas mixture.

pressure (6.1) a force per unit area.

standard temperature and pressure (STP) (6.1) 273 K and one atmosphere pressure.

surface tension (6.2) a measure of the strength of the attractive force at a liquid's surface.

surfactant (6.2) a substance that decreases the surface tension of a liquid.

van der Waals forces (6.2) a general term for intermolecular forces that include dipole-dipole and London forces.

vapor pressure of a liquid (6.2) the pressure exerted by the vapor at the surface of a liquid at equilibrium.

viscosity (6.2) a measure of the resistance to flow of a substance at constant temperature.

Self Test for Chapter Six

1. _____ is the conversion of a liquid to a gas below the boiling point of the liquid.

2. A(n) _____ gas is one in which gas particles do not interact and the volume of the individual gas particles is assumed to be negligible.

3. Convert 29.96 inches of mercury pressure into mm of Hg pressure.

4. A given mass of gas at 20.0°C occupies a volume of 250.0 mL at 1.00 atmosphere of pressure. What pressure must be applied to compress the gas to a volume of 50.0 mL, assuming no temperature change?

5. A balloon filled with air has a volume of 2.00 liters at 24.0°C. What would be the balloon's volume at 37.0°C if the pressure surrounding the balloon remained constant?

6. Calculate the number of moles of helium in a 0.500-liter balloon at 25.0°C and under 0.960 atmospheres of pressure.

7. Show the hydrogen bonds between hydrogen fluoride molecules. The polar structure of HF is: δ^+ H – F δ^-.

8. The total pressure of our atmosphere is primarily equal to the sum of the pressures of what two gases?

9. What law explains why oxygen is distributed to the cells and the waste product,

carbon dioxide, is expelled by the lungs?

10. What is defined as a measure of the attractive forces at the surface of a liquid?

11. What type of substances may be added to a liquid to decrease its surface tension?

12. What is defined as the pressure exerted by the vapor of a liquid at equilibrium?

13. What is defined as the temperature at which a solid is converted into a liquid?

14. Ionic solids have what kind of melting points?

15. If the temperature of 10.0 liters of gas is increased from 273 K to 546 K, what will be the new volume of that gas? Assume that all other conditions are held constant.

16. _____ is the conversion of a gas to a liquid.

17. _____ solids have no organized, regular structure.

18. A(n) _____ is a device for measuring gas pressure.

19. A(n) _____ is a device for measuring heat.

20. Ice is a _____ solid.

Vocabulary Quiz for Chapter Six

1. The _____ theory explains the behavior of an ideal gas.

2. The inverse relationship of pressure and volume in a gas is a statement of _____ law.

3. The direct proportionality of volume and temperature of a gas was first recognized by _____.

4. _____ decrease surface tension of a liquid.

5. An _____ solid gas has no regular, repeating pattern.

6. A _____ measures gas pressure.

7. _____ is a force per unit area.

8. The standard pressure of gas is _____ atm.

9. _____ solids are good electrical conductors.

10. A(n) _____ gas is one in which gas particles do not interact and the volume of the individual gas particles is assumed to be negligible.

7 *Solutions*

Learning Goals

1 Distinguish among the terms *solution*, *solute*, and *solvent*.
2 Describe the various kinds of solutions, and give examples of each.
3 Describe the relationship between solubility and equilibrium.
4 Calculate solution concentration in units of weight/volume percent, weight/weight percent, parts per thousand, and parts per million.
5 Calculate solution concentration using molarity.
6 Perform dilution calculations.
7 Describe and explain concentration-dependent solution properties.
8 Describe the ways in which the chemical and physical properties of water make it a truly unique solvent.
9 Explain the role of electrolytes in blood and their relationship to the process of dialysis.

Introduction

A **solution** is a homogeneous mixture of two or more substances. A solution is composed of one or more **solutes** dissolved in a **solvent.** When the solvent is water we refer to the homogeneous mixture as an **aqueous solution.**

7.1 Properties of Solutions

Learning Goal 1: Distinguish among the terms *solution, solute,* and *solvent.*

General Properties of Liquid Solutions

A **solution** is made up of a **solvent** (the substance in excess) and one or more **solutes** that are dissolved in the solvent.

Liquid solutions are clear and transparent. They may be colored or colorless, depending upon the properties of the solute and solvent.

Learning Goal 2: Describe the various kinds of solutions, and give examples of each.

Solutions may be classified as **electrolytic** or **nonelectrolytic.** Electrolytic solutions are formed from ionic compounds that dissociate in solution to produce ions.

Electrolytic solutions are good conductors of electricity. Nonelectrolytic solutions are formed from nondissociating molecular solutes (nonelectrolytes), and these solutions are nonconducting. Aqueous sodium chloride solution is an example of an electrolytic solution; sugar water is a common nonelectrolytic solution.

Solutions and Colloids

A **colloidal suspension** is not a true solution; the colloid particles are not identical in size nor homogeneously distributed throughout the solution. Particles with diameters between one nanometer and two hundred nanometers are colloids. Smaller particles form a true solution; larger particles are precipitates.

A colloidal suspension is characterized by its light-scattering property, the Tyndall effect.

Degree of Solubility

The degree of solubility, how much solute can dissolve in a given volume of solvent, is difficult to predict, but general trends are based upon the following considerations:

1. The magnitude of difference between polarity of solute and solvent (the rule of "like dissolves like").

2. Temperature. An increase in temperature usually, but not always, increases solubility.

3. Pressure. Pressure has little effect on the solubility of solids and liquids in liquids. However, the solubility of gases in solution is extremely pressure dependent.

When a solution contains all the solute that can be dissolved at a particular temperature, it is **saturated.** Cooling a saturated solution often results in the formation of a **precipitate.** Occasionally, however, on cooling, the excess solute may remain in solution for a period of time. Such a solution is described as a **supersaturated solution.**

Solubility and Equilibrium

Learning Goal 3: Describe the relationship between solubility and equilibrium.

When an excess of solute is brought into contact with a solvent, the resulting dissolution establishes a dynamic equilibrium.

Solubility of Gases: Henry's Law

Henry's law states that the number of moles of a gas dissolved in a liquid at a given temperature is proportional to the partial pressure of the gas. In other words, the gas solubility is directly proportional to the pressure of that gas in the atmosphere that is in contact with the liquid.

7.2 Concentration Based on Mass

Learning Goal 4:	Calculate solution concentration in units of weight/volume percent, weight/weight percent, parts per thousand, and parts per million.

The amount of solute dissolved in a given amount of solution is defined as the solution concentration. The concentration of a solution has a profound effect on the properties of a solution, both physical (melting and boiling points) and chemical (solution reactivity). The more widely used concentration units are considered below.

Weight/Volume Percent

$$\% \left(\frac{W}{V} \right) = \frac{\text{grams of solute}}{\text{mL of solution}} \times 10^2 \, \%$$

Example 1

Calculate the %(W / V) of a sodium chloride solution that was made by mixing 15.0 grams of NaCl with enough distilled water to prepare 500.0 mL of solution.

Work

1. Use the equation for % (W / V): $\qquad \% \left(\frac{W}{V} \right) = \frac{\text{grams of solute}}{\text{mL of solution}} \times 10^2$

2. The solute is NaCl, so grams of solute = 15.0 grams of NaCl.

3. $\% \left(\frac{W}{V} \right) = \frac{15.0 \text{ grams}}{500.0 \text{ mL}} \times 10^2$

   $\% \text{ (W / V)} = 3.00 \, \%$

Answer

The solution is a 3.00 % (W / V) sodium chloride solution.

Example 2

How many grams of glucose are found in 1.00 liter of 5.00% (W / V) glucose solution?

Work

1. The 5.00% (W / V) glucose solution tells us that there are 5.00 grams of glucose per 100.00 mL of total solution.

Example 2 continued

2. Use conversion factors to find the number of grams of glucose that are present.

3. $1.00 \text{ liter} \times \dfrac{1000 \text{ mL}}{1 \text{ liter}} \times \dfrac{5.00 \text{ grams glucose}}{100.00 \text{ mL solution}} = 50.0 \text{ mL}$

Answer
In 1.00 liter of 5.00% (W / V) glucose solution there are 50.0 grams of glucose.

Example 3
If a patient is to receive 2.00 grams of a specific drug, and the drug is supplied as a 5.00% (W / V) drug solution, how many milliliters of the drug solution must the patient be given?

Work
1. The 5.00% (W / V) drug solution means that there are 5.00 grams of the drug per 100.00 mL of the drug solution.
2. Use conversion factors to find the number of mL of drug solution needed.

3. $2.00 \text{ grams} \times \dfrac{100.00 \text{ mL of drug solution}}{5.00 \text{ grams of the drug}} = 40.0 \text{ mL}$

Answer
The patient must take 40.0 mL of the 5.00% (W / V) drug solution to receive 2.00 grams of the specific drug.

Weight/Weight Percent

$$\% \left(\frac{W}{W} \right) = \frac{\text{grams of solute}}{\text{grams of solution}} \times 10^2 \%$$

Example 4
Calculate the % (W / W) of gold in a wedding band that has a mass of 12.0 grams and was found to contain 8.97 g of gold.

Work
1. Use the equation for % (W / W): $\% \left(\dfrac{W}{W} \right) = \dfrac{\text{grams of solute}}{\text{grams of solution}} \times 10^2$

2. $\% \left(\dfrac{W}{W} \right) = \dfrac{8.97 \text{ grams}}{12.0 \text{ grams}} \times 10^2$

   $\% (W / W) = 74.8\%$

Example 4 continued

Answer
The wedding band is 74.8% (W / W) gold.

Parts per Thousand (ppt) and Parts Per Million (ppm)

$$\text{ppt} = \frac{\text{grams of solute}}{\text{grams of solution}} \times 10^3 \text{ ppt} \quad \text{and} \quad \text{ppm} = \frac{\text{grams of solute}}{\text{grams of solution}} \times 10^6 \text{ ppm}$$

Example 5
Recalculate Example 4 in (a) ppt and (b) ppm.

Work

1. Use the equation for ppt $\qquad\qquad \text{ppt} = \dfrac{\text{grams of solute}}{\text{grams of solution}} \times 10^3 \text{ ppt}$

$$\text{ppt} = \frac{8.97 \text{ g}}{12.0 \text{ g}} \times 10^3 \text{ ppt}$$

$$\text{ppt} = 748$$

2. Use the equation for ppm $\qquad\qquad \text{ppm} = \dfrac{\text{grams of solute}}{\text{grams of solution}} \times 10^6 \text{ ppm}$

$$\text{ppm} = \frac{8.97 \text{ g}}{12.0 \text{ g}} \times 10^6 \text{ ppm}$$

$$\text{ppm} = 7.48 \times 10^5$$

Answer
The wedding band is 748 ppt and 7.48×10^5 ppm gold.

7.3 Concentration Based on Moles

Learning Goal 5:   Calculate solution concentration using molarity.

Molarity

   Molarity, symbolized M, is defined as the number of moles of solute per liter of solution:

$$\text{Molarity} = M = \frac{\text{moles solute}}{\text{L solution}}$$

Example 6

Calculate the molarity of a sodium chloride solution made by mixing 3.51 grams of solid NaCl in enough water to prepare 500.0 mL of total solution.

Work

1. Use the molarity equation:
$$M = \frac{\text{moles solute}}{\text{L solution}}$$

2. **When using molarity, convert all volumes to liters!**
   500.0 mL of solution is the same as 0.5000 liter.

3. The number of moles of solute (NaCl) in the solution is found by first determining the weight of 1 mole of NaCl.
$$1 \text{ mole of NaCl} = 58.5 \text{ grams}$$

4. Since we have only 3.51 grams of NaCl, we use a conversion factor to find the number of moles of NaCl that is really in our solution:

$$(3.51 \text{ grams of NaCl})\left(\frac{1 \text{ mole NaCl}}{58.5 \text{ grams of NaCl}}\right) = 0.0600 \text{ moles NaCl}$$

5. We can now use the molarity equation, since we know all the needed values:
   The number of moles of NaCl in the solution = 0.0600 moles
   The number of liters of solution = 0.5000 liter

6. $M = \dfrac{\text{moles solute}}{\text{L solution}}$

   $M = \dfrac{0.0600 \text{ moles}}{0.5000 \text{ L}}$

   Molarity of our solution = 0.120 M

Answer

We have prepared a 0.120 molar NaCl solution.

Example 7

Calculate the number of grams of solid silver nitrate needed to prepare 250.0 mL of a 0.100 molar $AgNO_3$ solution.

Work

1. One mole of $AgNO_3$ is found to be 169.9 grams/mole (from the periodic table).

Example 7 continued

2. Convert 250.0 mL into liter units:

$$250.0 \text{ mL} \times \frac{1 \text{ liter}}{1000 \text{ mL}} = 0.250 \text{ liter solution}$$

3. 0.100 molar $AgNO_3$ solution means that there is 0.100 mole of $AgNO_3$ per 1 liter of solution.
4. Using suitable conversion factors:

$$0.250 \text{ L} \times \frac{0.100 \text{ mole of } AgNO_3}{1 \text{ liter of solution}} \times \frac{169.9 \text{ grams of } AgNO_3}{1 \text{ mole of } AgNO_3} = 4.25 \text{ grams of } AgNO_3$$

Answer
We need to add 4.25 grams of solid $AgNO_3$ to prepare 250.0 mL of a 0.100 molar $AgNO_3$ solution.

Dilution

Learning Goal 6:   Perform dilution calculations.

The technique of dilution is often used to prepare less-concentrated solutions. The approach to such a calculation is outlined below:

M_1 = molarity prior to dilution
M_2 = molarity after dilution
V_1 = volume prior to dilution
V_2 = volume after dilution
$(M_1)(V_1) = (M_2)(V_2)$

Knowing any three of these terms enables one to calculate the fourth.
The dilution equation is valid with any concentration units, such as % (W / V) or % (V / V), as well as molarity. Be certain to use the same units for both initial and final concentration values. Only in this way will proper unit cancellation occur.

Example 8
How many mL of 12.0 M HCl solution do we need to add to distilled water so we can prepare 500.0 mL of 2.50 M HCl solution?

Work
1.   Use $(M_1)(V_1) = (M_2)(V_2)$
       V_1 is the starting volume: we do not know this value yet.

Example 8 continued

M_1 is the molarity of the solution we have available to us. We will dilute some of it to make our final solution.

M_2 is the molarity of our final solution. It always will be the lower value.

V_2 is the final volume of our diluted solution.

2. From our problem:

$V_1 = ?$ (not known yet)

$M_1 = 12.0\ M$ HCl solution (the one we start with)

$M_2 = 2.50\ M$ HCl solution (the one we want to make)

$V_2 = 500.0$ mL (the volume we want to prepare)

3. Rearrange the equation to find V_1:

$$(M_1)(V_1) = (M_2)(V_2)$$

$$V_1 = \frac{(M_2)(V_2)}{M_1}$$

4. $$V_1 = \frac{(2.50\ M)(500.0\ \text{mL})}{(12.0\ M)}$$

$$V_1 = 104\ \text{mL}$$

Answer

If we measure out 104 mL of the 12.0 M HCl solution and then add distilled water to it in another container until we have made 500.0 mL of total solution, we will have prepared 500.0 mL of a 2.50 M HCl solution by dilution.

7.4 Concentration-Dependent Solution Properties

Learning Goal 7: **Describe and explain concentration-dependent solution properties.**

Solution properties that are dependent upon the concentration of the solute particles, rather than the identity of the solute, are referred to as colligative properties. There are four colligative properties of solutions.

Vapor Pressure Lowering

Raoult's law states that when a solute is added to a solvent, the vapor pressure of the solvent decreases in proportion to the concentration of solute particles.

Freezing Point Depression and Boiling Point Elevation

When a nonvolatile solid is added to a solvent, the freezing point of the resulting solution decreases (a lower temperature is required to form the solid state), and the boiling point increases (requiring a higher temperature to form the gaseous state).

The magnitude of the freezing point depression (ΔT_f) is proportional to the solute concentration over a limited range of concentration.

$$\Delta T_f = k_f \text{ (solute concentration)}$$

The boiling point elevation is also proportional to the solute concentration.

$$\Delta T_b = k_b \text{ (solute concentration)}$$

If the value of the proportionality factor is known, the magnitude of the freezing point depression or boiling point elevation may be calculated for a solution of known concentration of particles.

We have already worked with one mole-based unit, molarity, and this concentration unit can be used to calculate either the freezing point depression or boiling point elevation.

A second mole-based concentration unit is molality, which is more commonly used in these types of situations. **Molality** (symbolized by m) is defined as the number of moles of solute per kilogram of solvent in a solution:

$$m = \frac{\text{moles solute}}{\text{kg solvent}}$$

Molality does not vary with temperature, whereas molarity is temperature dependent.

Osmotic Pressure

Osmosis is the movement of solvent from a dilute solution to a more concentrated solution through a **semipermeable membrane.** Pressure must be applied to the more concentrated solution to stop this flow, and the magnitude of the pressure required to just stop the flow is termed the **osmotic pressure.** The "driving force" for the osmotic process is the need to establish an equilibrium between the solutions on either side of the membrane.

The osmotic pressure, not unlike the pressure exerted by a gas, may be treated quantitatively. Osmotic pressure, symbolized by π, follows the same form as the ideal gas equation:

Ideal gas		**Osmotic pressure**
$PV = nRT$		$\pi V = nRT$
or $\quad P = n\dfrac{RT}{V}$		or $\quad \pi = n\dfrac{RT}{V}$
and, since $\quad M = \dfrac{n}{V}$		and, since $\quad M = \dfrac{n}{V}$
then, $\quad P = MRT$		then, $\quad \pi = MRT$

The osmotic pressure may be calculated from the solution concentration at any given temperature. Osmosis is a colligative property; it is dependent on the concentration of solute particles.

By convention, the molarity of particles in solution is termed osmolarity, abbreviated Osm, for osmotic pressure calculations.

Osmosis is important as a transport mechanism to all living organisms. A living cell contains an aqueous solution, and material movement in and out of cells is based partly upon osmosis. If the concentration of the fluid surrounding red blood cells is higher than that inside the cell (a **hypertonic solution**), water flows from the cell, causing it to collapse. This process is known as crenation. On the other hand, too low a concentration of this fluid relative to the solution within the blood cell (a **hypotonic solution**) will cause water to flow into the cell, causing cell rupture, a process known as **hemolysis.**

Two solutions are **isotonic** if they have identical osmotic pressures. In that case, the osmotic pressure differential across the cell is zero, and no cell disruption occurs.

7.5 Water as a Solvent

Learning Goal 8:	Describe the ways in which the chemical and physical properties of water make it a truly unique solvent.

The role of water in the solution process deserves special attention because of its many unique characteristics:

1. It is often referred to as the universal solvent.

2. In view of its small size, it is remarkable that it exists as a liquid at room temperature.

3. Contrary to our expectations, liquid water is more dense than its solid form, ice.

4. Water is the principal biological solvent. Approximately 60% of the adult human body is water.

7.6   Electrolytes in Body Fluids

Learning Goal 9:	**Explain the role of electrolytes in blood and their relationship to the process of dialysis.**

Proper cell function in biological systems is critically dependent on the concentration of electrolytes. The maintenance of proper muscle and nervous system function depends on the sodium/potassium ratio inside and outside of the cell. A stable osmotic pressure in biological fluids depends on the electrolyte concentration inside and outside of the cell.

Glossary of Key Terms in Chapter Seven

aqueous solution (7.1) any solution in which the solvent is water.

colligative property (7.4) property of a solution that is dependent only on the concentration of solute particles.

colloidal suspension (7.1) a heterogeneous mixture of solute particles in a solvent; distribution of solute particles is not uniform because of the size of the particles.

concentration (7.2) a measure of the quantity of a substance contained in a specified volume of solution.

concentration gradient (7.4) region where concentration decreases over a distance.

dialysis (7.6) the removal of waste material via transport across a membrane.

diffusion (7.4) net movement of solute or solvent molecules from an area of high concentration to an area of low concentration.

electrolyte (7.1) a material that dissolves in water to produce a solution that conducts an electrical current.

equivalent (7.4) number of grams of an ion corresponding to Avogadro's number of electrical charges.

Henry's Law (7.1) a law stating that the number of moles of a gas dissolved in a liquid at a given temperature is proportional to the partial pressure of the gas.

hypertonic solution (7.4) the more concentrated solution of two separated by a semipermeable membrane.

hypotonic solution (7.4) the more dilute solution of two separated by a semipermeable membrane.

isotonic solution (7.4) a solution that has the same solute concentration as another solution with which it is being compared; a solution that has the same osmotic pressure as a solution existing within a cell.

molarity (7.3) the number of moles of solute per liter of solvent.

nonelectrolyte (7.1) a substance that, when dissolved in water, produces a solution that does not conduct an electrical current.

osmolarity (7.4) molarity of particles in solution; this value is used for osmotic pressure calculation.

osmosis (7.4) net flow of a solvent across a semipermeable membrane in response to a concentration gradient.

osmotic pressure (7.4) the net force with which water enters a solution through a semipermeable membrane; alternatively, the pressure required to stop net transfer of solvent across a semipermeable membrane.

precipitate (7.1) an insoluble substance formed and separated from a solution.

Raoult's law (7.4) a law stating that the vapor pressure of a compound is equal to its mole fraction times the vapor pressure of the pure component.

saturated solution (7.1) one in which undissolved solute is in equilibrium with the solution.

selectively permeable membranes (7.4) a membrane that restricts diffusion of some ions and molecules (based upon size and charge) across a membrane.

semipermeable membrane (7.4) a membrane permeable to the solvent but not the solute; a material that allows the transport of certain substance from one side of the membrane to the other.

solubility (7.1) the amount of a substance that will dissolve in a given volume of solvent at a specified temperature.

solute (7.1) a component of a solution that is present in lesser quantity than the solvent.

solution (7.1) a homogeneous (uniform) mixture of two or more substances.

solvent (7.1) the solution component that is present in the largest quantity.

supersaturated solution (7.1) a solution that is more concentrated than a saturated solution. (Note that such a solution is not at equilibrium.)

suspension (7.1) a heterogeneous mixture of particles; the suspended particles are larger than those found in a colloidal suspension.

weight/volume percent (% (W/V)) (7.2) the concentration of a solution expressed as a ratio of grams of solute to milliliters of solution multiplied by 100%.

weight/weight percent (% (W/W)) (7.2) the concentration of a solution expressed as a ratio of mass of solute to mass of solution multiplied by 100%.

Self Test for Chapter Seven

1. What is the % (W / V) of NaOH in 150.0 mL of a solution that contains 1.65 grams of sodium hydroxide?

2. How many milliliters of a 5.000% (W / V) glucose solution would contain 40.00 grams of glucose?

3. How many liters of D-5-W solution [5.00% (W / V) glucose solution] would need to be given to a patient by I.V. so that she would receive 250.0 food-calories of energy from the given glucose? One gram of glucose produces 4.00 food-calories of energy in the human body.

4. How many moles of KCl are present in 100.0 mL of a 0.552 molar potassium chloride solution?

5. How many grams of KCl are present in 50.0 mL of a 0.125 molar potassium chloride solution?

   KCl = 74.60 grams/mole.

6. If 20.80 grams of HCl are added to enough distilled water to form 3.0 liters of solution, what is the molarity of the solution? HCl = 36.46 grams/mole.

7. Calculate the molarity of a solution that contains 5.60 grams of KNO_3 in 300.0 mL of solution.

   KNO_3 = 101.1 grams/mole.

8. How many mL of 12.0 M HCl solution are required to prepare 250.0 mL of a 2.50 M HCl solution?

9. What term defines the amount of solute in grams or moles dissolved in a given amount of solution?

10. Normal physiological saline solution is a 0.900% (W / V) NaCl solution. How many grams of solid NaCl are needed to prepare 1.50 liters of this solution?

11. What term is defined as the number of moles of solute per liter of solution?

12. What term defines the movement of solvent through a semipermeable membrane from a dilute solution to a more concentrated solution?

13. Is a concentrated solution hypertonic or hypotonic when compared to a dilute solution?

14. What is the term used to describe solutions of exactly the same particle concentration?

Vocabulary Quiz for Chapter Seven

1. _____ is the shrinkage of red blood cells due to water loss to the surrounding medium.

2. A(n) _____ is a substance that ionizes in water.

3. A(n) _____ is the more concentrated solution of two separated by a semipermeable membrane.

4. A(n) _____ is the more dilute solution of two separated by a semipermeable membrane.

5. _____ is the number of moles of solute per liter of solution.

6. _____ is the net flow of solvent though a semipermeable membrane.

7. _____ is the pressure required to stop net transfer of solvent across a semipermeable membrane.

8. _____ describes a solution in which undissolved solute is in equilibrium with the solution.

9. The _____ is a component of a solution that is present in lesser quantity than the solvent.

10. A(n) _____ is a solution that is more concentrated than a saturated solution.

8 *Acids and Bases*

Learning Goals

1 Identify acids and bases and acid-base reactions.
2 Describe the role of the solvent in acid-base reactions.
3 Write equations describing acid-base dissociation and label the conjugate acid-base pairs.
4 Calculate pH from concentration data.
5 Calculate hydronium and/or hydroxide ion concentration from pH data.
6 Provide examples of the importance of pH in chemical and biochemical systems.
7 Describe the meaning and utility of neutralization reactions.
8 Describe the application of buffers to chemical and biochemical systems, particularly blood chemistry.

Introduction

A **solution** is a homogeneous (or uniform) mixture of two or more substances. A solution is composed of one or more **solutes,** dissolved in a **solvent.** When the solvent is water, we refer to the homogeneous mixture as an **aqueous solution.**

Two very important classes of solution reactions involve transfer of either positive or negative charge: acid-base reactions (proton transfer) and oxidation-reduction reactions (electron transfer). Each type is described in this chapter and is applied in numerous ways throughout the rest of your study of chemistry.

8.1 Acids and Bases

Learning Goal 1:   Identify acids and bases and acid-base reactions.

Acids have a sour taste, dissolve some metals, and cause vegetable dyes to change color. Release of the hydrogen ion (H+) is characteristic of acids. Bases have a bitter taste, a slippery feel, and accept hydrogen ions.

Arrhenius Theory of Acids and Bases

An Arrhenius acid dissociates to form hydrogen ions. An Arrhenius base dissociates to form hydroxide ions. The **Arrhenius theory** is less comprehensive than the Brønsted-Lowry theory, discussed below.

Brønsted-Lowry Theory of Acids and Bases

A **Brønsted-Lowry** acid is a proton donor and a Brønsted-Lowry base is a proton acceptor. (Remember that H^+ is simply a proton, so a *proton donor* is giving away an H^+.)
Examples include:

$$\text{acid} \quad HCl + H_2O \rightarrow H_3O^+ + Cl^-$$

$$\text{base} \quad NH_3 + H_2O \rightleftharpoons NH_4^+ + OH^-$$

(Note: The double arrow ($\rightleftharpoons$) implies that ammonia is a *weak* base. See the following discussion.)

Acid-Base Properties of Water

Learning Goal 2:	**Describe the role of the solvent in acid-base reactions, and explain the meaning of the term *pH*.**

Water is **amphiprotic,** meaning that it possesses *both* acid and base properties. The fact that water can either accept or donate protons (H^+ ions) makes it an excellent solvent for both acids and bases.

In this reaction, water acts a base, and accepts a proton from hydrochloric acid:

$$HCl + H_2O \rightarrow H_3O^+ + Cl^-$$

In this reaction, water acts as an acid, and donates a proton to ammonia:

$$NH_3 + H_2O \rightarrow NH_4^+ + OH^-$$

Acid and Base Strength

The terms acid or base *strength* and acid or base *concentration* are easily confused. Strength is a measure of the degree of dissociation of an acid or base in solution, independent of its concentration. Concentration refers to the amount of acid or base per quantity of solution.

The strength of acids and bases in water is dependent upon the extent to which they react with the solvent, water. Acids and bases are classified as *strong* when the reaction with water is virtually 100% complete and as *weak* when much less than 100% complete. All strong bases are metal hydroxides. Strong bases completely dissociate, or ionize, in aqueous solution to produce hydroxide ions and metal cations. For example, NaOH dissociates to form Na^+ and OH^- ions in water. Weak acids and weak bases dissolve in water principally in the molecular form. Only a small percentage of the molecules dissociate to form the **hydronium** or **hydroxide** ion. For example, the weak base ammonia, NH_3, dissolves in water primarily as NH_3, but a small percentage reacts with water to form NH_4^+ and OH^-.

The most fundamental chemical difference between strong and weak acids and bases is their equilibrium situation. A strong acid, such as HCl, does not exist in equilibrium with its ions, H_3O^+ and Cl^-. A weak acid, such as acetic acid, establishes a dynamic equilibrium with its ions, H_3O^+ and $C_2H_3O_2^-$. (Note that the acetate ion may be written as CH_3COO^- or as the more condensed form $C_2H_3O_2^-$.) This reaction is represented by the equilibrium constant expression:

$$CH_3COOH + H_2O \rightleftharpoons H_3O^+ + CH_3COO^-$$

$$K_a = \frac{[H_3O^+][CH_3COO^-]}{[CH_3COOH]}$$

where K_a is the equilibrium constant or acid-dissociation constant for acetic acid. The larger the dissociation constant, the more dissociated (hence, stronger) the acid will be.

The situation for bases is analogous. The equilibrium expression for the weak base ammonia in water is

$$NH_3 + H_2O \rightleftharpoons NH_4^+ + OH^-$$

$$K_b = \frac{[NH_4^+][OH^-]}{[NH_3]}$$

Note that in each of the above cases, the solvent, water, is not written as part of the equilibrium constant expression.

Conjugate Acids and Bases

Learning Goal 3: **Write equations describing acid-base dissociation and label the conjugate acid-base pairs.**

Any acid-base reaction can be represented by the general equation

$$HA + B \rightleftharpoons BH^+ + A^-$$

where HA represents an acid and B represents a base. In the forward reaction, the acid donates a proton to the base. The products of this reaction also have acid-base properties. The protonated base can, in fact, donate a proton to the anion, A^-, reforming HA and B. The product acids and bases are termed *conjugate acids and bases*.

A **conjugate acid** is the species formed when a base accepts a proton.

A **conjugate base** is the species formed when an acid donates a proton.

The acid and base on the opposite sides of the equation are collectively termed a **conjugate acid-base pair.** For our general expression, B and BH$^+$ are a conjugate acid-base pair. Similarly, HA and A$^-$ are a conjugate acid-base pair.

The Dissociation of Water

Although pure water is virtually 100% molecular, a small number of water molecules do ionize. This process occurs by the transfer of a proton from one water molecule to another, producing a hydronium and hydroxide ion. This equilibrium is shown below:

$$H_2O + H_2O \rightleftharpoons H_3O^+ + OH^-$$

This process is the **autoionization,** or self-ionization of water. Water is, therefore, a very weak electrolyte and a poor conductor of electricity. Water has both acid and base properties; the dissociation produces both the hydronium and hydroxide ion.

Water at room temperature has a hydronium ion concentration of 1.0×10^{-7} M. The hydroxide ion concentration is also 1.0×10^{-7} M.

The product of hydronium and hydroxide ion concentration is the **ion product for water,**

$$\text{ion product} = [H_3O^+][OH^-] = [1.0 \times 10^{-7}][1.0 \times 10^{-7}] = 1.0 \times 10^{-14} = K_w$$

The ion product is a constant because its value is not dependent upon the nature or concentration of the solute, as long as the temperature does not change. This relationship is the basis for the pH scale.

Example 1

If we have a 0.200 molar solution of sodium hydroxide, what are the hydroxide ion (OH$^-$) and hydronium ion (H$_3$O$^+$) concentrations in this solution?

Work
1. If we have a 0.200 molar NaOH solution, it means that we have 0.200 mole of Na$^+$ ions and 0.200 mole of OH$^-$ ions per liter of this solution.
2. Thus, [OH$^-$] = 0.200 moles/liter
3. Next use the ion product for water (K_w): ion product = $[H_3O^+][OH^-]$ = 1.0×10^{-14}
   The above equation will be used whenever we know the value of H$_3$O$^+$ or OH$^-$ and want to find the other one. The nature and concentration of the solutes added to water do alter the concentrations of H$_3$O$^+$ and OH$^-$ present, but the product $[H_3O^+][OH^-]$ always equals 1.0×10^{-14}.
4. Rearrange ion product = $[H_3O^+][OH^-]$ = 1.0×10^{-14} to find $[H_3O^+]$

$$[H_3O^+] = \frac{\text{ion product}}{[OH^-]}$$

Example 1 continued

5. $[H_3O^+] = \dfrac{\text{ion product}}{[OH^-]}$

$[H_3O^+] = \dfrac{1.0 \times 10^{-14}}{2.00 \times 10^{-1}}$

$[H_3O^+] = 0.50 \times 10^{-13} = 5.0 \times 10^{-14}$

Answer
In our solution, the $[OH^-] = 0.200$ moles/L and the $[H_3O^+] = 5 \times 10^{-14}$ moles/L.

8.2 pH: A Measurement Scale for Acids and Bases

Learning Goal 4:   **Calculate pH from concentration data.**

Learning Goal 5:   **Calculate hydronium and/or hydroxide ion concentration from pH data.**

A Definition of pH

The **pH scale** correlates the hydronium ion concentration with a number, the pH, which serves as a useful indicator of the degree of acidity or basicity of a solution. The pH scale specifies "how acidic" or "how basic" a solution is.

1. Addition of an acid (proton donor) to water *increases* $[H_3O^+]$ and decreases $[OH^-]$.

2. Addition of a base (proton acceptor) to water *decreases* $[H_3O^+]$ and increases $[OH^-]$.

3. $[H_3O^+] = [OH^-]$ when *equal* amounts of acid and base are present.

4. In all of the above cases, $[H_3O^+][OH^-] = 1.0 \times 10^{-14} = $ ion product for water at 25°C.

The pH of a solution is defined as the negative logarithm of the molar concentration of the hydronium ion.

$$pH = -\log [H_3O^+]$$

Measuring pH
The pH of a solution may be measured by a pH meter. Also, indicating paper (pH paper) can give an approximate value for the solution pH.

Calculating pH
The pH of a solution can be calculated if the concentration of either H_3O^+ or OH^- is known. Also, $[H_3O^+]$ or $[OH^-]$ can be calculated from the pH.

103

Example 2

Calculate the pH of a 0.010 molar HCl solution.

Work

1. Since HCl is a strong acid, if 1 mole of HCl dissociates, it produces 1 mole of H_3O^+ ions.

   Therefore a 1.0×10^{-2} M HCl solution has $[H_3O^+] = 1.0 \times 10^{-2}$ M.

2. Use the pH equation:

   pH = –logarithm of the molar concentration of the hydronium ion

   or

   pH = –log $[H_3O^+]$ (Note: [ ] means molar.)

3. pH = –log $[H_3O^+]$

   pH = –log (1.0×10^{-2})

   pH = –(log 1.0 + log 10^{-2})   (Note that log 1.0 = zero; log $10^{-2} = -2$)

   pH = –[0 + (–2)]

   pH = –(–2)

   pH = 2.0

Answer

The pH of a 0.010 molar HCl solution is pH = 2.0

Example 3

Calculate the pH of a 1.0×10^{-4} molar solution of NaOH.

Work

1. NaOH is a strong base; thus, in our problem $[OH^-] = 1.0 \times 10^{-4}$ molar.

2. Use ion product = $[H_3O^+][OH^-]$ to find $[H_3O^+]$, so we can use the pH equation.

3. $[H_3O^+] = \dfrac{\text{ion product}}{[OH^-]}$

   $[H_3O^+] = \dfrac{1.0 \times 10^{-14}}{1.0 \times 10^{-4}}$ $[H_3O^+] = 1.0 \times 10^{-10}$

4. Next, use the pH equation:

   pH = –log$[H_3O^+]$

   pH = –log (1×10^{-10})

   pH = –(log 1.0 + log 10^{-10})

   pH = –[0 + (–10)]

   pH = 10.0

Answer

The pH of a 1.0×10^{-4} M NaOH solution is pH = 10.0

104

The Importance of pH and pH Control

Learning Goal 6:	Provide examples of the importance of pH in chemical and biochemical systems.

Solution pH and pH control play a major role in many facets of our everyday lives:

1. Agriculture (soil pH)
2. Physiology (blood pH)
3. Industry
4. Municipal services (purification of drinking water; sewage treatment)
5. Acid rain

8.3 Reactions Between Acids and Bases

Neutralization

Learning Goal 7:	Describe the meaning and utility of neutralization reactions.

The reaction of an acid with a base to produce a salt and water is a **neutralization reaction.** In the strictest sense, neutralization requires equal numbers of moles of H_3O^+ and OH^- to produce a neutral solution (no excess acid or base).

A neutralization reaction may be used to determine the concentration of an unknown acid or base solution. The technique of **titration** involves the addition of measured amounts of a standard solution (one whose concentration is known) to neutralize the second, unknown solution. From the volumes of the two solutions and the concentration of the **standard solution,** the concentration of the unknown solution may be determined.

Example 4

A 5.00 g sample of vinegar is titrated with 0.100 *M* NaOH. If the vinegar requires 40.0 mL of the NaOH for a complete reaction,
   a. Calculate the *M* of acetic acid in the vinegar.
   b. Calculate the % (W / V) of acetic acid in the vinegar.
Assume that the density of the vinegar solution is 1.00 g/mL, and that the only acidic component of the vinegar solution is acetic acid (CH_3COOH). The reaction is:

$$CH_3COOH(aq) + NaOH(aq) \rightarrow CH_3COONa(aq) + H_2O(l)$$

Work, Part A
1. Since the coefficients of both acetic acid and sodium hydroxide (the reactants) in the equation are 1, moles acid = moles base at the equivalence point.

Example 4 continued

2. moles acid = M_{acid} x $V_{liters\ acid}$

   moles base = M_{base} x $V_{liters\ base}$

3. Therefore, M_{acid} x $V_{liters\ acid}$ = M_{base} x $V_{liters\ base}$
   and,

   $$M_{acid} = M_{base} \ x \ \frac{(V_{liters\ base})}{(V_{liters\ acid})}$$

4. M_{base} = 0.100 M

   $$V_{liters\ base} \ = \ 40.0 \ mL \ x \ \frac{1\ L}{10^3\ mL} \ = 0.0400 \ L$$

   $$V_{liters\ acid} = 5.00 \ g \ \ x \ \frac{1\ mL}{1.00\ g} \ = 5.00 \ mL$$

   $\longleftarrow$————— Density conversion

   and

   $$V_{liters\ acid} = 5.00 \ mL \ x \ \frac{1\ L}{10^3\ mL} = 0.00500 \ L$$

5. Substituting,

   $$M_{acid} \ = \ M_{base} \ x \ \frac{(V_{liters\ base})}{(V_{liters\ acid})}$$

   $$M_{acid} \ = \ 0.100 \ M \ x \ \frac{0.0400\ L}{0.00500\ L}$$

   M_{acid} = 0.800 M acetic acid

Work, Part B

1. % (W/V) = $\dfrac{grams\ acetic\ acid}{milliliters\ of\ solution}$ x 100%

2. grams acetic acid =

   $$\frac{0.800\ mL\ acetic\ acid}{1\ L} \ x \ \frac{60.0\ g\ acetic\ acid}{1\ mol\ acetic\ acid} \ = \ \frac{48.0\ g\ acetic\ acid}{1\ L}$$

Example 4 continued

3. Convert L to mL:
$$\frac{48.0 \text{ g acetic acid}}{1 \text{ L} \times \dfrac{10^3 \text{ mL}}{1 \text{ L}}} = \frac{0.0480 \text{ g acetic acid}}{1 \text{ mL}}$$

4. Substitute into the initial equation for % (W / V)

$$\% \text{ (W/V)} = \frac{0.0480 \text{ g}}{1 \text{ mL}} \times 100\% = 4.80\%$$

Answer
a. The vinegar solution is 0.800 M acetic acid
b. The vinegar solution is 4.80% (W / V)

Polyprotic Substances

Polyprotic substances can donate or accept more that one proton per formula unit. Sulfuric acid, H_2SO_4, is a diprotic acid, and phosphoric acid, H_3PO_4, is a triprotic acid, and barium hydroxide, $Ba(OH)_2$, is a diprotic base. Acid-base reactions may occur in various combining ratios if they involve a polyprotic substance.

One mole of a diprotic acid, such as H_2SO_4, can react with two moles of NaOH.

$$H_2SO_4 + 2\,NaOH \rightarrow Na_2SO_4 + 2\,H_2O$$

Dissociation of a diprotic acid takes place in two steps.

$$H_2SO_4 + H_2O \rightarrow HSO_4^- + H_3O^+$$

$$HSO_4^- + H_2O \rightleftharpoons HSO_4^{2-} + H_3O^+$$

8.4 Acid-Base Buffers

Learning Goal 8:   **Describe the application of buffers to chemical and biochemical systems, particularly blood chemistry.**

A buffer solution contains components that enable the solution to resist large changes in pH when acids or bases are added.

The Buffer Process

The basis of buffer action is the establishment of an equilibrium between either a weak acid and its conjugate base, or a weak base and its conjugate acid.

A buffer solution functions in accordance with LeChatelier's principle (Section 5.4), which states that an equilibrium system, when stressed, will shift its equilibrium to alleviate that stress.

Addition of Base (OH⁻) to a Buffer Solution

For the acetic acid/sodium acetate system:

$$CH_3COOH + H_2O \rightleftharpoons H_3O^+ + CH_3COO^-$$

OH⁻ added, equilibrium shifts to the right

$\longrightarrow$

Addition of Acid (H₃O⁺) to a Buffer Solution

For the acetic acid/sodium acetate system:

$$CH_3COOH + H_2O \rightleftharpoons H_3O^+ + CH_3COO^-$$

H₃O⁺ added, equilibrium shifts to the left

$\longleftarrow$

Higher-than-normal CO_2 levels shift the above equilibrium to the right increasing $[H_3O^+]$ and lowering the pH. A situation of high blood CO_2 levels and low pH is **acidosis.**

Lower-than-normal CO_2 levels shift the equilibrium to the left, decreasing $[H_3O^+]$ and making the pH more basic; this condition is termed **alkalosis** (from alkali, implying basic in nature).

Glossary of Key Terms in Chapter Eight

acid (8.1) a substance that behaves as a proton donor.
amphiprotic (8.1) a substance that can behave either as a Brønsted acid or a Brønsted base.
Arrhenius theory (8.1) describes an acid as a substance that dissociates to produce H+ and a base as a substance that dissociates to produce OH⁻.
autoionization (8.1) or self-ionization, the reaction of a substance, such as water, with itself to produce a positive and a negative ion.
base (8.1) a substance that behaves as a proton acceptor.
Brønsted-Lowry theory (8.1) describes an acid as a proton donor and a base as a proton acceptor.
buffer capacity (8.4) a measure of the ability of a solution to resist large changes in pH when a strong acid or strong base is added.
buffer solution (8.4) a solution containing a weak acid or base and its salt; it is resistant to large changes in

pH upon addition of strong acid or bases.

buret (8.3) a device calibrated to deliver accurately know volumes of liquid, as in a titration.

conjugate acid (8.1) substance that has one more proton than the base from which it is derived.

conjugate acid-base pair (8.1) two species related to each other through the gain or loss of a proton.

conjugate base (8.1) substance that has one proton fewer than the acid from which it is derived.

equivalence point (8.3) describes the situation in which reactants have been mixed in the molar ratio corresponding to the balanced equation.

hydronium ion (8.1) a protonated water molecule.

indicator (8.3) a solute that shows some condition of a solution (such as acidity or basicity) by its color.

ion product for water (8.1) the product of the hydronium and hydroxide ion concentrations in pure water at a specified temperature, $25^{\circ}C$; it has a value of 1.0×10^{-14}

neutralization (8.3) the reaction between an acid and a base.

pH scale (8.2) a numerical representation of acidity or basicity of a solution (pH = -log [H+]).

polyprotic substance (8.3) a substance that can donate or accept more than one proton per molecule.

standard solution (8.3) a solution whose concentration is accurately known.

titration (8.3) the process of adding a solution from a buret to a sample until a reaction is completed, at which time the volume is accurately measured and the concentration of the sample is calculated.

Self Test for Chapter Eight

1. Calculate the pH of a solution that contains 6.50×10^{-5} moles of hydronium ion per liter.

2. If 14.8 mL of 0.100 M NaOH solution are needed to completely react with 25.0 mL of an unknown HCl solution, what is the molarity of the unknown HCl solution?

3. If we have a 0.200 molar solution of NaOH, what is the molarity of H_3O^+ ions present in the solution?

4. Calculate the pH of a 1.0×10^{-3} molar solution of HCl.

5. What is the term for the reaction of an acid plus a base to yield a salt plus water?

6. Which theory defines an acid as a proton donor and a base as a proton acceptor?

7. What is the measure of the strength of an acid in water?

8. Which one or ones of the following are not strong acids?

    a. HCl

    b. CH_3COOH

    c. H_2SO_4

    d. HNO_3

9. Which one or ones of the following are not strong bases?

    a. NaOH

    b. NH_4OH

    c. $Ca(OH)_2$

    d. KOH

Vocabulary Quiz for Chapter Eight

1. The _____ theory describes the behavior of bases such as ammonia.

2. _____ is the reaction of a substance, such as water, with itself to produce a positive and negative ion.

3. _____ is a solution that contains an acid-base pair; that acid-base pair is resistant to large changes in pH upon addition of strong acids or bases.

4. The _____ theory describes an acid as a proton donor and a base as a proton acceptor.

5. A _____ substance can donate or accept more than one proton per molecule.

6. _____ is a numerical representation of acidity or basicity of a solution.

7. A(n) _____ solution is a solution whose concentration is accurately known.

8. _____ is the process of adding from a buret a solution to a sample until a reaction is completed, at which time the volume is accurately measured.

9 The Nucleus and Radioactivity

Learning Goals

1 Enumerate the characteristics of alpha, beta, and gamma radiation.
2 Write balanced equations for nuclear processes.
3 Calculate the amount of radioactive substance remaining after a specified period of time has elapsed.
4 Describe how nuclear energy can generate electricity: fission, fusion, and the breeder reactor.
5 Cite examples of the use of radioactive isotopes in medicine.
6 Describe the use of ionizing radiation in cancer therapy.
7 Discuss the preparation and use of radioisotopes in diagnostic imaging studies.
8 Explain the difference between natural and artificial radioactivity.
9 Describe the characteristics of radioactive materials that relate to radiation exposure and safety.
10 Be familiar with common techniques for the detection of radioactivity.
11 Know the common units of radiation intensity: the curie, roentgen, rad, and rem.

Introduction

This chapter considers the nucleus, nuclear properties, and their applications.

9.1 Natural Radioactivity

Radioactivity is the process by which atoms emit high-energy particles or rays. These particles or rays are termed radiation. Nuclear radiation occurs as a result of an alteration in nuclear composition or structure, which happens because the nucleus is unstable and hence radioactive.

Nuclear symbols are analogous to atomic symbols. The nuclear symbols consist of the elemental symbol, the atomic number (equivalent to the number of protons in the nucleus), and the mass number, which is defined as the sum of the neutrons and protons in the nucleus. Note: Be careful not to confuse the *mass number* (neutrons and protons) with the *atomic mass*, which includes the contribution of electrons, and is a true mass figure.

Only unstable nuclei undergo change and produce radioactivity. Not all atoms of a particular element undergo radioactive decay. When writing the symbols for a nuclear process, it is important to designate the particular isotope involved.

Three types of natural radiation emitted by unstable nuclei are alpha particles, beta particles, and gamma rays.

| Learning Goal 1: | Enumerate the characteristics of alpha, beta, and gamma radiation. |

111

Alpha Particles

Alpha particles (α) contain two protons and two neutrons. An alpha particle is identical to the helium ion (He^{2+}) and is represented as:

$$^4_2 He$$

or

$$\alpha$$

Alpha particles emitted by radioisotopes move relatively slowly (approximately 10% of the speed of light), and they are stopped by very little mass.

Beta Particles and Positrons

The beta particle (β) is a fast-moving electron traveling at approximately 90% of the speed of light as it leaves the nucleus. The beta particle is represented as

$$^0_{-1} e$$

or

$$\beta$$

the subscript -1 is written in the same position as the atomic number and, like the atomic number (number of protons), indicates the charge of the particle.

Beta particles are smaller, faster, and more energetic and penetrating than alpha particles. A positron has the same mass as a beta particle but carries a positive charge:

$$^0_{+1} e$$

Gamma Rays

Gamma rays (γ) are pure energy. Since energy has no mass or charge, the symbol for a gamma ray is simply:

$$\gamma$$

Gamma radiation is highly energetic and is the most penetrating form of nuclear radiation.

Example 1
List the names, symbols, and characteristics of the three basic types of radiation.

Example 1 continued

Answer
1. The alpha particles have the lowest velocity of the three types of radiation. They have a mass of about 4.0 amu and are composed of two protons and two neutrons. Notice that they have the same structure as the nucleus of a helium atom. Since they have two protons present, they have a charge of +2. The alpha particle may be symbolized as $^{4}_{2}He$.
2. The beta particle is just a fast-moving electron that has been produced during a nuclear reaction. It has a speed of about 90% the speed of light. Since it is only an electron, it has a charge of -1 and may be symbolized as $^{0}_{-1}e$.
3. The gamma rays are pure energy, in contrast to alpha and beta radiation, which are particles. Gamma radiation is highly energetic and is the most penetrating and deadly to living things. It is represented by the gamma symbol, γ.

Properties of Alpha, Beta, and Gamma Radiation

Important properties of α, β, and γ radiation are summarized in Table 9.1 of the textbook.

9.2 Writing a Balanced Nuclear Equation

Learning Goal 2:   Write balanced equations for nuclear processes.

A **nuclear equation** represents a nuclear process such as radioactive decay. To write a balanced equation, we must remember the following:

1. The total mass on each side of the reaction arrow must be identical.

2. The sum of the charges of the reactant nuclei must be equal to the sum of the charges of the product nuclei.

Alpha Decay

A reaction involving alpha decay is shown below:

$$^{226}_{88}Ra \quad \rightarrow \quad ^{222}_{86}Rn \quad + \quad ^{4}_{2}He$$
Radium-226 $\qquad$ Radon-222 $\qquad$ Helium-4

Beta Decay

Beta decay may be illustrated by the following:

$$^{234}_{90}Th \quad \rightarrow \quad ^{234}_{91}Pa \quad + \quad ^{0}_{-1}e$$
Thorium-234 $\qquad$ Protactinium-234 $\qquad$ β particle

Positron Emission

An example of positron emission is the conversion of a proton to a neutron:

$$_1^1H \rightarrow \ _0^1n \ + \ _1^0e$$

Proton   neutron   positron

Gamma Production

If gamma radiation were the only product of nuclear decay, there would be no change in the mass or identity of the radioactive nuclei, since a gamma ray is pure energy, possessing no mass or charge. The gamma emitter has simply gone to a lower energy state. The decay of the **metastable isotope** technetium-99m is shown:

$$_{43}^{99m}Tc \quad \rightarrow \quad _{43}^{99}Tc \quad + \quad \gamma$$

Predicting Products of Nuclear Decay

It is possible to use a nuclear equation to predict one of the products of a nuclear reaction if the others are known. We know that the mass and charge of the total of all products and reactants must be equal. By difference, we can compute the missing charge and mass that represent the unknown product and deduce its identity.

Example 2
Complete the following nuclear equations:

1. $_{92}^{238}U \rightarrow \ _{90}^{234}Th \ + \ ?$

2. $_7^{16}N \rightarrow \ _8^{16}O \ + \ ?$

3. $_{19}^{40}K \rightarrow \ _{-1}^0e \ + \ ?$

4. $_{79}^{197}Au \ + \ ? \rightarrow \ _{79}^{198}Au$

Work
Use the following rules:
1. The total mass on each side of the equation arrow must be identical.
2. The sum of the atomic numbers on each side of the reaction arrow must be identical.

Answer
The missing parts are given below:
1. $_2^4He$ 2. $_{-1}^0e$ 3. $_{20}^{40}Ca$ 4. $_0^1n$

Example 3

Show how molybdenum-98 can be converted by neutron bombardment into technetium-99 plus beta particles. The technetium produced is unstable and can be used as a gamma source for tracer applications.

Work

1. From the periodic table, we find that molybdenum has an atomic number of 42, and technetium's is 43.
2. Next we set up the equation:

$$^{98}_{42}\text{Mo} + {}^{1}_{0}\text{n} \rightarrow {}^{99m}_{43}\text{Tc} + {}^{0}_{-1}\text{e}$$

Notice that the "m" following the mass number means that the isotope is metastable.

Answer

$$^{98}_{42}\text{Mo} + {}^{1}_{0}\text{n} \rightarrow {}^{99m}_{43}\text{Tc} + {}^{0}_{-1}\text{e}$$

9.3 Properties of Radioisotopes

Nuclear Structure and Stability

The energy that holds the protons, neutrons, and other particles together in the nucleus is the **binding energy** of the nucleus. When an isotope decays, some of this binding energy is released.

Factors related to nuclear stability include the following:

1. Nuclear stability correlates with the ratio of neutrons to protons in the isotope.
2. Nuclei with large numbers of protons (84 or more) tend to be unstable.
3. Isotopes containing 2, 8, 20, 50, 82, or 126 protons or neutrons are stable. These "**magic numbers**" seem to indicate the presence of stable energy levels in the nucleus.
4. Isotopes with even numbers of protons or neutrons are generally more stable than those with odd numbers of protons or neutrons.

Example 4

Explain why some isotopes are more stable than others.

Answer

Nuclear stability correlates with the ratio of neutrons to protons in a specific isotope. For elements with atomic numbers less than 84, a neutron/proton ratio of 1 usually means a stable isotope of a specific element. Elements that have an atomic number over 83 usually form only unstable isotopes. Isotopes containing the magic numbers of 2, 8, 20, 50, 82, or 126 protons or neutrons are stable. Isotopes with an even number of protons or neutrons are generally more stable than those with odd numbers of protons or neutrons.

Half-Life

Learning Goal 3: Calculate the amount of radioactive substance remaining after a specified period of time has elapsed.

The **half-life,** $t_{1/2}$, is the time required for one-half of a given quantity of a substance to undergo change. Each isotope has its own characteristic half-life.

The degree of stability of an isotope is indicated by the isotope's half-life. Isotopes with short half-lives decay rapidly; they are very unstable.

Decay of a radioisotope that has a reasonably short $t_{1/2}$ is experimentally determined by following its activity as a function of time. Graphing the results produces a radioactive decay curve (see Figure 9.2 in the textbook).

Example 5

Radium-226 ($^{226}_{88}Ra$) has a half-life of 1620 years. If a watch's hands are coated with paint containing Ra-226 in 1920, why is it still radioactive?

Answer

Since only 72 years (a small fraction of one half-life, ($\frac{72 \text{ years}}{1620 \text{ years}}$) have passed, nearly all the radium-226 used in the paint is still radioactive.

Example 6

The following half-lives are known for different isotopes of radium:

Ra-223 ....11 days
Ra-224 ....3.64 days
Ra-226 ....1620 years
Ra-228 ....6.7 years

Why is there such a difference in their half-lives?

Answer

It is primarily due to the proton/neutron ratio in the isotope, but other as yet undiscovered reasons may play a part as well.

Example 7

How much of a 1.00-gram sample of Ra-226 would remain after three half-lives? The half-life of Ra-226 is 1620 years.

Example 7 continued

Work

	first		second		third	
1.00g	$\longrightarrow$	0.50 g	$\longrightarrow$	0.25 g	$\longrightarrow$	0.13 g
	half-life		half-life		half-life	

Answer

0.13 grams of radium-226 remain.

9.4 Nuclear Power

Learning Goal 4: Describe how nuclear energy can generate electricity: fission, fusion, and the breeder reactor.

Energy Production

Einstein's equation relating mass and energy predicts that a small amount of nuclear mass is converted to a very large amount of energy when the nuclear particle breaks apart.

Einstein's equation is as follows:

$$E = mc^2$$

where E = energy

m = mass

c = speed of light

Three major modes of generating electrical power through a nuclear reaction involve the processes of fission, fusion, and the use of what is termed a breeder reactor.

Nuclear Fission

Fission (splitting) occurs when a heavy nuclear particle is split into smaller nuclei and large amounts of energy by a smaller nuclear particle (such as a neutron). The fission process intensifies, producing very large amounts of energy. This process of intensification is referred to as a chain reaction.

Nuclear Fusion

Fusion (meaning to join together) results from the combination of two small nuclei to form a larger nucleus with the concurrent release of large amounts of energy. The best example of a fusion reactor is the sun. Continuous fusion processes furnish our solar system with light and heat.

$$^2_1H + {}^3_1H \rightarrow {}^4_2He + {}^1_0n + \text{Energy}$$

Breeder Reactors

A **breeder reactor** literally manufactures its own fuel. A perceived shortage of fissionable isotopes makes the breeder an attractive alternative to fission reactors. A breeder reactor uses uranium-238, which is abundant but nonfissionable. In a series of steps, the uranium-238 is converted to plutonium-239, which is fissionable and undergoes a fission chain reaction, producing energy.

9.5 Medical Applications of Radioactivity

Learning Goal 5:	Cite examples of the use of radioactive isotopes in medicine.

The use of radiation in the treatment of various forms of cancer, as well as the newer area of **nuclear medicine,** has become widespread in the past quarter-century.

Cancer Therapy Using Radiation

Learning Goal 6:	Describe the use of ionizing radiation in cancer therapy.

When high-energy radiation, such as gamma radiation, passes through a cell, it may collide with one of the molecules in the cell and cause it to lose one or more electrons, producing an ion pair. For this reason, such radiation is termed **ionizing radiation.** Ions produced in this way may cause subtle changes in cellular biochemical processes, which may result in diminished or altered cell function, or in extreme cases the death of the cell.

An organ that is cancerous is composed of both healthy cells and malignant cells. Cells undergoing division are particularly sensitive to gamma radiation. Therefore, exposing the tumor area to controlled dosages of high-energy gamma radiation from cobalt-60 (a high-energy gamma ray source) will generally kill a higher percentage of abnormal cells than normal cells.

Nuclear Medicine

The diagnosis of a host of biochemical irregularities or diseases of the human body has been made routine through the use of *radioactive tracers.* Tracers are small amounts of radioactive substances used as probes to study internal organs. Because the isotope is radioactive, its path may be followed using suitable detection devices. A "picture" of the organ is obtained, often far more detailed than is possible with conventional X-rays. Such techniques are noninvasive; that is, surgery is not required to investigate the condition of the internal organ, eliminating the risk associated with an operation. These techniques are successful because the radioactive isotope of an element has exactly the same chemical behavior as any other isotope of the same element.

Making Isotopes for Medical Applications

Learning Goal 7:	Discuss the preparation and use of radioisotopes in diagnostic imaging studies.

Learning Goal 8:	Explain the difference between natural and artificial radioactivity.

The radioactivity produced by unstable isotopes is described as **natural radioactivity.** If a normally stable, nonradioactive nucleus is made radioactive through bombardment with protons, neutrons, or alpha particles, the resulting radioactivity is termed **artificial radioactivity.**

The bombardment process is often accomplished in the core of a **nuclear reactor,** where an abundance of small nuclear particles, particularly neutrons, are available. Alternately, extremely high-velocity charged particles (such as α and β) may be produced in a **particle accelerator,** such as a cyclotron.

These product isotopes are often used in hospital laboratories as tracers in nuclear medicine.

9.6 Biological Effects of Radiation

Learning Goal 9: Describe the characteristics of radioactive materials that relate to radiation exposure and safety.

Radiation Exposure and Safety

Safety considerations are based on the following:

1. The magnitude of the half-life
2. Shielding
3. Distance from the radioactive source
4. Time of exposure
5. Types of radiation emitted

Virtually all applications of nuclear chemistry create radioactive waste and, along with it, the problems of safe handling and disposal. Most disposal sites at present are considered temporary, until a long-term, safe solution can be found.

9.7 Measurement of Radiation

Learning Goal 10: Be familiar with common techniques for the detection of radioactivity.

The changes that take place when radiation interacts with matter provide the basis of operation for various radiation detection devices.

Nuclear Imaging

An isotope is administered to a patient, and the isotope begins to concentrate in the organ of interest. Photographs of that region of the body are taken at periodic intervals using a special type of film. Upon development of the series of photographs, a record of the organ's uptake of the isotope as a function of time enables the radiologist to assess the condition of the organ.

Computer Imaging

A specialized television camera, sensitive to emitted radiation from a radioactive substance administered to a patient, develops a continuous and instantaneous record of the voyage of the isotope throughout the body.

The Geiger Counter

A **Geiger counter** is an instrument capable of detecting ionizing radiation.

Film Badges

A **film badge** is merely a piece of photographic film that is sensitive to energies corresponding to radioactive emissions.

Units of Radiation Measurement

Learning Goal 11:	Know the common units of radiation intensity: the curie, roentgen, rad, and rem.

The Curie
The **curie** is a measure of the amount of radioactivity in a radioactive source. The curie is independent of the nature of the radiation (α, β, or γ) as well as its effect on biological tissue. A curie is defined as the amount of radioactive material that produces 3.7×10^{10} atomic disintegrations per second.

The Roentgen
The **roentgen** is a measure of ionizing radiation (X-ray and gamma ray) only. The roentgen is defined as the amount of radioactive isotope needed to produce 2×10^9 ion pairs when passing through 1 cubic centimeter of air at 0°C. The roentgen is a measure of radiation's interaction with air, and gives no information regarding its effect on biological tissue.

The Rad
The **rad,** or radiation absorbed dosage, provides more meaningful information than either of the previous units of measure. It takes into account the nature of the absorbing material. It is defined as the dosage of radiation able to transfer 2.4×10^{-3} calories of energy to 1 kilogram of matter.

The Rem
The **rem,** or roentgen equivalent for man, describes the biological damage caused by the absorption of different kinds of radiation by the body. The rem is obtained by multiplication of the rad by a factor called the relative biological effect, or **RBE.** The RBE is a function of the type of radiation (α, β, or γ). Although a beta particle is more energetic than an alpha particle, an alpha particle is approximately ten times more damaging to biological tissue. As a result, the RBE is 10 for alpha particles and 1 for beta particles.

An estimated lethal dose, symbolized by **LD$_{50}$,** is 500 rems. The lethal dose, LD$_{50}$, is defined as the dose that would be fatal for 50% of the exposed population within 30 days. Some biological effect, however, is detectable at a level as low as 25 rem.

Glossary of Key Terms in Chapter Nine

alpha particle (9.1) consists of two protons and two neutrons (the alpha particle is identical to helium nucleus).

artificial radioactivity (9.5) radiation that results from the conversion of a stable nucleus to another, unstable nucleus.

background radiation (9.6) the radiation that emanates from natural sources.

beta particle (9.1) an electron formed in the nucleus by the conversion of a neutron into a proton.

binding energy (9.3) the energy required to break down the nucleus into its component parts.

breeder reactor (9.4) a nuclear reactor that produces its own fuel in the process of providing electrical energy.

chain reaction (9.4) in a fission reactor, involves neutron production causing subsequent reactions accompanied by the production of more neutrons in a continuing process.

curie (9.7) the quantity of radioactive material that produces 3.7×10^{10} nuclear disintegrations per second.

fission (9.4) involves heavy nuclei split into lighter nuclei accompanied by the release of large quantities of energy.

fusion (9.4) the joining of light nuclei to form heavier nuclei accompanied by the release of large quantities of energy.

gamma ray (9.1) high-energy emission from nuclear processes, traveling at the speed of light.

half-life (9.3) of an isotope, the length of time required for one-half of the initial mass of the isotope to decay to products.

ionizing radiation (9.1) radiation that is sufficiently high in energy to cause ion formation upon impact.

lethal dose (9.7) the dosage of a toxic material, such as radiation, that would be lethal to 50% of a test population.

metastable isotope (9.2) an isotope that will give up some energy to produce a more stable form of the same isotope.

natural radioactivity (9.5) the spontaneous decay of a nucleus to produce high-energy particles or rays.

nuclear equation (9.2) a balanced equation accounting for the products and reactants in a nuclear reaction.

nuclear imaging (9.5) the generation of images of components of the body (organs, tissues) by using techniques based on the measurement of radiation.

nuclear medicine (9.5) a field of medicine that uses radioisotopes for diagnostic and therapeutic purposes.

nuclear reactor (9.5) a device for conversion of nuclear energy into electrical energy.

nuclide (9.1) nuclear species with specified numbers of protons and neutrons.

particle accelerator (9.5) a device for production of high-energy nuclear particles based on the interaction of charged particles with magnetic and electrical fields.

positron (9.2) particle emitted from the nucleus that has the same mass as an electron but has a positive charge.

rad (9.7) or radiation absorbed dosage, the absorption of 2.4×10^{-3} calories of energy per kilogram of absorbing tissue.

radioactivity (9.1) the process by which atoms emit high-energy particles or rays.

rem (9.7) or roentgen equivalent for man, the product of rad and RBE.

roentgen (9.7) the dose of radiation producing 2.1×10^{9} ions in 1 cm^3 of air at 0°C and 1 atmosphere of pressure.

shielding (9.6) material used to provide protection from radiation.

tracer (9.5) a radioisotope that is rapidly and selectively transmitted to the part of the body for which diagnosis is desired.

Self Test for Chapter Nine

1. Complete the following nuclear reactions by supplying the missing part:

    a. $^{27}_{13}\text{Al} + ^{4}_{2}\text{He} \rightarrow ^{30}_{15}\text{P} + ?$

    b. $^{210}_{82}\text{Pb} \rightarrow ^{0}_{-1}\text{e} + ?$

2. How many mg of a 100.0 mg sample of Tc-99m will remain after 30 hours? The half-life of Tc-99m is 6 hours.

3. When using the following nuclear symbols for an isotope of fluorine, which number is equal to the atomic number: $^{19}_{9}\text{F}$?

4. Name the three types of natural radiation emitted by unstable nuclei.

5. Which of the types of radiation is really pure energy?

6. Which of the types of radiation is the least-penetrating form of nuclear radiation?

7. What term means that an isotope is unstable?

8. What is defined as the time period required for one-half of a given quantity of a substance to undergo a nuclear change?

9. Nuclear power plants use what nuclear process to produce energy?

10. What term do we use for cells that undergo unnaturally rapid cell division?

11. Where does iodine tend to concentrate in the human body?

12. What is the term that describes the amount of radiation attributable to our surroundings on a day-to-day basis?

13. The intensity of radiation varies in what way with the square of the distance from the source?

14. What do we call the specialized system that is sensitive to emitted radiation from a radioactive substance administered to a patient and produces a continuous picture of the voyage of the isotope throughout the body?

15. What badge is often worn by staff members that shows their exposure to radiation?

16. Which type of radiation travels at the speed of light?

17. What is defined as the spontaneous decay of a nucleus to produce high-energy particles or rays?

18. Which type of radiation is similar to a helium atom?

19. Which radioactive element is found in some homes?

Vocabulary Quiz for Chapter Nine

1. The _____ consists of two protons and two neutrons, carries a +2 charge, and results from nuclear decay.

2. The _____ is an electron formed in the nucleus by the conversion of a neutron into a proton.

3. The energy required to break down the nucleus into its component parts is called the _____.

4. _____ is the name of the equation that represents the energy equivalent of mass being equal to the mass times the square of the speed of light.

5. _____ is the process of joining of light nuclei to form heavier nuclei accompanied by the release of large amounts of energy.

6. A form of electromagnetic radiation from nuclear processes is termed _____.

7. The length of time required for one-half of the initial mass of an isotope to decay to products is referred to as the _____.

8. _____ is radiation that is sufficiently high in energy to cause ion formation upon impact.

9. The _____ is the dosage of radiation that would be fatal to 50% of the exposed population.

10. A (n) _____ is a radioisotope that is selectively transmitted to the part of the body for which diagnosis is desired.

10 *An Introduction to Organic Chemistry: The Saturated Hydrocarbons*

Learning Goals

1. Compare and contrast organic and inorganic compounds.
2. Draw structures that represent each of the families of organic compounds.
3. Write the names and draw the structures of the common functional groups.
4. Write condensed and structural formulas for saturated hydrocarbons.
5. Describe the relationship between the structure and physical properties of saturated hydrocarbons.
6. Use the basic rules of the I.U.P.A.C. Nomenclature System to name alkanes and substituted alkanes.
7. Draw constitutional (structural) isomers of simple organic compounds.
8. Write the names and draw the structures of simple cycloalkanes.
9. Write equations for combustion reactions of alkanes.
10. Write equations for halogenation reactions of alkanes.

Introduction

Until 1828 it was thought that all organic compounds were derived from natural sources. All early attempts to synthesize these compounds in the laboratory failed, and it was proposed that a vital force was necessary for their formation. In 1828, Wöhler synthesized urea (an organic compound) from potassium cyanate and ammonium sulfate (two inorganic compounds). Wöhler's experiment is now recognized as the beginning of modern organic chemistry.

Example 1

Draw the dash structures of ammonium cyanate and urea. Why are these two compounds important to the beginning of modern organic chemistry?

Answer

1. Dash structure of ammonium cyanate:

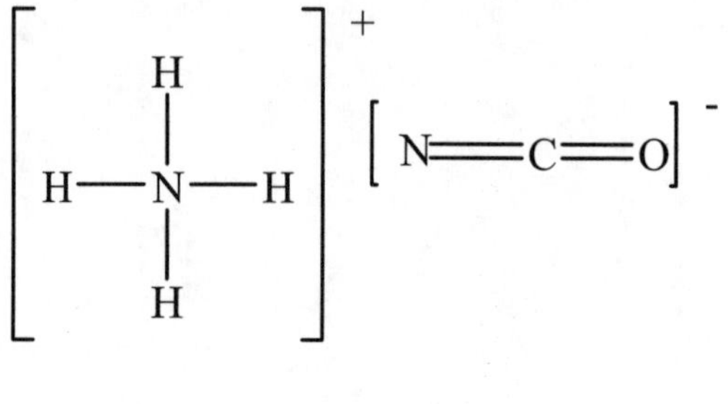

Example 1 continued

2. Dash structure of urea:

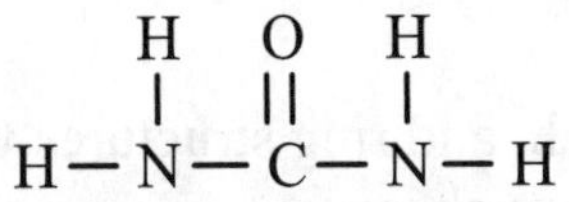

$$\begin{array}{ccccc} H & & O & & H \\ | & & || & & | \\ H-N & - & C & - & N-H \end{array}$$

3. In 1828, the German chemist, Friedrich Wöhler was trying to produce ammonium cyanate, but due to experimental conditions, the compound urea was produced. The two compounds have the molecular formula, CON_2H_4, but they have different structures and properties. The "vital force theory" was laid to rest. Up until his discovery, it was believed that only living things could make compounds like urea, which is found as a waste product in the urine of animals.

10.1 The Chemistry of Carbon

There are several reasons for the existence of hundreds of thousands of organic compounds:

1. Carbon atoms are able to form stable, covalent bonds with other carbon atoms. Consider the family of alkanes as an example. Each molecule in this family contains only carbon and hydrogen, but each is a different chemical substance with unique chemical and physical properties. There are an infinite number of possible alkanes; alkanes with hundreds or even thousands of carbon and hydrogen atoms may be easily synthesized. Even elemental carbon exists in different forms, allotropic forms, which have different properties.

2. Carbon can form stable bonds with other elements, for example nitrogen, oxygen, sulfur, and the halogens.

3. The number of ways that these elements may combine to form unique structures is practically limitless. Compounds having the same molecular formulas but different structures (hence, different properties) are termed **constitutional** or **structural isomers.**

Example 2
List three reasons why there are so many organic compounds and provide examples.

Answer
1. Carbon can form chain, branch, and ring structures with other carbon atoms:
   a. Chain structures with no branches:

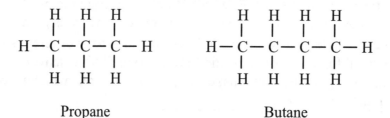

   Propane   Butane

   b. Chain structure that also has branch structures:

   Isobutane
```
            H   H   H
            |   |   |
        H — C — C — C — H
            |   |   |
            H   |   H
                |
            H — C — H
                |
                H
```

   c. Ring structures:

   Cyclobutane
```
            H   H
            |   |
        H — C — C — H
            |   |
        H — C — C — H
            |   |
            H   H
```

2. Carbon can form compounds that also contain other elements, such as oxygen, sulfur, and nitrogen, in their structure:

```
            H                       H   H   H
            |                       |   |   |
        H — C — O — H           H — C — C — N — H
            |                       |   |
            H                       H   H
```

   Methyl alcohol   Ethyl amine

3. Carbon can form compounds that have the same molecular formula but a different structure and a different set of properties. Urea and ammonium cyanate are one example of such a set of isomeric compounds.

Important Differences Between Organic and Inorganic Compounds

> **Learning Goal 1:** Compare and contrast organic and inorganic compounds.

The bonds in organic molecules are almost always covalent bonds, while those found in many inorganic substances are ionic bonds. When comparing organic and inorganic compounds it helps to remember the differences between ionic and covalent bonds.

1. Ionic bonds result from a transfer of one or more electrons. Covalent bonds are formed by a sharing of electrons to form a stable orbital containing two electrons.
2. The ionic bond is electrostatic in nature. It is formed by the attraction of positive and negative ions resulting from the electron transfer process.
3. Ions are arranged in large, three-dimensional crystals, consisting of many positive and negative ions. Covalently bonded substances exist as discrete units: molecules.
4. Ionic compounds often dissociate in solution (electrolytes), whereas most covalently bonded molecules retain their identity in solution (nonelectrolytes).

Families of Organic Compounds

> **Learning Goal 2:** Draw structures that represent each of the families of organic compounds.

Organic compounds are classified according to groups, or families. The two most general classifications are **hydrocarbons** and **substituted hydrocarbons.** Compounds that contain only carbon and hydrogen are classified as hydrocarbons: hydrogen (hydro) and carbon (carbon). Hydrocarbons are subdivided into two principal classes: **aliphatic** and **aromatic hydrocarbons.** Aliphatic hydrocarbons are further subdivided into three families: alkanes, alkenes, and alkynes.

The **alkanes** (and cycloalkanes) are termed **saturated hydrocarbons.** They are composed solely of C—C and C—H single bonds. The **unsaturated hydrocarbons,** hydrocarbons that contain at least one carbon-carbon double bond or carbon-carbon triple bond, are referred to as **alkenes** and **alkynes,** respectively. Aromatic compounds contain a benzene ring or a derivative of the benzene ring.

Substitution of one or more functional groups for hydrogens in a hydrocarbon brings about major changes in properties. A **functional group** is an atom or group of atoms in a molecule principally responsible for the chemical and physical properties of that molecule. All compounds that contain a particular functional group, for example, the hydroxyl group (—OH), are classified as being in the same family. The hydroxyl group is the functional group of the alcohols.

> **Learning Goal 3:** Write the names and draw the structures of the common functional groups.

Example 3
To which family of organic compounds do each of the following belong? (Refer to Table 10.2 in the text.)

1. CH_3CH_3

2. $CH_3CH=CHCH_3$

3. CH_3OH

4.
$$H_3C - \overset{\overset{\displaystyle O}{\|}}{C} - CH_3$$

5. CH_3Br

Answer
1. alkane
2. alkene
3. alcohol
4. ketone
5. alkyl halide

10.2 Alkanes

Structure and Physical Properties

Learning Goal 4:   Write condensed and structural formulas for saturated hydrocarbons.

Alkanes are hydrocarbons that contain only carbon and hydrogen bonded together through carbon-hydrogen and carbon-carbon single bonds. They have the general formula C_nH_{2n+2}.

Several different types of formulas are used to describe organic molecules. The **structural formula** shows all of the atoms in a molecule and shows all bonds as lines. The **molecular formula** provides the atoms and number of each type of atom in a molecule, but gives no information regarding the bonding pattern. The **condensed formula** shows all of the atoms in a molecule and places them in a sequential arrangement that details which atoms are bonded to each other. The simplest is the **line formula.** In this formula, we assume that there is a carbon at any location where two or more lines meet. We also assume that there is a carbon at the end of any line and that each carbon is bonded to the correct number of hydrogen atoms.

A molecular formula tells us the atomic composition of a molecule, but provides no information about the structure of the molecule. For instance, there are three possible structures for a compound with the molecular formula of C_5H_{12}:

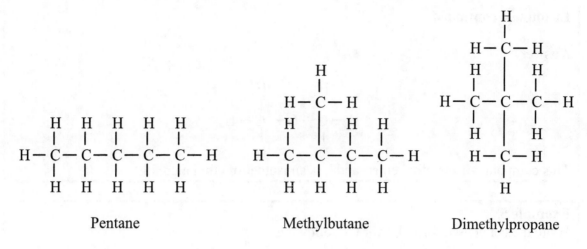

Pentane   Methylbutane   Dimethylpropane

These are the three possible isomers of pentane. These structural formulas provide more information about a molecule.

A suitable compromise between the convenience of the molecular formula and the detail of the structural formula is the condensed formula.

Examples of condensed formulas:

$CH_3CH_2CH_2CH_2CH_3$ $CH_3CHCH_2CH_3$ $H_3C - C - CH_3$

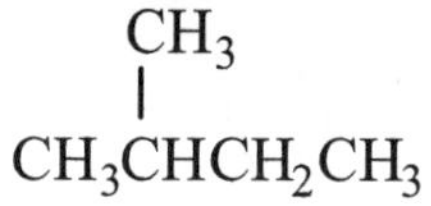

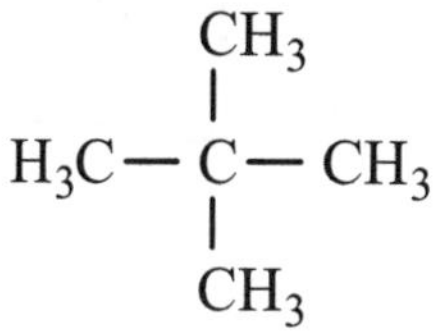

Pentane   Methylbutane   Dimethylpropane

Example of line formulas:

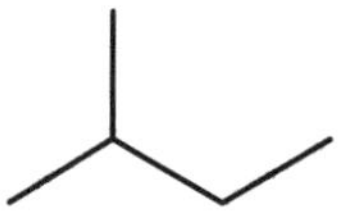

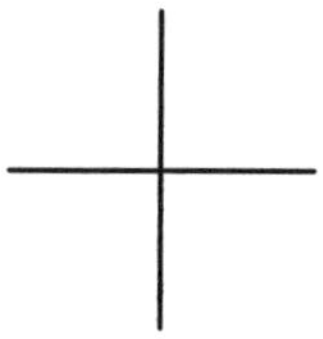

Pentane   Methylbutane   Dimethylpropane

Example 4

Draw the structural formula of an isomer of ethyl alcohol (CH_3CH_2OH), in which an oxygen atom is placed between two carbon atoms.

Work

Hydrogen always forms one bond; carbon forms four bonds and oxygen forms two bonds.

129

Example 4 continued

Answer

$$
\begin{array}{ccc}
& \text{H} & & & & \text{H} & \\
& | & & & & | & \\
\text{H}- & \text{C} & -\text{O}- & \text{C} & -\text{H} \\
& | & & & & | & \\
& \text{H} & & & & \text{H} &
\end{array}
$$

This compound is dimethyl ether, and it is an isomer of ethyl alcohol.

Example 5
Write the condensed formulas for the following:

1. Propane

$$
\begin{array}{ccccc}
\text{H} & \text{H} & \text{H} \\
| & | & | \\
\text{H}-\text{C}-\text{C}-\text{C}-\text{H} \\
| & | & | \\
\text{H} & \text{H} & \text{H}
\end{array}
$$

2. Methylbutane

$$
\begin{array}{c}
\text{H} \\
| \\
\text{H}-\text{C}-\text{H} \\
\text{H} \quad | \quad \text{H} \quad \text{H} \\
| \quad | \quad | \quad | \\
\text{H}-\text{C}-\text{C}-\text{C}-\text{C}-\text{H} \\
| \quad | \quad | \quad | \\
\text{H} \quad \text{H} \quad \text{H} \quad \text{H}
\end{array}
$$

3. Methyl alcohol

$$
\begin{array}{c}
\text{H} \\
| \\
\text{H}-\text{C}-\text{O}-\text{H} \\
| \\
\text{H}
\end{array}
$$

4. Acetone

$$
\begin{array}{c}
\text{H} \quad \text{O} \quad \text{H} \\
| \quad || \quad | \\
\text{H}-\text{C}-\text{C}-\text{C}-\text{H} \\
| \qquad | \\
\text{H} \qquad \text{H}
\end{array}
$$

5. 2-Chloropropane

$$
\begin{array}{c}
\text{H} \quad \text{H} \quad \text{H} \\
| \quad | \quad | \\
\text{H}-\text{C}-\text{C}-\text{C}-\text{H} \\
| \quad | \quad | \\
\text{H} \quad \text{Cl} \quad \text{H}
\end{array}
$$

Answer
1. $CH_3CH_2CH_3$
2. $CH_3CH(CH_3)CH_2CH_3$
3. CH_3OH
4. CH_3COCH_3
5. $CH_3CH(Cl)CH_3$

Example 6
Write the molecular formulas of the alkanes that contain 4, 8, and 40 carbon atoms.

Work
Use the general formula for finding the molecular formula of any alkane, provided you are given the number of carbon atoms.

C_nH_{2n+2} where n = number of carbon atoms

130

Example 6 continued

Answer

4 carbons: C_4H_{10}; 8 carbons: C_8H_{18}; 40 carbons: $C_{40}H_{82}$

Learning Goal 5:   Describe the relationship between the structure and physical properties of saturated hydrocarbons.

Since all the hydrocarbons are composed of nonpolar carbon-carbon and carbon-hydrogen bonds, hydrocarbons are nonpolar molecules. As a result, they are not water soluble, but are readily soluble in nonpolar solvents. Furthermore, virtually all of the hydrocarbons are less dense than water and have relatively low melting points and boiling points.

Alkyl Groups

Alkyl groups result when a hydrogen atom is removed from an alkane. The name is derived from the name of the corresponding alkane by removing the -*ane* ending and replacing it with -*yl*.

Carbon atoms are classified according to the number of carbon atoms to which they are attached. A **primary (1°) carbon** is directly bonded to one other carbon. A **secondary (2°) carbon** is bonded to two other carbon atoms. A **tertiary (3°) carbon** is bonded to three other carbon atoms. Similarly, alkyl groups are classified according to the number of carbon atoms attached to the carbon atom that joins the alkyl group to a molecule.

Nomenclature

Learning Goal 6:   Use the basic rules of the I.U.P.A.C. Nomenclature System to name alkanes and substituted alkanes.

The basic rules used for naming compounds in the I.U.P.A.C. Nomenclature System follow:

1. The name of the compound is defined by the longest continuous carbon chain in the compound. This chain is the **parent compound.** (Refer to Table 10.3 in the text for the names of parent compounds.)

2. Each substituent attached to the parent compound is given a name and a number. The number designates the position of the substituent on the main chain, and the name tells what type of substituent is present at that position. If the location of the substituent is unambiguous, no number is needed (see Examples 7 and 9).

3. The chain must be numbered from one end to the other in order to provide the lowest position number for each substituent. If more than one substituent is present, number from the end that gives the lowest number of the first substituent encountered regardless of the numbers that result for the other substituents.

4. If the same substituent occurs more than once in the compound, a separate position number is supplied for each substituent, and the prefixes *di-*, *tri-*, *tetra-*, *penta-*, *hexa-*, *hepta-*, and so forth are used.

5. Place the names of the substituents in alphabetical order before the name of the parent compound. Numbers are separated from each other by commas, and numbers are separated from names by hyphens. By convention halogen substituents are placed before alkyl substituents in this priority sequence regardless of the alphabetization. Names of some common alkyl groups are listed in Tables 10.5 and 10.6.

Example 7
Give the I.U.P.A.C. names of the following branched-chain alkanes:

$$CH_3$$
$$|$$
1. $CH_3CHCH_2CH_3$ 2. $CH_3CH_2CCH_2CH_2CH_3$
$$|$$ $$|$$
$$CH_3$$ $$CH_3$$

Work
1. Determine the longest continuous carbon chain. This is called the parent compound.
2. Number the parent compound chain so that the branches have the lowest possible numbers.

Answer
1. The parent chain is butane. Since the only possible position for the methyl group is carbon-2, no number is needed. The correct name of this branched alkane is methylbutane.
2. The parent chain is hexane. We will number from left to right. The correct name of this branched alkane is 3,3-dimethylhexane.

Example 8
Give the I.U.P.A.C. name of the following:

$$CH_3$$
$$|$$
$$CH_3CHCH_2CCH_2Br$$
$$| |$$
$$Br   CH_3$$

Work
1. The parent compound is pentane.
2. Number the parent compound from right to left:

$$CH_3$$
$$|$$
$$C-C-C-C-C-Br$$
$$| |$$
$$Br   CH_3$$

5   4   3   2   1

132

Example 8 continued

Answer
1,4-dibromo-2,2-dimethylpentane

Constitutional or Structural Isomers

Learning Goal 7:	Draw constitutional (structural) isomers of simple organic compounds.

Two molecules having the same molecular formulas but different structures are called **constitutional** or **structural isomers.** Isomers having the same molecular formula are unique compounds due to their structural differences. They may have similar physical and chemical properties, but in many cases their properties are quite dissimilar.

Example 9
Draw and give the I.U.P.A.C. names of all the alkane isomers of C_5H_{12}.

Work
1. The first isomer you should always draw is the one that has all the carbons in a continuous chain structure:

$$CH_3CH_2CH_2CH_2CH_3 \qquad \text{I.U.P.A.C. name} = \text{pentane}$$

2. Next draw four carbons in a continuous chain, and connect the fifth carbon as a methyl group to this continuous chain:

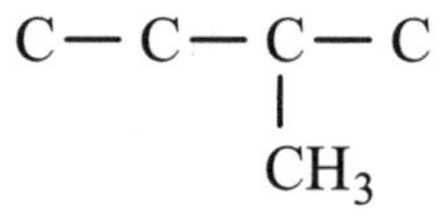

$$
\begin{array}{cccc}
C & - C & - C & - C \\
& & | & \\
& & CH_3 &
\end{array}
$$

Fill in the structure using hydrogen atoms, so that each carbon atom has four bonds.

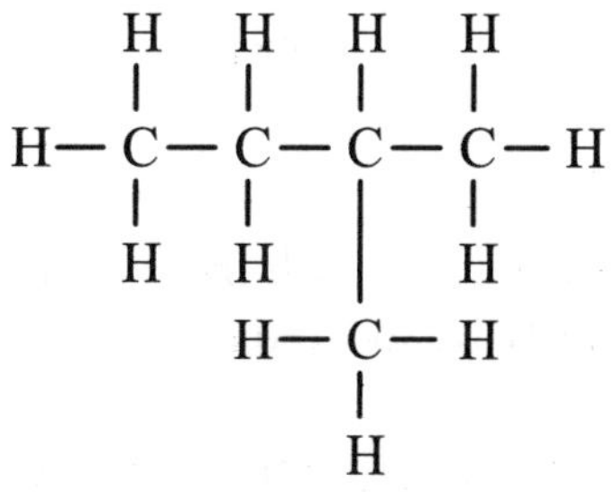

$$
\begin{array}{ccccc}
& H & H & H & H \\
& | & | & | & | \\
H- & C & - C & - C & - C - H \\
& | & | & | & | \\
& H & H & | & H \\
& & & H - C - H & \\
& & & | & \\
& & & H &
\end{array}
$$

Then write the condensed structure:

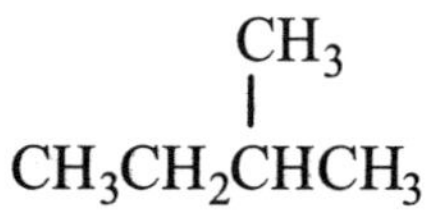

$$
\begin{array}{c}
CH_3 \\
| \\
CH_3CH_2CHCH_3
\end{array}
$$

Example 9 continued

I.U.P.A.C. name = methylbutane. Note that you can move the methyl group one carbon to the left. However, if you apply the I.U.P.A.C. nomenclature test, the product is still methylbutane. Similarly, you can move the methyl group one carbon to the right. The I.U.P.A.C. nomenclature test reveals that you have drawn pentane, the linear isomer.

3. The last isomer of C_5H_{12} is found by drawing three carbons in a continuous chain and connecting the other two carbons as methyl groups to the middle carbon of the continuous chain:

$$
\begin{array}{c}
CH_3 \\
| \\
C - C - C \\
| \\
CH_3
\end{array}
$$

Fill in the structure using hydrogen atoms as before:

$$
\begin{array}{c}
H \\
| \\
H - C - H \\
 | \\
H \quad | \quad H \\
| \quad | \quad | \\
H - C - C - C - H \\
| \quad | \quad | \\
H \quad | \quad H \\
H - C - H \\
| \\
H
\end{array}
$$

Then write the condensed structure:

$$
\begin{array}{c}
CH_3 \\
| \\
CH_3CCH_3 \\
| \\
CH_3
\end{array}
$$

Answer
I.U.P.A.C. name = dimethylpropane. As with methylbutane, no numbers are needed because the methyl groups can only be located on carbon-2.

Remember that all isomers must have different I.U.P.A.C. names. To check whether you have drawn duplicate isomers, name them using the I.U.P.A.C. nomenclature system.

10.3 Cycloalkanes

The **cycloalkanes** are another family of hydrocarbons closely related to the alkanes. Cycloalkanes have the general molecular formula C_nH_{2n}. Note that they contain two fewer hydrogens than the corresponding alkane.

Cycloalkanes are named by adding the prefix *cyclo-* before the name of the alkane with the same number of carbon atoms. For example, cyclobutane is a cyclic alkane that has four carbon atoms. Substituted cycloalkanes are named by placing the name of the substituent before the name of the cycloalkane. No number is needed if only a single substituent is present. If more than one substituent is present, then the numbers that result in the lowest possible position numbers for the substituents are used.

10.4 Reactions of Alkanes and Cycloalkanes

Combustion

Alkanes and other hydrocarbons may be oxidized (burned in air), producing carbon dioxide and water; this reaction is called **combustion.** During combustion, a large amount of heat energy is released.

Example 10

Show the complete balanced equation for the combustion of octane.

Work

1. Octane: The *-ane* ending tells us that the compound is an alkane. The *oct-* means that it has eight carbon atoms. By using the general molecular formula C_nH_{2n+2}, we can derive the molecular formula of octane. Thus, the molecular formula is C_8H_{18}.
2. The complete combustion of an alkane requires O_2 to react with the alkane to produce $CO_2 + H_2O$.
3. $C_8H_{18} + O_2 \rightarrow CO_2 + H_2O$ This equation is not balanced.

Answer

The balanced equation for the complete combustion of octane is
$$2\,C_8H_{18} + 25\,O_2 \rightarrow 16\,CO_2 + 18\,H_2O$$

Halogenation

Learning Goal 10:   Write equations for halogenation reactions of alkanes.

Halogenation of an alkane or cycloalkane is a substitution reaction in which a halogen atom (usually bromine or chlorine) replaces a hydrogen atom.

Typically, when an alkane reacts with certain halogens in the presence of heat and/or light, a substitution reaction results. One of the C–H bonds of the alkane is broken and replaced with a C–X bond in which a halogen atom (X = Br or Cl) has substituted for a hydrogen atom. The products of this reaction are an **alkyl halide** and a hydrogen halide. In more complex alkanes, substitution can occur to some extent at all positions to give a mixture of products.

Example 11

Show the reaction for the stepwise substitution of chlorine atoms for the hydrogen atoms on methane; also, determine the I.U.P.A.C. name for each organic compound produced.

Answer

$$CH_4 + Cl_2 \xrightarrow{\text{UV}} CH_3Cl + HCl$$

$$CH_3Cl + Cl_2 \xrightarrow{\text{UV}} CH_2Cl_2 + HCl$$

$$CH_2Cl_2 + Cl_2 \xrightarrow{\text{UV}} CHCl_3 + HCl$$

$$CHCl_3 + Cl_2 \xrightarrow{\text{UV}} CCl_4 + HCl$$

The products of these reactions are chloromethane, dichloromethane, trichloromethane, and tetrachloromethane, respectively.

Glossary of Key Terms in Chapter Ten

aliphatic hydrocarbon (10.1) any member of the alkanes, alkenes, and alkynes or the substituted alkanes, alkenes, and alkynes.

alkane (10.2) saturated hydrocarbon; a hydrocarbon that contains only carbon and hydrogen and that is bonded together through carbon-hydrogen and carbon-carbon single bonds. Alkanes have the general molecular formula C_nH_{2n+2}.

alkyl group (10.2) a simple hydrocarbon group that results from the removal of one hydrogen from the original hydrocarbon (e.g., methyl, -CH_3; ethyl, -CH_2CH_3).

alkyl halide (10.6) a substituted hydrocarbon that has the general structure R-X, where R- represents any alkyl group and X is one of the halogens (F-, Cl-, Br-, or I-).

aromatic hydrocarbon (10.1) an organic compound that contains the benzene ring or a derivative of the benzene ring.

combustion (10.4) oxidation of hydrocarbons by burning in the presence of air to produce carbon dioxide and water.

condensed formula (10.2) a formula that shows all of the atoms in a molecule and places them in a sequential arrangement that details which atoms are bonded to each other; the bonds themselves are not shown.

constitutional isomers (10.2) two molecules having the same molecular formulas but different chemical structures.

cycloalkane (10.3) cyclic alkanes; saturated hydrocarbons, that have the general formula C_nH_{2n}.

functional group (10.1) an atom (or group of atoms and their bonds) that imparts specific chemical and physical properties to a molecule.

halogenation (10.4) a reaction in which one of the C-H bonds of a hydrocarbon is replaced with a C-X bond of a halogen atom (X = Br or Cl, generally).

hydrocarbon (10.1) a compound composed solely of the elements carbon and hydrogen.

I.U.P.A.C. Nomenclature System (10.2) The International Union of Pure and Applied Chemistry (I.U.P.A.C.) standard, universal system for the nomenclature of organic compounds.

line formula (10.2) the simplest representation of a molecule in which it is assumed that there is a carbon atom at any location where two or more lines intersect, there is a carbon at the end of any line, and each carbon is bonded to the correct number of hydrogen atoms.

molecular formula (10.2) a formula that provides the atoms and number of each type of atom in a molecule but gives no information regarding the bonding pattern involved in the molecule's structure.

parent compound (10.2) In the I.U.P.A.C. Nomenclature System, the parent compound is the longest chain containing the principal functional group (e.g., the hydroxyl group) in the molecule that is being named.

primary (1⁰) carbon (10.2) a carbon atom that is bonded to only one other carbon atom.

quaternary (4⁰) carbon (10.2) a carbon atom that is bonded to four other carbon atoms.

saturated hydrocarbon (10.1) alkane; a hydrocarbon that contains only carbon and hydrogen and that is bonded together through carbon-hydrogen and carbon-carbon single bonds. Alkanes have the general molecular formula C_nH_{2n+2}.

secondary (2⁰) carbon (10.2) a carbon atom that is bonded to two other carbon atoms.

structural formula (10.2) a formula that shows all of the atoms in a molecule and exhibits all bonds as lines.

structural isomers (10.2) molecules having the same molecular formula but different chemical structures.

substituted hydrocarbon (10.1) a hydrocarbon in which one or more hydrogen atoms is replaced by another atom or group of atoms.

substitution reaction (10.4) a reaction that results in the replacement of one group by another.

tertiary (3⁰) carbon (10.2) a carbon atom that is bonded to three other carbon atoms.

unsaturated hydrocarbon (10.1) a hydrocarbon containing at least one multiple (double or triple) bond.

Self Test for Chapter Ten

1. Which of the following are isomers of C_5H_{12}? Give letter(s) as your answers.

    a. $CH_3CH_2CH_2CH_3$

    b. $CH_3CH_2CH_2CH_2CH_3$

    c. $CH_3CH(CH_3)CH_2CH_3$

d.

$$CH_3$$
$$|$$
$$CH_3CCH_3$$
$$|$$
$$CH_3$$

e. $CH_3CH(CH_3)_2$

2. What is the geometry of a carbon surrounded by four single bonds?

3. What is the angle around carbon surrounded by a double bond and two single bonds?

4. What is the hydrogen-carbon-carbon bond angle in ethyne, HCCH?

5. What is the term used to describe the group in an organic compound that is primarily responsible for the chemical and physical properties of that compound?

6. Name the functional group found in all alcohols.

7. Name the functional group found in all aldehydes.

8. Name the family whose simplest member is benzene.

9. Which of the following is an isomer of 1,2-dibromoethane?

a. Br_2CHCH_3 b. CH_2BrCH_2Br c. $BrCH_2CH_2CH_3$

10. Which of the following is an isomer of C_4H_{10}?

a. benzene d. $CH_3CHBrCH_2CH_3$

b. cyclobutane e. dimethylpropane

c.

$$CH_3CHCH_3$$
$$|$$
$$CH_3$$

11. Which of the following is an isomer of CH_3OCH_3?

a. CH_3CH_2OH d. oxymethylurea

b. methylbutane e. dimethylether

c. $CH_3CH_2OCH_3$

12. Which of the following is the simplest organic compound?

a. ethyl alcohol d. urea

b. methane e. acetylene

c. ammonium cyanate

13. Which of the following is not a normal property of an inorganic compound?

a. flammable d. reactions are fast

b. high boiling point e. high melting point

c. soluble in water

14. Which of the following formulas have limited usage in organic chemistry?

    a. space filling formula   c. molecular formula

    b. condensed formula   d. structural formula

15. Compounds that contain only carbon and hydrogen atoms are classified in what major subdivision?

    a. alkanes   d. heterocyclic

    b. aliphatics   e. hydrocarbons

    c. aromatics

16. What field of study is defined as the chemistry of all the elements except carbon?

17. What organic compound did Friedrich Wöhler synthesize from inorganic reactants?

18. Alkanes contain only carbon and what other element?

19. Name the functional group found in all alkenes.

20. Give the molecular formula of the alkane that contains nine carbon atoms.

21. Give the I.U.P.A.C. names of the following branched-chain alkanes:

    a. $CH_3CHCH_2CH_3$
        |
        CH_3

    b. CH_3
           |
      $CH_3CCH_2CH_2CH_3$
           |
           CH_3

    c. CH_3 CH_3
           | |
      $CH_3 - C - CH - CH_2CH_3$
            |
         CH_3

22. Give the I.U.P.A.C. name of the following:

            H CH_3
            | |
       $CH_3CCH_2CCH_2Br$
           | |
          Cl   H

23. Give the I.U.P.A.C. name of the following:

          CH_3
          |
       CH_3CCH_3
          |
         Br

24. Give the I.U.P.A.C. name of the following:

            CH_2CH_3
            |
      $CH_3 - C - CH_3$
            |
           CH_2CHCH_3
               |
              Br

25. Write an equation representing the complete combustion of propane.

26. Write the names of the monosubstitution products that are produced when Cl_2 and propane react under UV light.

27. Describe the solubility of alkanes in water.

28. Are the melting and boiling points of alkanes generally higher or lower than those of other organic compounds?

29. In the I.U.P.A.C. system of naming compounds, the name of the compound is defined by the longest continuous carbon chain in the compound. What do we call this continuous chain?

30. What is the I.U.P.A.C. name of $CHCl_3$?

31. The general formula C_nH_{2n} can be used to find the molecular formula of any member of what alkane like class of organic compounds?

32. What does the prefix *cyclo-* mean?

33. What is the term for the oxidation of alkanes and other hydrocarbons to carbon dioxide and water?

34. How many moles of water are produced by the complete combustion of one mole of methane?

35. For methane to react with chlorine, what catalyst is needed?

36. What type of organic reaction involves the replacement of one or more atoms in a molecule by new atoms?

Vocabulary Quiz for Chapter Ten

1. A(n) _____ is a hydrocarbon that contains only carbon and hydrogen bonded together through carbon-hydrogen and carbon-carbon single bonds. They are saturated hydrocarbons that have the general molecular formula C_nH_{2n+2}.

2. Two molecules having the same molecular formulas but different chemical structures are called _____.

3. A(n) _____ is a simple hydrocarbon group that results from the removal of one hydrogen from the original hydrocarbon (e.g., methyl, $-CH_3$; ethyl, $- CH_2CH_3$).

4. In the I.U.P.A.C. Nomenclature System, the _____ is the longest chain containing the major functional group (e.g., the hydroxyl group) in the molecule.

5. A(n) _____ is any hydrocarbon that contains one or more carbon-carbon multiple bonds.

6. _____ is a reaction in which one of the C-H bonds of a hydrocarbon is replaced with a C-X bond of a halogen atom (X = Br or Cl, generally).

7. A(n) _____ is a compound containing only hydrogen and carbon.

8. An atom or group of atoms that imparts specific chemical and physical properties to a molecule is called a(n) _____.

9. A(n) _____ is a reaction that results in the replacement of one group by another.

11 The Unsaturated Hydrocarbons
Alkenes, Alkynes, and Aromatics

Learning Goals

1   Describe the physical properties of alkenes and alkynes.
2   Draw the structures and write the I.U.P.A.C. names of simple alkenes and alkynes.
3   Write the names and draw the structures of simple geometric isomers of alkenes.
4   Write equations predicting the products of the simple addition reactions of alkenes and alkynes: hydrogenation, halogenation, and hydration.
5   Apply Markovnikov's rule to predict the major and minor products of hydration reactions of unsymmetrical alkenes.
6   Write equations representing the formation of addition polymers of alkenes.
7   Draw the structures and write the names of common aromatic hydrocarbons.
8   Describe heterocyclic aromatic compounds and list several biological molecules in which they are found.

11.1 Alkenes and Alkynes: Structure and Physical Properties

Learning Goal 1:   Describe the physical properties of alkenes and alkynes.

    Alkenes, alkynes, and **aromatic compounds** contain at least one carbon-carbon double or triple bond. As a result of multiple bonding, these compounds contain fewer hydrogens than alkanes with the same number of carbon atoms. They are referred to as unsaturated because they do not contain as many hydrogens as their carbon skeleton will allow. The structures seen on the next page (page 142) reveal the structural differences among alkanes, alkenes, and alkynes.
    Each of the compounds in the diagram has the same number of carbon atoms, but they differ in the number of hydrogen atoms. Alkynes have the general formula C_nH_{2n-2} and thus contain two fewer hydrogens than the corresponding alkene (general formula C_nH_{2n}). Alkenes have two fewer hydrogens than the corresponding alkane (general formula C_nH_{2n+2}).
    Alkanes contain only single bonds; alkenes have at least one carbon-to-carbon double bond; and alkynes contain at least one carbon-to-carbon triple bond. The differences in bond order result in variation in molecular geometry and chemical reactivity among these three families. Alkanes have tetrahedral carbon atoms. When carbon is bonded by one double bond and two single bonds, as in alkenes, the molecule is planar and each bond angle is 120°. When two carbons are bonded by a triple bond, the molecule is linear and the bond angles are 180°.

	Ethane (Ethane)	Ethene (Ethylene)	Ethyne (Acetylene)
Structural Formulas	H–C–C–H with H atoms above and below each carbon	C=C with H atoms attached	H–C≡C–H
Molecular Formulas	C_2H_6	C_2H_4	C_2H_2
Condensed Formulas	CH_3CH_3	$CH_2=CH_2$	$HC≡CH$

Alkenes and alkynes are nonpolar. As a result they are not water-soluble but are readily soluble in nonpolar solvents and in many low-polarity organic solvents such as ether or chloroform. They are also less dense than water and have relatively low boiling points.

11.2 Alkenes and Alkynes: Nomenclature

Learning Goal 2: Draw the structures and write the I.U.P.A.C. names of simple alkenes and alkynes.

The nomenclature of alkenes and alkynes is analogous to that of the alkanes, with the following exceptions. For alkenes, the parent name is derived from the longest continuous carbon chain containing the double bond. Then the *-ane* ending of the alkane is replaced with the *-ene* ending of an alkene. The chain is numbered to give the lowest numbers for the two carbons containing the double bond.

Alkynes are named in the same way as the alkenes, except that the *-ane* ending of the corresponding alkane is replaced with the *-yne* ending of alkynes. The rules used in numbering the alkene chain are also used in alkyne nomenclature.

Example 1
Draw and name the alkane, alkene, and alkyne that contain three carbon atoms.

Work
1. To find the molecular formula of an alkane, use the general formula C_nH_{2n+2}
2. To find the molecular formula of an alkene, use the general formula C_nH_{2n}
3. To find the molecular formula of an alkyne, use the general formula C_nH_{2n-2}

Answer
1. The alkane with three carbon atoms has the molecular formula C_3H_8:

$$CH_3CH_2CH_3 \qquad \text{I.U.P.A.C. name: Propane}$$

2. The alkene with three carbon atoms has the molecular formula C_3H_6:

$$CH_3CH=CH_2 \qquad \text{I.U.P.A.C. name: Propene}$$

3. The alkyne with three carbon atoms has the molecular formula C_3H_4:

$$CH_3C\equiv CH \qquad \text{I.U.P.A.C. name: Propyne}$$

Example 2
Give the I.U.P.A.C. name of the following molecule:

$$
\begin{array}{c}
CH_2CH_2CH_2CH_3 \\
| \\
CH_3-CH=C-CH_2-CH_3
\end{array}
$$

Work
1. First determine the parent compound (longest continuous chain) that contains the double bond. The parent compound is heptene. Since the *hept-* means seven carbons and the *-ene* means double bond, we have

$$
\begin{array}{c}
C-C-C-C \\
| \\
C-C=C
\end{array}
$$

2. Next, label the parent to give the double bond carbons the lowest possible numbers:

$$
\begin{array}{cccc}
4 & 5 & 6 & 7 \\
C-C-C-C \\
| \\
C-C=C \\
1 \quad 2 \quad 3
\end{array}
$$

2-heptene; the 2 tells us that the double bond is between carbons 2 and 3.

Example 2 continued

3. Finally, list the attached groups, as in naming alkanes.

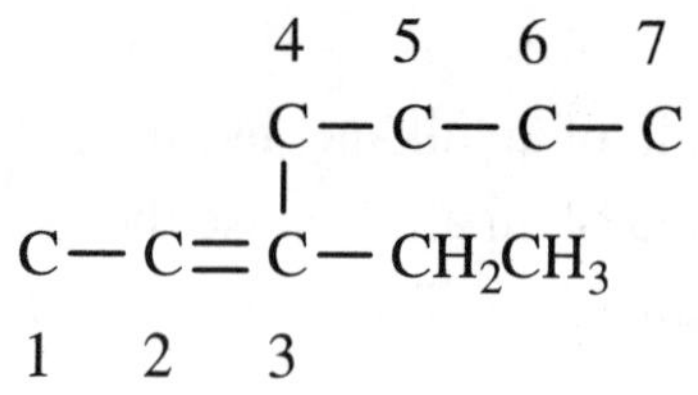

$$
\begin{array}{cccc}
4 & 5 & 6 & 7 \\
C-C-C-C \\
| \\
C-C=C-CH_2CH_3 \\
1 \quad 2 \quad 3
\end{array}
$$

Answer
3-Ethyl-2-heptene

Example 3
Determine the I.U.P.A.C. name of the following molecule:

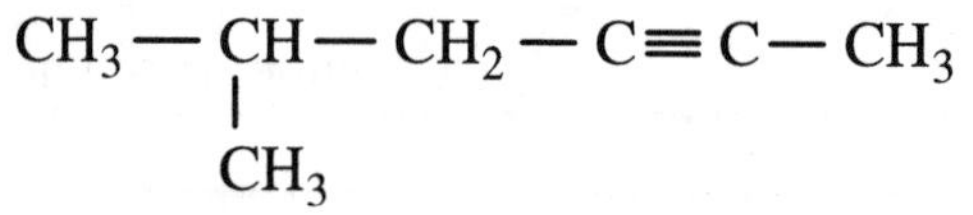

$$CH_3 - CH - CH_2 - C \equiv C - CH_3$$
$$|$$
$$CH_3$$

Work
1. First determine the parent compound that contains the triple bond. The parent compound is hexyne. The *hex-* means six carbons, and the *-yne* means a triple bond.
2. Label the compound so that the triple bond carbons have the lowest numbers possible.

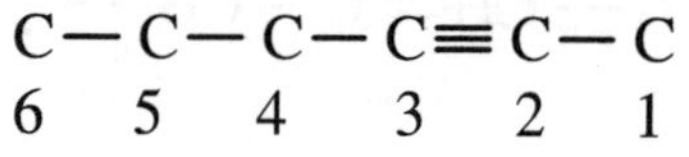

$$
\begin{array}{cccccc}
C-C-C-C\equiv C-C \\
6 \quad 5 \quad 4 \quad 3 \quad 2 \quad 1
\end{array}
$$

Thus, it is 2-hexyne.
3. Finally, list the attached groups as before:

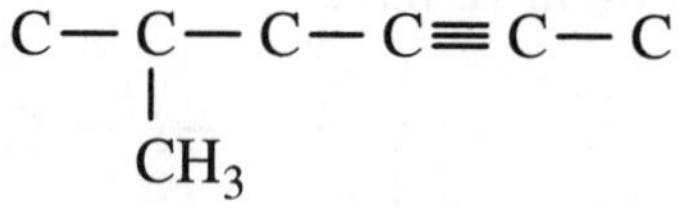

$$
\begin{array}{c}
C-C-C-C\equiv C-C \\
| \\
CH_3
\end{array}
$$

Answer
5-Methyl-2-hexyne

Example 4
Draw each of the following using structural formulas:

1. 1-Bromo-2-hexyne   2. 2-Butene

Work
1. 1-Bromo-2-hexyne: there are 6 carbons (*hex-*) in the parent compound and a triple bond between carbons 2 and 3 (*2-yne*). Also, there is a bromine (Br–) bonded to carbon 1.

144

Example 4 continued

$$\underset{\underset{\displaystyle |}{\overset{\displaystyle Br}{}}}{C} - C \equiv C - C - C - C$$

Next, fill in the other bonds with hydrogen atoms.

$$H - \underset{\underset{\displaystyle H}{\displaystyle |}}{\overset{\overset{\displaystyle Br}{\displaystyle |}}{C}} - C \equiv C - \underset{\underset{\displaystyle H}{\displaystyle |}}{\overset{\overset{\displaystyle H}{\displaystyle |}}{C}} - \underset{\underset{\displaystyle H}{\displaystyle |}}{\overset{\overset{\displaystyle H}{\displaystyle |}}{C}} - \underset{\underset{\displaystyle H}{\displaystyle |}}{\overset{\overset{\displaystyle H}{\displaystyle |}}{C}} - H \qquad \text{or} \qquad BrCH_2 - C \equiv C - CH_2CH_2CH_3$$

2. 2-Butene: There are four carbons (*but-*) in the parent compound and a double bond between carbons 2 and 3.

$$C - C = C - C$$

Next, fill in the other bonds with hydrogen atoms.

$$H - \underset{\underset{\displaystyle H}{\displaystyle |}}{\overset{\overset{\displaystyle H}{\displaystyle |}}{C}} - \overset{\overset{\displaystyle H}{\displaystyle |}}{C} = \overset{\overset{\displaystyle H}{\displaystyle |}}{C} - \underset{\underset{\displaystyle H}{\displaystyle |}}{\overset{\overset{\displaystyle H}{\displaystyle |}}{C}} - H \qquad \text{or} \qquad CH_3CH = CHCH_3$$

11.3 Geometric Isomers

Learning Goal 3: **Write the names and draw the structures of simple geometric isomers of alkenes.**

The carbon-carbon double bond is rigid because there is no free rotation around the double bond. The rigidity is caused by the shapes of the orbitals involved in the double bond.

The rigidity of the carbon-carbon double bond in alkenes produces another class of isomers: **geometric isomers.** Geometric isomers are described using the prefixes *cis* and *trans*, which provide an easy method for naming and distinguishing between the two isomeric forms.

Geometric isomers are different molecules with different physical and chemical properties. Although their properties may be similar, they are never identical.

The prefixes *cis* and *trans* refer to the placement of the substituents attached to the carbon-carbon double bond. When identical groups are on the same side of the double bond, the prefix *cis* is used; when identical groups are on opposite sides of the double bond, *trans* is the appropriate prefix.

Example 5

Draw the *cis-* and *trans-* isomers of 2-butene.

Work

1. The carbon-carbon double bond is rigid as a result of the shapes of the orbitals involved in its formation. The specific atoms that are bonded to the two carbon atoms are locked into specific arrangements around the double bond.

2. The *cis* arrangement means that the two reference groups (say A and A) are on the same side of the double bond. The *trans* arrangement means the two A groups are across from each other.

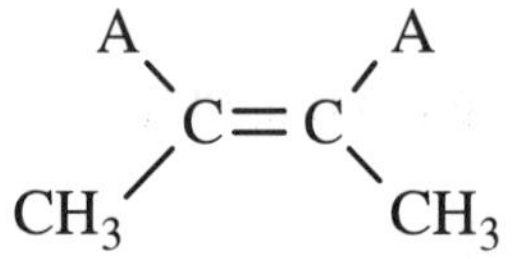

$$\underset{CH_3}{\overset{A}{\diagdown}}C=C\underset{CH_3}{\overset{A}{\diagup}} \qquad \underset{CH_3}{\overset{A}{\diagdown}}C=C\underset{A}{\overset{CH_3}{\diagup}}$$

cis-isomer 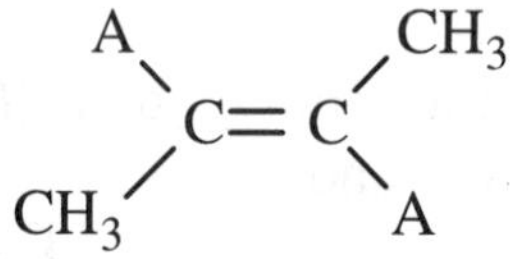   *trans*-isomer

Answer

1. First we draw 2-butene in a straight form, as before:

$$CH_3CH=CHCH_3$$

2. Next, locate the double bond and place the remaining groups either *cis* or *trans* from one another:

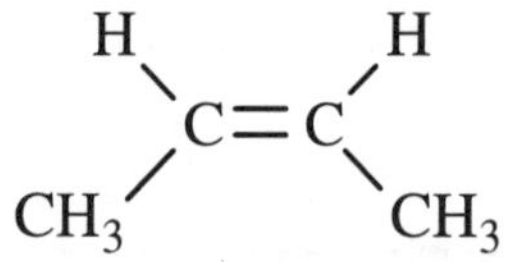

$$\underset{CH_3}{\overset{H}{\diagdown}}C=C\underset{CH_3}{\overset{H}{\diagup}} \qquad \underset{CH_3}{\overset{H}{\diagdown}}C=C\underset{H}{\overset{CH_3}{\diagup}}$$

cis-2-Butene 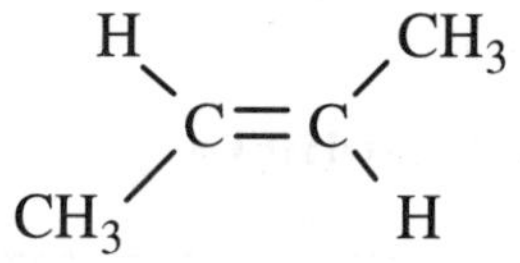   *trans*-2-Butene

11.4 Reactions Involving Alkenes and Alkynes

Learning Goal 4: **Write equations predicting the products of the simple addition reactions of alkenes and alkynes: hydrogenation, halogenation, and hydration.**

The major reactions of alkenes and alkynes involve the addition of atoms or molecules to the carbon-carbon double or triple bond. The principal kinds of addition reactions are hydrogenation, halogenation, and hydration.

146

Hydrogenation: Addition of H_2

Hydrogenation is the addition of a molecule of hydrogen (H_2) to a carbon-carbon double bond to produce an alkane. Two new C-H single bonds are formed as the double bond is broken.

Hydrogenation generally requires heat and/or pressure. The reaction always requires a metal catalyst, such as platinum or nickel, to allow the reaction to occur at a reasonably rapid rate.

The food industry takes advantage of the hydrogenation reaction. Vegetable oils are liquid because the fat molecules have many double bonds. When these oils are hydrogenated, many of the double bonds become carbon-carbon single bonds. This increases the rigidity of the fat molecules and solid fats are formed.

Example 6

Write an equation representing the hydrogenation reaction for 2-butene. Indicate the conditions needed and name all reactants and products.

Work

1. The term *hydrogenation* means the reaction of gaseous H_2 with a specific reactant. A catalyst such as Pt, Pd, or Ni is needed to speed up the reaction. Elevated temperature and pressure are also required.

Answer

$$CH_3CH = CHCH_3 \quad + \quad H_2 \quad \xrightarrow[\text{pressure}]{\text{Ni, heat}} \quad CH_3CH_2CH_2CH_3$$

$$\text{2-Butene} \hspace{6cm} \text{Butane}$$

Hydrogenation of an alkyne requires two moles of H_2.

Halogenation: Addition of X_2

Chlorine (Cl_2) or bromine (Br_2) can be added to a double bond. This reaction, called **halogenation,** proceeds readily and does not require a catalyst.

Example 7

Write an equation representing the halogenation of ethene to produce 1,2-dichloroethane.

Work

1. The term *halogenation* means the addition of a halogen molecule (X_2) across the double bond of an alkene. The double bond is broken and two new C–X bonds are formed.

Example 7 continued

Answer

Ethene   + Chlorine ——————▶ 1,2-Dichloroethane

Alkynes also react with halogens. Two moles of halogen are required for halogenation of an alkyne.

Halogenation involving bromine (Br_2) can be used to show the presence of double or triple bonds in an organic compound. The reaction mixture is red due to the presence of dissolved bromine. If the red color is lost, it indicates that bromine was consumed, bromination occurred, and the compound was unsaturated. One mole of Br_2 is consumed per mole of alkene, and two moles of Br_2 are consumed per mole of alkyne.

Hydration: Addition of H_2O

Learning Goal 5:   **Apply Markovnikov's rule to predict the major and minor products of hydration reactions of unsymmetrical alkenes.**

A **hydration** reaction is the addition of a water molecule to a carbon-carbon double bond. When an alkene is reacted with water containing a trace of strong acid, an –OH group bonds to one carbon of the carbon-carbon double bond and an –H atom bonds to the other carbon. The product of the hydration of an alkene is an alcohol.

When an alkene is unsymmetrical (carries two different groups on the double bond carbons) hydration can yield two different products. One is usually favored over the other, as explained by **Markovnikov's rule,** which tells us that the "rich get richer." This means that the carbon of the carbon-carbon double bond that carries the greater number of hydrogen atoms most often will receive the hydrogen atom being added to the double bond. The other carbon becomes bonded to the –OH group.

Example 8
Write an equation representing the hydration of ethene to produce ethanol (ethyl alcohol).

Work
1. The term *hydration* means the addition of a water molecule (HOH) across the double bond of an alkene. The –H bonds to one carbon of the double bond and the –OH

Example 8 continued

attaches to the other carbon of the double bond. The double bond is broken, and in its place are two new single bonds.

Answer

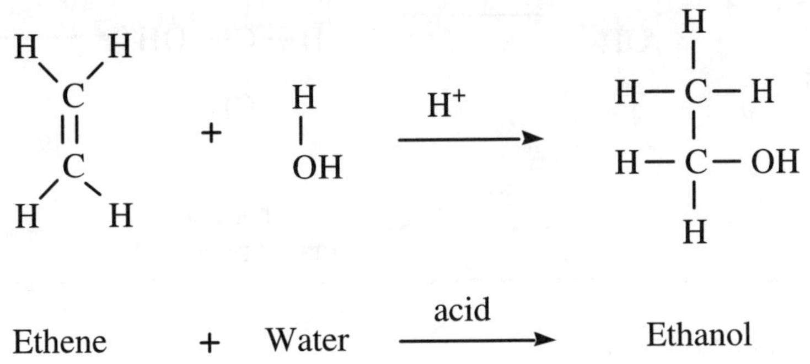

$$
\begin{array}{ccc}
\underset{\substack{| \\ H \quad H}}{\overset{H \quad H}{\diagdown C \diagup}} & & \underset{\substack{| \\ H}}{\overset{H}{\diagup}} \\
\parallel & + & | \\
C & & OH \\
\end{array}
\xrightarrow{\;H^+\;}
\begin{array}{c}
H \\
| \\
H-C-H \\
| \\
H-C-OH \\
| \\
H
\end{array}
$$

$$
\text{Ethene} \quad + \quad \text{Water} \quad \xrightarrow{\;\text{acid}\;} \quad \text{Ethanol}
$$

Example 9
Write an equation representing the hydration of propene, and name and draw the major product.

Work
The hydration of an unsymmetrical alkene (like propene—having different groups on each side of the double bond) favors one product over the other.

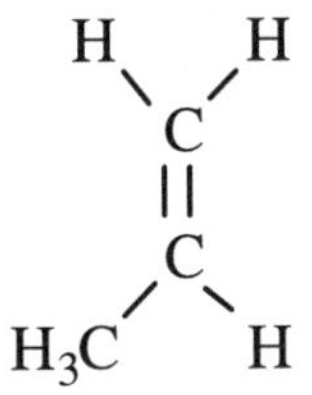

$$
\begin{array}{c}
H \quad\quad H \\
\diagdown C \diagup \\
\parallel \\
C \\
\diagup \quad \diagdown \\
H_3C \quad\quad H
\end{array}
$$

The carbon atom of the double bond that has the greater number of directly attached hydrogen atoms most often will receive the hydrogen atom being added to the double bond. The remaining carbon atom bonds to the –OH group. This is known as Markovnikov's rule.

Note that propene is unsymmetrical, since one carbon of the double bond has 2 –H atoms bonded to it while the other carbon of the double bond has only one –H atom and one –CH$_3$ group attached to it.

Example 9 continued

Answer

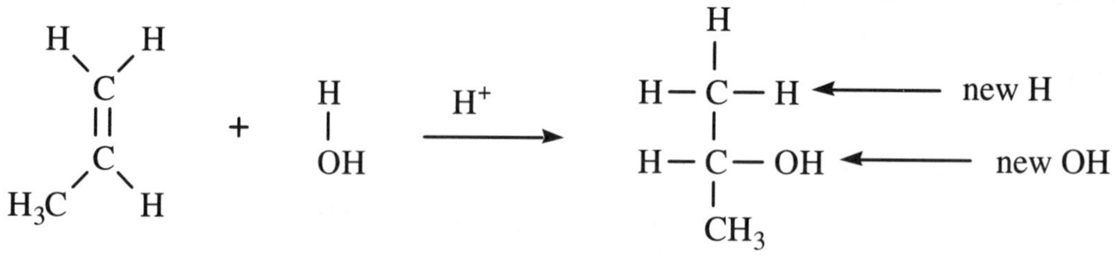

Propene	+	Water	$\xrightarrow{\text{acid}}$	2-Propanol
				(major product)

Addition Polymers of Alkenes

Learning Goal 6:   Write equations representing the formation of addition polymers of alkenes.

Addition polymers are produced by the sequential addition of an alkene monomer to produce the polymer, which is a macromolecule composed of many repeating structural units (monomers). Many useful plastics are addition polymers produced from alkene monomers. Table 11.1 of the text presents a number of common addition polymers.

Example 10
Write an equation representing the reaction that produces the addition polymer polyvinyl chloride (PVC) from vinyl chloride monomers. What is the I.U.P.A.C. name for vinyl chloride?

Work
1. Begin by determining the structure of vinyl chloride (Table 11.1):

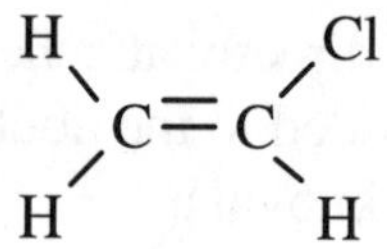

2. Now determine the structure of the addition polymer:

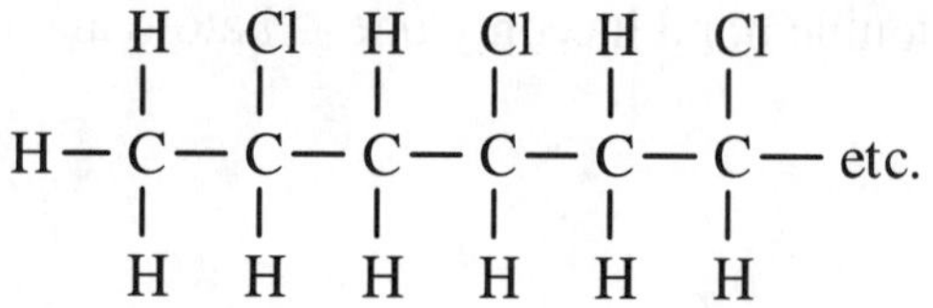

150

Example 10 continued

Using the simplified general form for the polymer, write the equation representing the reaction.

Answer

$$n \begin{bmatrix} \begin{array}{ccc} H & & Cl \\ & C = C & \\ H & & H \end{array} \end{bmatrix} \xrightarrow[\text{pressure}]{\text{heat}} \begin{bmatrix} \begin{array}{cc} H & Cl \\ | & | \\ C - C \\ | & | \\ H & H \end{array} \end{bmatrix}_n$$

The I.U.P.A.C. name for vinyl chloride is chloropropene.

11.5 Aromatic Hydrocarbons

Structure and Properties

Learning Goal 7:   **Draw the structures and write the names of common aromatic hydrocarbons.**

   Aromatic compounds are all characterized by the presence of an aromatic ring within the structure. The simplest, and most common, aromatic compound is benzene, which consists of a six carbon ring in a planar hexagonal arrangement. Each carbon is bonded to two other carbon atoms and to a single hydrogen atom.
   In 1865, Friedrich Kekulé proposed that single and double bonds alternated around the hexagonal ring. Since benzene did not decolorize bromine, he further proposed that the double and single bonds shifted position rapidly.
   The current model of benzene structure proposes that each carbon atom is bonded to two other carbon atoms and to a hydrogen atom. The remaining six electrons are located in a cloud of electrons above and below the ring. Because the electrons are delocalized within the cloud, benzene is unusually stable and resists addition reactions typical of alkenes.

Nomenclature

   Most simple aromatic compounds are named as derivatives of benzene. Others, for example phenol and toluene, have common historical names that must simply be memorized.
   When there are two or more groups on the ring, the ring carbons are numbered to give the lowest numbers to the groups.

Example 11
Name the following aromatic compounds:

1.   2.   3.   4.

   Cl   Cl   Br   Br

                                      Br

                       Br   Br

          Cl

Work
The term *aromatic* means that there is a benzene ring present.
If only one group is bonded to the benzene structure, no number is needed in its name.
If two or more groups are bonded to the benzene ring, the ring carbons are numbered to give the lowest numbers to the groups.

Answer
1. Chlorobenzene
2. 1,4-Dichlorobenzene
3. 1,3,5-Tribromobenzene
4. 1,2-Dibromobenzene

    In I.U.P.A.C. nomenclature, the group that results from the removal of a single hydrogen atom from the benzene ring is called the **phenyl group.** Aromatic hydrocarbons having long aliphatic side chains are often named as phenyl-substituted hydrocarbons.

Polycyclic Aromatic Hydrocarbons

    Polynuclear aromatic hydrocarbons are composed of two or more aromatic rings joined to one another. Naphthalene has a distinctive aroma and is used as mothballs. Benzopyrene, found in cigarette smoke, is one of the most potent carcinogens known.

11.6 Heterocyclic Aromatic Compounds

Learning Goal 8:   Describe heterocyclic aromatic compounds and list several biological molecules in which they are found.

    Heterocyclic aromatic compounds are those that have at least one atom other than carbon as part of the structure of the aromatic ring. Molecules such as pyridine, pyrimidines, purines, and pyrrole are heterocyclic aromatic compounds that are found in nicotine, DNA, RNA, and the porphyrin ring found in chlorophyll and hemoglobin.

Glossary of Key Terms in Chapter Eleven

addition polymer (11.4) a polymer prepared by sequential addition of monomers.

addition reaction (11.4) a reaction in which two molecules add together to form a new molecule; often involves the addition of one molecule to a double or triple bond in an unsaturated molecule.

alkene (11.1) a hydrocarbon that contains one or more carbon-carbon double bonds; an unsaturated hydrocarbon with the general formula C_nH_{2n}.

alkyne (11.1) a hydrocarbon that contains one or more carbon-carbon triple bonds; an unsaturated hydrocarbon with the general formula C_nH_{2n-2}.

aromatic compound (11.5) a hydrocarbon that contains a benzene ring or that has properties that are similar to those exhibited by benzene.

geometric isomers (11.3) isomers that differ from one another in the placement of substituents on a double bond.

halogenation (11.4) For an alkane, a reaction in which one of the C-H bonds of a hydrocarbon is replaced with a C-X bond of a halogen atom. For an alkene or alkyne, the addition of a halogen (Cl_2 or Br_2) to a double bond.

heterocyclic aromatic compound (11.6) a compound having at least one atom other than carbon in the structure of the aromatic ring.

hydration (11.4) a reaction in which water is added to a molecule.

hydrogenation (11.4) a reaction in which hydrogen (H_2) is added to a double or a triple bond.

Markovnikov's rule (11.4) the rule that states that a hydrogen atom, adding to a carbon-carbon double bond, is more likely to add to the carbon having the larger number of hydrogens attached to it already.

monomer (11.4) the individual molecules from which a polymer is formed.

phenyl group (11.5) a benzene ring that has had a hydrogen atom removed.

polymer (11.4) a very large molecule formed by the combination of many small molecules (called monomers).

unsaturated compound (11.1) any hydrocarbon that contains one or more carbon-carbon double or triple bonds.

Self Test for Chapter Eleven

1. Write the molecular formula for and name the straight chain alkane, alkene, and alkyne that contain four carbon atoms.

2. Write the I.U.P.A.C. name of the following compound:

$$CH_3 - CH = C - CH_2CH_3$$
$$|$$
$$CH_2CH_3$$

3. Write the I.U.P.A.C. name of the following compound:

$$Br$$
$$|$$
$$CH_3CHCHC \equiv CCH_3$$
$$|$$
$$Br$$

4. Draw the structure of 2-methyl-2-pentene.

5. Draw the structure of *cis*-3-methyl-2-pentene.

6. Name the following compound using the prefixes *cis-* or *trans-*:

$$\underset{CH_3}{\overset{Br}{\diagdown}}C=C\underset{H}{\overset{CH_3}{\diagup}}$$

7. Write an equation for the hydrogenation reaction for 1-butene.

8. Write an equation for the hydration of propene. Indicate both the major and minor products.

9. Name the following aromatic compounds:

a.   b.   c.   d.   e.

10. What is the common name of ethyne?

11. What is the I.U.P.A.C. name of the following molecule?

$$\underset{CH_3}{\overset{H}{\diagdown}}C=C\underset{Br}{\overset{CH_3}{\diagup}}$$

12. What is the I.U.P.A.C. name of the following molecule?

$$\underset{Br}{\overset{CH_3}{\diagdown}}C=C\underset{H}{\overset{Br}{\diagup}}$$

13. What reaction is used to convert alkenes into alkanes?

14. What two catalysts are often used in the hydrogenation of alkenes?

15. Provide the product of the reaction represented by the following equation:

$$CH_2 = CH_2 \ + \ H_2O \ \xoverset{acid}{\longrightarrow} \ ?$$

16. What compound makes up the basic unit of aromatic compounds?

17. What is the common name of trichloromethane?

18. In ethene, the two carbon atoms are joined by what type of bond?

19. Alkenes are characterized by what type of reaction?

20. What does the ending *-ane* mean?

21. What does the ending *-yne* mean?

22. Write an equation showing the reaction of 1-butyne with 2 moles of hydrogen gas.

23. Provide the product of the reaction represented by the following equation:

$$CH_3CH_2CH{=}CH_2 \ + \ H_2 \ \xrightarrow{\text{Pd, heat}} \ ?$$

24. What materials are hydrogenated to produce a product such as Crisco?

Vocabulary Quiz for Chapter Eleven

1. A(n)_____ is a hydrocarbon that contains at least one carbon-carbon double bond. They are also described as unsaturated hydrocarbons with the general formula C_nH_{2n}.

2. A large molecule formed by the sequential addition of small units is called a (an) _____.

3. _____ are isomers (which by definition have the same molecular formula) that differ from one another due to the placement of substituents on a double bond.

4. A(n)_____ is a benzene ring that has had a hydrogen atom removed.

5. In the I.U.P.A.C. Nomenclature System, the _____ is the longest chain containing the major functional group in the molecule that is being named.

6. A(n)_____ is any hydrocarbon that contains one or more carbon-carbon triple bonds.

7. A(n) _____ is a reaction in which two molecules add together to form a new molecule. This often involves the addition of one molecule to a double or triple bond in an unsaturated molecule.

8. A(n) _____ is a compound having at least one atom other than carbon in the structure of the aromatic ring.

9. A(n) _____ is a reaction that results in the replacement of one group by another.

12 *Oxygen- and Sulfur-Containing Organic Compounds*

Learning Goals

1 Draw the structures and be able to rank selected alcohols by relative water solubility, boiling points, or melting points.
2 Write the common and I.U.P.A.C. names for common alcohols.
3 Discuss the biological, medical, or environmental significance of several alcohols.
4 Classify alcohols as primary, secondary, or tertiary.
5 Write equations representing the preparation of alcohols by the hydration of an alkene.
6 Write equations showing the dehydration of an alcohol.
7 Write equations representing the oxidation of alcohols.
8 Discuss the use of phenols as germicides.
9 Write names and draw structures for common ethers and discuss their use in medicine.
10 Write equations representing formation of an ether by a dehydration reaction between two alcohol molecules.
11 Write names and draw structures for simple thiols and discuss their biological significance.
12 Draw the structures and discuss the physical properties of aldehydes and ketones.
13 Write the common and I.U.P.A.C. names for common aldehydes and ketones.
14 Discuss the biological, medical, or environmental significance of several aldehydes and ketones.
15 Write equations representing the oxidation and reduction of aldehydes and ketones.

Introduction

The functional group of the **alcohols** and **phenols** is the **hydroxyl group** (–OH). Alcohols have the general formula R–OH, in which R represents an alkyl group. Thus, the simplest alcohol is methyl alcohol: a methyl group bonded to a hydroxyl group, CH_3OH. Phenols have an aryl group in place of the alkyl group of the alcohols, and thus have the general formula Ar–OH. Phenol, the simplest member of this family, consists of a phenyl group (a benzene ring missing one hydrogen atom) bonded to a hydroxyl group.

Ethers contain two alkyl or aryl groups attached to an oxygen atom. Thus, the functional group of the ethers is R–O–R or Ar–O–Ar. **Thiols** are a family of compounds in which the sulfur atom has been substituted for the oxygen atom of an alcohol.

The **carbonyl group** (–C = O) is characteristic of many groups of organic compounds, including the **aldehydes** and **ketones,** carboxylic acids and esters. The carbonyl group and the two atoms attached to it are coplanar.

Aldehydes and ketones are carbonyl-containing compounds that differ from one another in the type of atom or atoms attached to the carbonyl carbon. In ketones, the carbonyl carbon is attached to two carbon atoms, whereas in aldehydes the carbonyl carbon is attached to at least one hydrogen atom; the second atom attached to the carbonyl carbon in aldehydes may be another hydrogen or a carbon atom.

12.1 Alcohols

Learning Goal 1: **Draw the structures and be able to rank selected alcohols by relative water solubility, boiling points, or melting points.**

We can think of alcohols (and phenols) as substituted water molecules: an H–O–H molecule in which one hydrogen atom has been replaced by an alkyl or aryl group. The R–O–H portion of an alcohol is planar, as is H–O–H. In addition, the bond angles of both R–O–H and H–O–H are 104.5°.

Alcohols are polar molecules owing to the hydroxyl group. The oxygen and hydrogen atoms have very different electronegativities (oxygen is 3.5 and hydrogen is 2.1). Consequently, the oxygen atom displays a partial negative charge and the hydrogen atom bears a partial positive charge. Alcohol molecules can form hydrogen bonds with other alcohol molecules and with the molecules of a polar solvent.

Both the polarity of alcohol molecules and the ability to form intermolecular hydrogen bonds exert a strong influence on the physical properties of alcohols. They have higher boiling points than hydrocarbons or ethers of similar molecular weight and show greater solubility in water. In fact, the smallest alcohols (1–4 carbons) are highly soluble in water; those having 5 or 6 carbons have decreased solubility; and those with more than 6 carbons are insoluble in water.

Example 1
For the following pairs of compounds, indicate the one that has the higher boiling point:

1. CH_4 (methane) or CH_3OH (methanol)
2. $CH_3CH_2CH_2CH_3$ (butane) or $CH_3OCH_2CH_3$ (ethyl methyl ether)
3. ethyl methyl ether or 1-propanol

Work
When comparing the boiling points of alkanes, ethers, and alcohols of similar molecular weight, remember that the alcohols have the highest boiling points. The ethers are found to have higher boiling points than the alkanes: alcohol > ethers > alkanes
Answer
1. Methanol
2. Ethyl methyl ether
3. 1-Propanol

Example 2
Predict the solubility of each of the following alcohols in water solutions:

1. $CH_3CH_2CH_2CH_2CH_2CH_2OH$ 2. CH_3CH_2OH 3. $CH_3(CH_2)_8CH_2OH$

Work
1. Alcohols of 1–4 carbons are very soluble in water. As the carbon chain length increases, the alcohol becomes less soluble. Those with more than 6 carbons are insoluble in water.

Answer
1. Not very soluble   2. Very soluble   3. Insoluble in water

Nomenclature

Learning Goal 2:   **Write the common and I.U.P.A.C. names for common alcohols.**

Determine the *parent compound*, in this case the longest continuous carbon chain containing the –OH group. Drop the *-e* ending of the alkane chain and replace it with *-ol*. Number the chain so that the hydroxyl group has the lowest possible number. Add substituents, named and numbered appropriately, as prefixes to the alcohol name.

If two hydroxyl groups are present, the suffix *-diol* is used; if three hydroxyl groups are present, the suffix used is *-triol*. The positions of hydroxyl groups are numbered as usual.

Example 3
Name the following compounds using the I.U.P.A.C. method:

a.
CH_3CHCH_3
   |
   OH

b.
$CH_3CHCH_2CHCH_3$
   | |
   OH CH_3

c.
CH_2CH_2
 | |
OH OH

d.
$CH_3CHCHCH_3$
   | |
   Br OH

Work
1. Determine the parent compound that contains the –OH bonded to it.
2. Change the *-e* ending of the parent alkane's name to *-ol*.
3. Next, number the parent carbon chain so that the –OH has the lowest possible number.
4. If the alcohol has two or more –OH groups, add *-diol* or *-triol* to the parent's name.
5. Then use all the other rules we have used previously to finish naming the other groups that may be present.

Example 3 continued

Answer
a. 2-Propanol
b. 4-Methyl-2-pentanol
c. 1,2-Ethanediol
d. 3-Bromo-2-butanol

    The common names for the alcohols are derived from the name of the corresponding alkyl group. The name of the alkyl group is followed by the word *alcohol*.

Example 4
Give the common names of each of the following alcohols:

1. CH_3OH

2. CH_3CH_2OH

3. 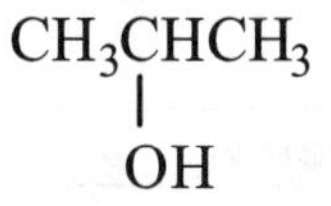
$$CH_3CHCH_3$$
$$|$$
$$OH$$

4. $CH_3CH_2CH_2CH_2OH$

5. 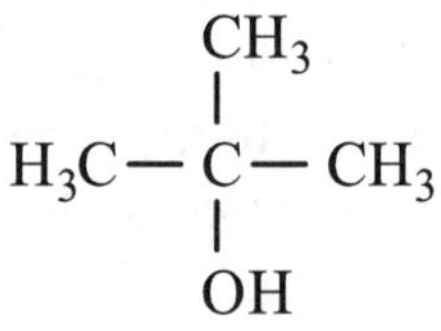
$$CH_3$$
$$|$$
$$H_3C - C - CH_3$$
$$|$$
$$OH$$

Work
To determine the common names for the alcohols, first find which alkyl group is bonded to the –OH group. Next, write down the alkyl group name and add the word alcohol to complete the common name of that alcohol.

Answer
1. Methyl alcohol
2. Ethyl alcohol
3. Isopropyl alcohol
4. Butyl alcohol
5. *t*-Butyl alcohol

159

Medically Important Alcohols

Several of the small alcohols are important in medicine and industry. The smallest, *methanol* (methyl alcohol, wood alcohol) is a common solvent. It is extremely toxic if ingested.

Ethanol (ethyl alcohol, grain alcohol) is a solvent and disinfectant. Ethanol for human consumption is produced by fermentation of sugars by yeast. Denatured alcohol is ethanol to which a denaturing agent has been added to make it undrinkable.

2-Propanol (isopropyl alcohol, rubbing alcohol) is commonly used as a disinfectant and solvent. Like methanol, it is very toxic if ingested.

1,2-Ethanediol (ethylene glycol) is a common antifreeze for cars.

1,2,3-Propanetriol (glycerol) is a component of stored fats in the body. It is used in cosmetics and pharmaceuticals.

Classification of Alcohols

Alcohols are classified as methyl, **primary (1°), secondary (2°),** or **tertiary (3°)** depending on the number of alkyl groups attached to the **carbinol carbon,** the carbon that bears the hydroxyl group. If there is one alkyl group attached to the carbinol carbon, the alcohol is a primary alcohol. An alcohol with two alkyl groups bonded to the carbinol carbon is a secondary alcohol. A tertiary alcohol has three alkyl groups bonded to the carbinol carbon.

Example 5

Classify the following alcohols as either primary, secondary, or tertiary:

1. CH_3CH_2OH

$$3. \quad CH_3CH_2CHCH_3$$
$$| $$
$$OH$$

2. CH_3CHCH_3
$| $
OH

$$4. \quad CH_3CH_2CH_2\overset{\overset{\textstyle CH_3}{|}}{\underset{\underset{\textstyle OH}{|}}{C}}CH_3$$

Answer
1. Primary
2. Secondary
3. Secondary
4. Tertiary

160

Reactions Involving Alcohols

Learning Goal 5:	Write equations representing the preparation of alcohols by the hydration of an alkene.

Addition of a water molecule to the carbon-carbon double bond of an alkene produces an alcohol. This reaction, called hydration, requires a trace of acid (H^+) as a catalyst, as shown in the following example:

$$CH_2{=}CH_2 \;\; + \;\; H_2O \;\; \xrightarrow{H^+} \;\; CH_3CH_2OH$$

$$\text{Ethene} \qquad\qquad \text{Water} \qquad\qquad \text{Ethanol}$$

Dehydration of Alcohols

Learning Goal 6:	Write equations showing the dehydration of an alcohol.

Alcohols undergo **dehydration** (loss of water) when heated in the presence of concentrated sulfuric or phosphoric acids. The products of dehydration are an alkene and a water molecule. Quite simply, a dehydration reaction is the reverse of the hydration reactions that produce an alcohol from an alkene and water.

$$CH_3CH_2OH \;\; \xrightarrow{H^+} \;\; CH_2{=}CH_2 \;\; + \;\; H_2O$$

$$\text{Ethanol} \qquad\qquad \text{Ethene} \qquad\qquad \text{Water}$$

In some cases, dehydration of an alcohol may produce a mixture of products. **Zaitsev's rule** tells us that the more highly substituted alkene will be the major product.

Example 6

Write an equation representing the dehydration reactions for the following alcohols:

1. 1-Propanol
2. Ethanol
3. 2-Butanol

Work

The dehydration of an alcohol produces an alkene plus water. Remember that in some cases the dehydration of an alcohol yields a mixture of products. In those instances, it is the more highly branched alkene that is the major product.

Example 6 continued

Answer

1. $CH_3CH_2CH_2OH \xrightarrow[\text{heat}]{\text{acid}} CH_3CH=CH_2 + H_2O$

   1-Propanol   Propene

2. $CH_3CH_2OH \xrightarrow[\text{heat}]{\text{acid}} CH_2=CH_2 + H_2O$

   Ethanol   Ethene

3. $CH_3CH_2 - \overset{\displaystyle OH}{\underset{\displaystyle H}{C}} - CH_3 \xrightarrow[\text{heat}]{\text{acid}} CH_3CH_2CH=CH_2 + H_2O$

                                                     1-Butene

   2-Butanol

                                           $CH_3CH=CHCH_3 + H_2O$

                                                 2-Butene

Two products are formed. The major product is 2-butene.

Oxidation Reactions

Learning Goal 7:   **Write equations representing the oxidation of alcohols.**

       Some alcohols can be **oxidized** to produce aldehydes, ketones, or carboxylic acids. The most commonly used oxidizing agents are basic potassium permanganate ($KMnO_4/OH^-$) and chromic acid (H_2CrO_4).

       Methanol and all primary alcohols produce aldehydes, while secondary alcohols form ketones. Tertiary alcohols cannot be oxidized. This is because the carbon bonded to the hydroxyl group must contain at least one C–H bond in order for oxidation to occur. Since tertiary alcohols contain three C–C bonds to the carbon bonded to the hydroxyl group they cannot be oxidized. Aldehydes can undergo further oxidation to produce carboxylic acids.

Example 7
Write an equation representing the oxidation of methanol and ethanol by the liver. Name all products.

Example 7 continued

Work
The oxidation of an alcohol first produces an aldehyde or ketone. If an aldehyde is formed, it can be further oxidized to produce a carboxylic acid.

Answer

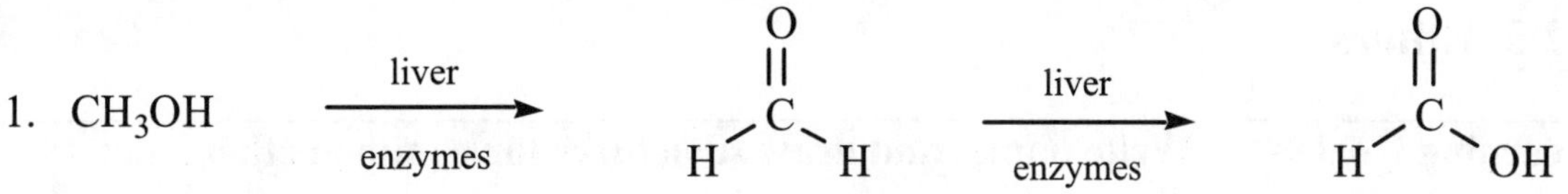

1. CH$_3$OH $\xrightarrow[\text{enzymes}]{\text{liver}}$

    Methanol

    (Formaldehyde structure: H–C(=O)–H)

    Formaldehyde
    (very toxic)

    $\xrightarrow[\text{enzymes}]{\text{liver}}$

    (Formic acid structure: H–C(=O)–OH)

    Formic acid
    (very toxic)

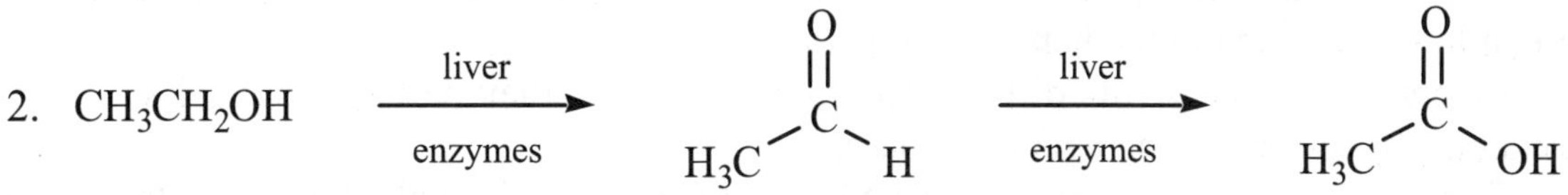

2. CH$_3$CH$_2$OH $\xrightarrow[\text{enzymes}]{\text{liver}}$

    Ethanol

    (Acetaldehyde structure: H$_3$C–C(=O)–H)

    Acetaldehyde

    $\xrightarrow[\text{enzymes}]{\text{liver}}$

    (Acetic acid structure: H$_3$C–C(=O)–OH)

    Acetic Acid

Example 8
Provide the oxidation products for the following compounds.
    1. primary alcohol   4. aldehyde
    2. secondary alcohol   5. 2-propanol
    3. tertiary alcohol   6. 2-methyl-2-propanol

Answer
1. aldehyde, then to a carboxylic acid
2. ketone
3. no reaction
4. carboxylic acid
5. ketone—called acetone
6. no reaction—a tertiary alcohol

12.2 Phenols

Learning Goal 8:   Discuss the use of phenols as germicides.

Phenols are compounds in which the hydroxyl group is attached to an aryl group. Owing to the polar hydroxyl group, the phenols are also polar compounds.

The simplest member of this family is known by the common name phenol. This compound is of interest in the history of medicine. Joseph Lister, a British physician, observed that the incidence of post-surgical infections could be radically decreased if the surgical instruments and the incision were treated with an antimicrobial chemical. The agent he used was carbolic acid, a dilute solution of phenol. As a result of his observations, the use of antiseptics and disinfectants has become routine medical practice.

12.3 Ethers

Learning Goal 9:	Write names and draw structures for common ethers and discuss their use in medicine.

Ethers are structurally related to alcohols. However, a quick look at the geometry of the functional group characteristic of the ethers (R–O–R) reveals that these compounds are much less polar than alcohols. Indeed, they are much less water-soluble and have much lower boiling points than the comparable alcohols.

Ethers are chemically inert. Under normal conditions, they will not react with oxidizing agents, reducing agents, or bases.

Common names for ethers are derived from the names of the alkyl groups attached to the ether oxygen. The alkyl group names are listed as prefixes before the word ether and may be ordered by size (small to large) or alphabetically. For instance, the compound CH_3–O–CH_2CH_3 can be called methyl ethyl ether (prefixes arranged by size of the alkyl group) or ethyl methyl ether (prefixes ordered alphabetically).

This same compound would be named methoxyethane using the I.U.P.A.C. Nomenclature System. With this system, an ether is named as a substituted hydrocarbon, in this case an alkane. In this system, the –OR group is named as an alkoxy group.

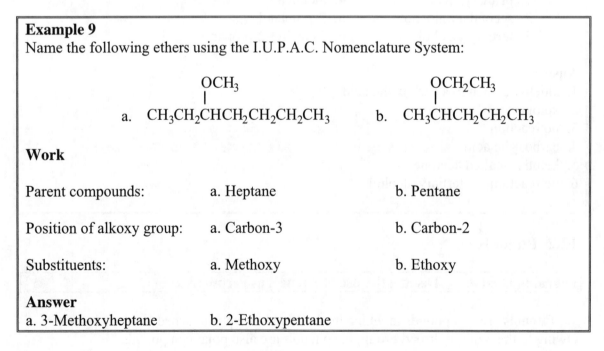

Example 9
Name the following ethers using the I.U.P.A.C. Nomenclature System:

$$\text{a.} \quad \underset{\overset{|}{OCH_3}}{CH_3CH_2CHCH_2CH_2CH_2CH_3} \qquad \text{b.} \quad \underset{\overset{|}{OCH_2CH_3}}{CH_3CHCH_2CH_2CH_3}$$

Work

Parent compounds:	a. Heptane	b. Pentane
Position of alkoxy group:	a. Carbon-3	b. Carbon-2
Substituents:	a. Methoxy	b. Ethoxy

Answer
a. 3-Methoxyheptane   b. 2-Ethoxypentane

Learning Goal 10:	Write equations representing formation of an ether by a dehydration reaction between two alcohol molecules.

Ethers may be prepared by a dehydration reaction between two alcohol molecules.

Example 10
Write an equation showing the synthesis of diethyl ether.

Work
Diethyl ether would be prepared by a dehydration reaction between two molecules of ethanol.

Answer

$$\underset{\text{Ethanol}}{CH_3CH_2OH} + \underset{\text{Ethanol}}{CH_3CH_2OH} \xrightarrow{H^+} \underset{\text{Diethyl ether}}{CH_3CH_2{-}O{-}CH_2CH_3} + H_2O$$

Diethyl ether was the first general anesthetic used in medical practice. However, since ethers are highly flammable and can form explosive peroxides upon storage, diethyl ether has largely been replaced by halogenated ethers, such as Penthrane and Enthrane.

12.4 Thiols

Learning Goal 11:	Write names and draw structures for simple thiols and discuss their biological significance.

Compounds that contain the –SH group are known as **thiols.** Although similar to alcohols in structure, thiols generally have lower boiling points than corresponding alcohols, though they are higher in molecular weight. Furthermore, thiols, and many other sulfur compounds, have nauseating aromas.

Thiols are also involved in protein structure and conformation. It is the ability of two thiol groups to easily undergo oxidation to a –S–S– (disulfide) bond that helps maintain the correct shape of a protein. The amino acid cysteine has a thiol group and can participate in disulfide bond formation in proteins.

Coenzyme A is a thiol that serves as a "carrier" of acetyl groups and fatty acids in cellular metabolic reactions.

12.5 Aldehydes and Ketones

Structure and Physical Properties

Learning Goal 12:	Draw the structures and discuss the physical properties of aldehydes and ketones.

Owing to the polar carbonyl group, aldehydes and ketones are moderately polar compounds. As a result, they boil at higher temperatures than hydrocarbons or ethers having the same number of carbon atoms, but at temperatures lower than comparable alcohols.

 Aldehydes and ketones composed of five or fewer carbon atoms are reasonably soluble in water because of the hydrogen bonding between the carbonyl group and water molecules. Larger members of these carbonyl-containing compounds are less polar, more hydrocarbon-like, and thus more soluble in nonpolar organic solvents.

Example 11

Which member of each of the following pairs of compounds that have similar molecular weights has the higher boiling point?

1. butane or methoxyethane
2. methoxyethane or propanol
3. propanone or methoxyethane

Work

Aldehydes and ketones are moderately polar compounds and boil at higher temperatures than hydrocarbons or ethers of similar molecular weights. The aldehydes and ketones, however, boil at lower temperatures than similar alcohols.

Answer
1. methoxyethane
2. propanol
3. propanone

Example 12

Show the polar attraction that is present between two acetone molecules.

Answer

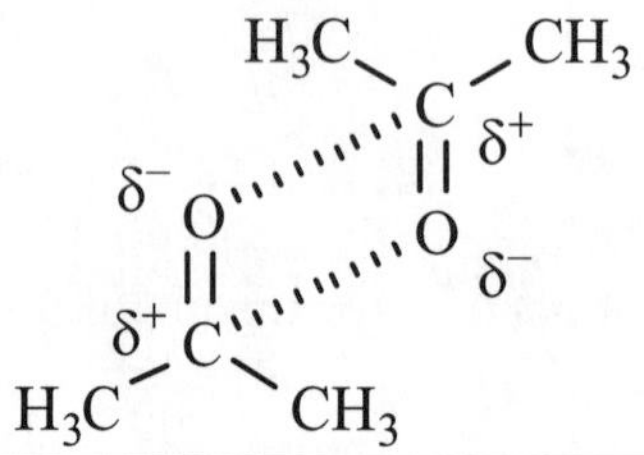

Nomenclature

Naming Aldehydes

Determine the parent compound, that is, the longest continuous carbon chain containing the carbonyl group. Drop the final *-e* of the parent alkane and replace it with *-al*. The parent chain is always numbered beginning with the carbonyl carbon as carbon-1. All other substituents are named and numbered as usual.

Common names for aldehydes are derived from the same Latin root as the corresponding carboxylic acids (see Tables 12.1 and 13.1). Substituted aldehydes are named as derivatives of the straight-chain parent compound, using Greek letters to indicate the positions of substituents. The carbon atom bonded to the carbonyl carbon is referred to as the α-carbon.

Naming Ketones

Determine the parent compound. Replace the *–e* ending of the parent alkane with the *-one* suffix of the ketone family. The longest carbon chain is numbered to give the carbonyl carbon the lowest possible number. All other substituents are named and numbered as usual.

The common names for ketones are derived by naming the alkyl groups that are bonded to the carbonyl carbon. These are used as prefixes followed by the word *ketone*. Alkyl groups may be arranged alphabetically or by size.

Example 13
Determine the I.U.P.A.C. and common names of the following aldehydes and ketones:

1. $H-\overset{\overset{\displaystyle O}{\|}}{C}-H$

3. $CH_3CH_2-\overset{\overset{\displaystyle O}{\|}}{C}-CH_3$

2. $CH_3CH_2-\overset{\overset{\displaystyle O}{\|}}{C}-H$

4. $CH_3CH_2CH_2-\overset{\overset{\displaystyle O}{\|}}{C}-CH_3$

Work
Aldehydes are named according to the longest continuous carbon chain (parent) that contains the carbonyl group. The final *-e* of the parent alkane is dropped and replaced by *-al* for aldehyde. Numbers are not needed for unsubstituted aldehydes, since the carbonyl group is always at the end of the molecule. For ketones, the final *-e* of the alkane name is replaced with *-one*, and numbers may be needed to show the position of the carbonyl group in the compound.

Example 13 continued

Answer
1. methanal (common name: formaldehyde)
2. propanal (common name: propionaldehyde)
3. butanone (common name: ethyl methyl ketone)
4. 2-pentanone (common name: methyl propyl ketone)

Example 14
Write structural formulas for each of the following:

1. 5-bromohexanal
2. 3-methylheptanal
3. 3-bromo-2-pentanone
4. cyclohexanone

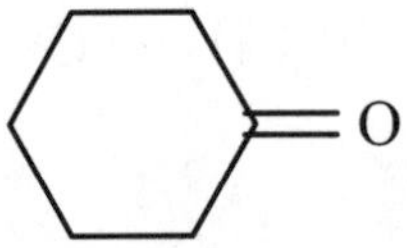

Answer

1. 5-bromohexanal:

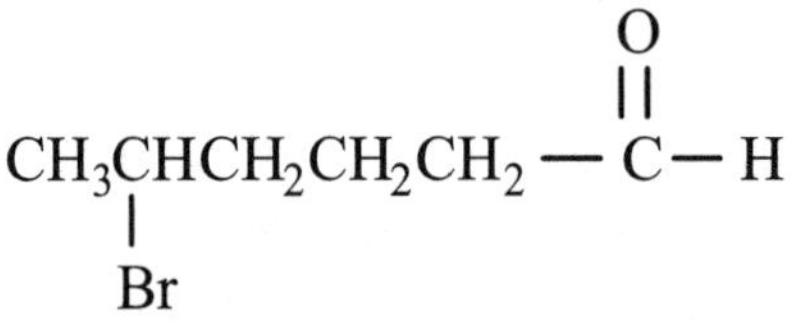

$$CH_3CHCH_2CH_2CH_2 - \overset{\overset{\displaystyle O}{\|}}{C} - H$$
$$\underset{Br}{|}$$

2. 3-methylheptanal:

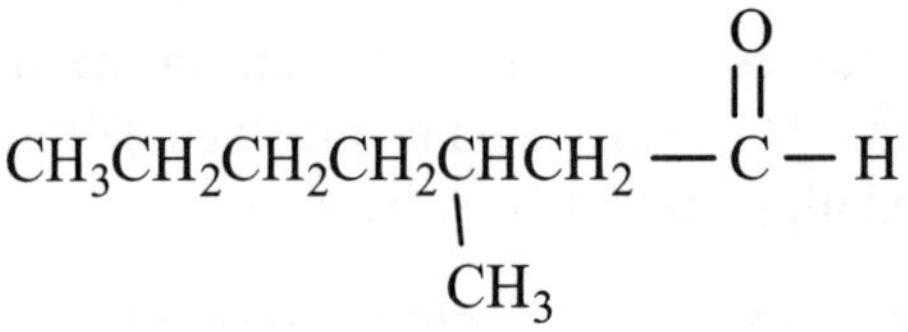

$$CH_3CH_2CH_2CH_2CHCH_2 - \overset{\overset{\displaystyle O}{\|}}{C} - H$$
$$\underset{CH_3}{|}$$

3. 3-bromo-2-pentanone:

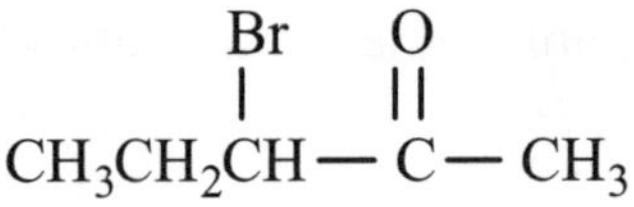

$$CH_3CH_2\overset{\overset{\displaystyle Br}{|}}{CH} - \overset{\overset{\displaystyle O}{\|}}{C} - CH_3$$

4. cyclohexanone:

Important Aldehydes and Ketones

Learning Goal 14:   Discuss the biological, medical, or environmental significance of several aldehydes and ketones.

Methanal (formaldehyde) is a gas. It is available commercially as an aqueous solution (formalin) that is used to preserve tissue samples.

Ethanal (acetaldehyde) is produced from ethanol in the liver and is responsible for the symptoms of a hangover.

168

Propanone (acetone) is the simplest ketone. It is an important solvent because it can dissolve organic compounds and is also miscible (mixes) with water. It is found as a solvent in adhesives, paints, and nail polish remover. Many complex members of the ketone family are important in the food industry as food additives. Others are useful as medicinals and agricultural chemicals.

Reactions Involving Aldehydes and Ketones

In the laboratory, aldehydes and ketones are often prepared by the oxidation of the corresponding alcohol. Any aldehyde or ketone can be prepared if the correct alcohol is available.

Oxidation of methyl alcohol gives methanal. Oxidation of a primary alcohol produces an aldehyde; oxidation of a secondary alcohol yields a ketone. Tertiary alcohols do not undergo oxidation. The conclusion we can draw from this information is that the carbonyl carbon must have at least one hydrogen substituent for oxidation to occur.

Example 15

Provide the I.U.P.A.C. names of the oxidation products for the following alcohols:
1. methanol
2. 2-propanol

Answer
1. methanal (or methanoic acid, if methanal is further oxidized)
2. propanone

Oxidation Reactions

Learning Goal 15:	**Write equations representing the oxidation and reduction of aldehydes and ketones.**

Aldehydes are very easily oxidized further to carboxylic acids. In fact, they are so easily oxidized that it is often very difficult to prepare or store them. Oxidation of an aldehyde yields a carboxylic acid:

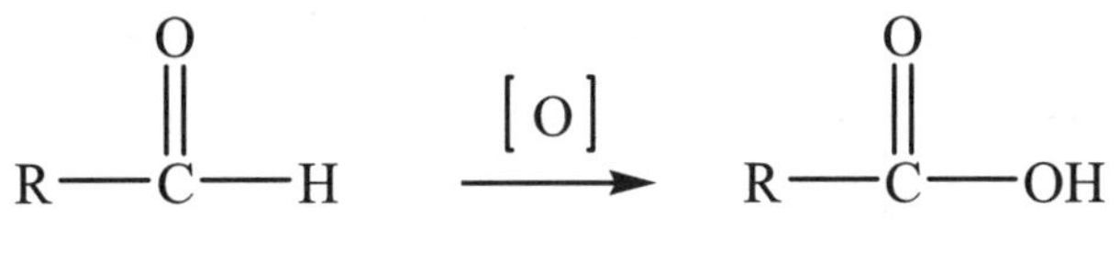

Ketones do not undergo further oxidation reactions because a carbon-hydrogen bond to the carbonyl carbon is necessary for the reaction to occur.

Aldehydes and ketones can be distinguished from one another based on their ability to undergo oxidation reactions. The **Tollens' test,** or Tollens' silver mirror test, is the most common such test. Tollens' reagent consists of a basic solution of $Ag(NH_3)_2^+$. An aldehyde will undergo an oxidation-reduction reaction in which the silver ion (Ag^+) is reduced to silver metal (Ag^0) as the aldehyde is oxidized to a carboxylic acid. The silver metal precipitates from solution and coats the vessel, giving a smooth silver mirror. Because ketones cannot undergo further oxidation, they do not react with the Tollens' reagent.

Example 16

Write the major products for each of the following reactions, using Tollens' reagent as the oxidizing agent:

$$1. \quad CH_3 - \overset{\displaystyle O}{\overset{\displaystyle \|}{C}} - H \ + \ Ag(NH_3)_2^+ \longrightarrow \ ?$$

$$2. \quad CH_3CH_2 - \overset{\displaystyle O}{\overset{\displaystyle \|}{C}} - H \ + \ Ag(NH_3)_2^+ \longrightarrow \ ?$$

$$3. \quad CH_3 - \overset{\displaystyle O}{\overset{\displaystyle \|}{C}} - CH_3 \ + \ Ag(NH_3)_2^+ \longrightarrow \ ?$$

Work

Treatment of an aldehyde with Tollens' reagent gives an oxidation-reduction reaction. The aldehyde is oxidized to the carboxylic acid, and the silver ion (Ag^+) is reduced to silver metal (Ag^0).

Answer

1. $CH_3COOH + Ag^0$
2. $CH_3CH_2COOH + Ag^0$
3. No reaction—Tollens' reagent can not oxidize ketones.

Another test used to distinguish between aldehydes and ketones is **Benedict's test.** In this test, Cu^{2+} is reduced to Cu^+. Cu^{2+} is soluble and gives a blue solution, while the Cu^+ precipitates as the red solid, Cu_2O. Benedict's test has been used to determine the concentrations of glucose in urine.

Reduction Reactions

Aldehydes and ketones are easily reduced to the corresponding alcohol by a large number of different reducing agents, designated [H]. The general reduction reaction is shown:

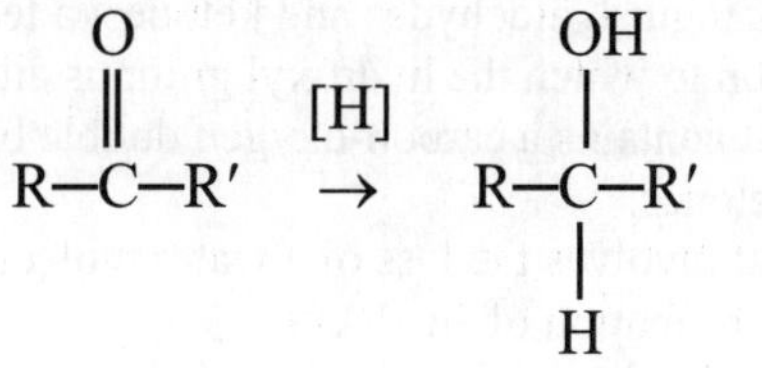

$$
\underset{\text{Aldehyde or ketone}}{R-\overset{\displaystyle O}{\overset{\|}{C}}-R'} \quad \xrightarrow{[H]} \quad \underset{\text{Alcohol}}{R-\overset{\displaystyle OH}{\underset{\underset{\displaystyle H}{|}}{\overset{|}{C}}}-R'}
$$

The classical reaction for aldehyde and ketone reduction is **hydrogenation,** in which the carbonyl compound is reacted with hydrogen gas. This reaction requires a metal catalyst, pressure and/or heat. The carbon-oxygen double bond (the carbonyl group) is reduced to a carbon-oxygen single bond.

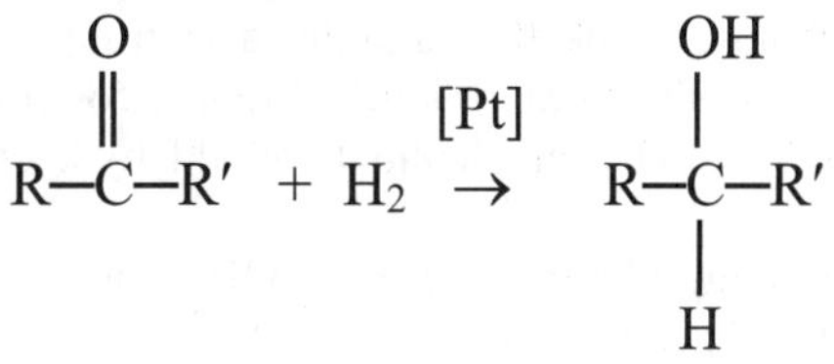

$$
\underset{\text{Aldehyde or ketone}}{R-\overset{\displaystyle O}{\overset{\|}{C}}-R'} + H_2 \quad \xrightarrow{[Pt]} \quad \underset{\text{Alcohol}}{R-\overset{\displaystyle OH}{\underset{\underset{\displaystyle H}{|}}{\overset{|}{C}}}-R'}
$$

Example 17

Label the following as oxidation or reduction reactions:

1. methanol to methanoic acid
2. methanol to formaldehyde
3. benzoic acid to benzaldehyde
4. 2-pentanol to 2-pentanone
5. acetone to 2-propanol

Answer
1. oxidation
2. oxidation
3. reduction
4. oxidation
5. reduction

Glossary of Key Terms in Chapter Twelve

alcohol (12.1) an organic compound that contains a hydroxyl group (-OH) attached to an alkyl group.

aldehyde (12.5) a class of organic molecules characterized by a carbonyl group; the carbonyl carbon is bonded to a hydrogen atom and to another hydrogen or an alkyl or aryl group.

Benedict's test (12.5) a test reagent used to distinguish aldehydes and ketones; a test for reducing sugars.

carbinol carbon (12.1) in an alcohol, the carbon to which the hydroxyl group is attached.

carbonyl group (12.5) the functional group that contains a carbon-oxygen double bond; -C=O; the functional group found in aldehydes and ketones.

dehydration (of alcohols) (12.1) a reaction that involves the loss of a water molecule; e.g., the loss of water from an alcohol and the concomitant formation of an alkene.

disulfide (12.4) an organic compound that contains the disulfide group (-S-S-).

ether (12.3) an organic compound that contains two alkyl and/or aryl groups attached to an oxygen atom; R-O-R, Ar-O-R, and Ar-O-Ar.

fermentation (12.1) the anaerobic (in the absence of oxygen) metabolism or degradation of glucose by microorganisms.

hydration (of alkenes) (12.1) a reaction in which water is added to a molecule; e.g., the addition of water to an alkene to form an alcohol.

hydrogenation (12.5) a reaction in which hydrogen (H_2) is added to a double or a triple bond.

hydroxyl group (12.1) the -OH functional group.

ketone (12.5) a family of organic molecules characterized by a carbonyl group; the carbonyl carbon is bonded to two alkyl groups, two aryl groups, or one alkyl and one aryl group.

oxidation (12.1) the loss of electrons by a molecule, atom, or ion; in organic compounds, the gain of oxygen or loss of hydrogen; e.g., the conversion of an alcohol to an aldehyde or ketone via the use of an oxidizing agent.

phenol (12.2) an organic compound that contains a hydroxyl group (-OH) attached to a benzene ring.

primary (1°) alcohol (12.1) an alcohol with the general formula RCH_2OH.

reduction (12.1) a gain of electrons by a molecule, ion, or atom; in organic compounds, the loss of oxygen or gain of hydrogen; e.g., the conversion of a carbonyl compound to an alcohol via the use of a reducing agent.

secondary (2°) alcohol (12.1) an alcohol with the general formula R_2CHOH.

tertiary (3°) alcohol (12.1) an alcohol with the general formula R_3COH.

thiol (12.4) an organic compound that contains a thiol group (-SH).

Tollens' test (12.5) a test reagent (silver nitrate in ammonium hydroxide) used to distinguish between aldehydes and ketones; also called the Tollens' silver mirror test.

Zaitsev's rule (12.1) a rule that states that in the dehydration of an alcohol, the alkene with the greatest number of alkyl groups on the double bonded carbon (the more highly substituted alkene) is the major product of the reaction.

Self Test for Chapter Twelve

1. Which of the following compounds of each set has the highest boiling point?

    a. 1-nonanol or 1-hexanol

    b. ethane or methanol

    c. methoxyethane or 1-propanol

    d. methane or dimethyl ether

    e. 1-propanol or propane or propanone

    f. pentane or methoxyethane or propanal

2. Which of the following alcohols has the greatest solubility in water?

    a. 1-propanol   c. 1-heptanol

    b. 1-butanol   d. 1-decanol

3. Provide the I.U.P.A.C. name of the following compounds:

    a.   b.

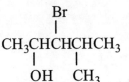

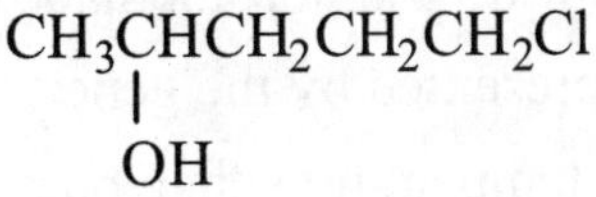

$CH_3CHCH_2CH_2CH_2Cl$
        |
       OH

4. Determine the I.U.P.A.C. name of the trialcohol called glycerol:
       $HOCH_2CHOHCH_2OH$

5. Provide the common names of the following:

    a. $CH_3CH_2CH_2OH$

    b. $CH_3CHOHCH_3$

    c. $C(CH_3)_3OH$

6. Complete the following chemical reactions:

    a.   (formaldehyde) $+ H_2 \xrightarrow{\text{catalyst}} ?$

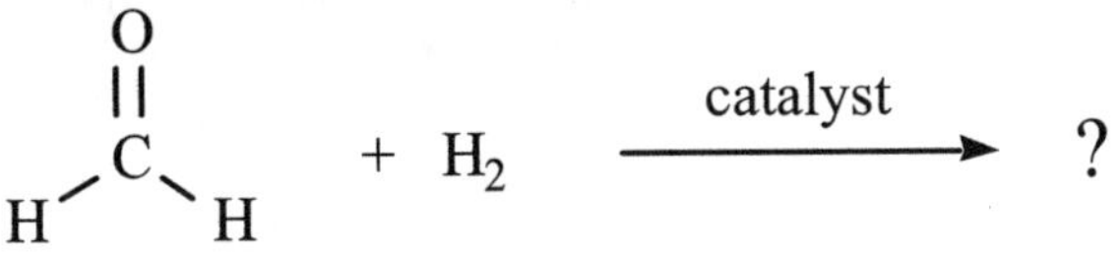

    b. $CH_2 = CH_2 + H_2O \xrightarrow[\text{catalyst}]{H^+} ?$

7. Complete the oxidation reactions of each of the following alcohols to form aldehydes:

    a. $CH_3OH \xrightarrow{\text{oxidation}}$

    b. $CH_3CH_2OH \xrightarrow{\text{oxidation}}$

8. What are the products of the dehydration of each of the following alcohols?

    a. ethanol

    b. 1-propanol

    c. 2-butanol

9. Provide the common names of each of the following ethers:

    a. CH_3OCH_3 c. $CH_3OCH_2CH_3$

    b. $CH_3CH_2OCH_2CH_3$

10. Which of the following compounds is the most oxidized?

    a. propane   d. acetic acid

    b. 1-propanol   e. propanone

    c. 2-propanol

11. Which of the following compounds is the most reduced?

    a. methane   c. dichloromethane

    b. chloromethane   d. trichloromethane

12. What class of compounds is represented by the general formula RSH?

13. Why do alcohols boil at higher temperatures than pure hydrocarbons or ethers of similar molecular weights?

14. In a secondary alcohol, how many hydrogen atoms are directly attached to the carbinol carbon atom?

15. Which kind of alcohol will not undergo oxidation under normal conditions?

16. Potassium permanganate and potassium dichromate are used with alcohols in what kind of chemical reaction?

17. Which class of compounds with a strong odor is found in the warning scent of skunks?

18. Which of the following is the least polar? Give the letter as the answer.

    a. alcohols

    b. ethers

    c. water

19. Which of the following cannot hydrogen bond to other molecules?

    a. alcohols   b. ethers   c. water

20. Give the I.U.P.A.C. names of the following:

$$\text{a.} \quad CH_3 - \overset{\displaystyle O}{\overset{\displaystyle \|}{C}} - H \qquad\qquad \text{d.} \quad CH_3CH_2 - \overset{\displaystyle O}{\overset{\displaystyle \|}{C}} - CH_2CH_3$$

$$\text{b.} \quad CH_3 - \overset{\displaystyle O}{\overset{\displaystyle \|}{C}} - CH_3 \qquad\qquad \text{e.} \quad CH_3CH_2\overset{}{\underset{\displaystyle \underset{Br}{|}}{C}}HCH_2CH_2 - \overset{\displaystyle O}{\overset{\displaystyle \|}{C}} - H$$

$$\text{c.} \quad CH_3CH_2 - \overset{\displaystyle O}{\overset{\displaystyle \|}{C}} - H$$

21. Write structural formulas for the following:

    a. 2-bromobutanal

    b. 2-methyl-3-pentanone

22. Complete the following oxidation reactions:

a. $H-\overset{\overset{\displaystyle O}{\|}}{C}-H \xrightarrow{[O]}$?

b. $CH_3-\overset{\overset{\displaystyle O}{\|}}{C}-H \xrightarrow{[O]}$?

c. $CH_3-\overset{\overset{\displaystyle O}{\|}}{C}-CH_3 \xrightarrow{[O]}$?

d. $CH_3CH_2-\overset{\overset{\displaystyle O}{\|}}{C}-H \xrightarrow{[O]}$?

e. $CH_3-\overset{\overset{\displaystyle OH}{|}}{\underset{\underset{\displaystyle CH_3}{|}}{C}}-CH_3 \xrightarrow{[O]}$?

23. Give the I.U.P.A.C. names of the oxidation products for the following alcohols:

a. CH_3CH_2OH

b. 1-propanol

c. 2-propanol

d. 2-methyl-2-propanol

24. Show the major products for each of the following oxidation reactions:

a. $CH_3-\overset{\overset{\displaystyle O}{\|}}{C}-H \; + \; Ag(NH_3)_2^+ \longrightarrow$?

b. $CH_3-\overset{\overset{\displaystyle O}{\|}}{C}-CH_3 \; + \; Ag(NH_3)_2^+ \longrightarrow$?

c. $CH_3CH_2-\overset{\overset{\displaystyle O}{\|}}{C}-CH_3 \; + \; Ag(NH_3)_2^+ \longrightarrow$?

25. The general formula RCOR represents what class of compounds?

26. Aldehydes are easily further oxidized to what compounds?

Vocabulary Quiz for Chapter Twelve

1. Organic compounds that contain hydroxyl groups attached to alkyl groups are _____.

2. A family of organic compounds that contain a sulfhydryl group is the _____.

3. The carbon atom bonded to the hydroxyl group is called the _____.

4. A(n) _____ is an organic compound that contains two alkyl and/or aryl groups attached to an oxygen atom: R-O-R, Ar-O-R, and Ar-O-Ar.

5. A(n) _____ is an organic compound that contains a hydroxyl group (-OH) attached to a benzene ring.

6. _____ states that in the dehydration of an alcohol, the alkene with the greatest number of alkyl groups on the double bonded carbon (the more highly substituted alkene) is the major product of the reaction.

7. In inorganic chemistry, _____ is the loss of electrons accompanied by a change in charge. In organic chemistry, this is more generally seen as a gain of oxygen or loss of hydrogen.

8. The functional group that consists of a carbon double bonded to an oxygen is the _____.

9. The conversion of an alcohol to an aldehyde or ketone is a(n) _____ reaction.

10. _____ is a test reagent (silver nitrate in ammonium hydroxide) used to test for the carbonyl functional group.

13 *Carboxylic Acids, Esters, Amines, and Amides*

Learning Goals

1 Write structures and describe the physical properties of carboxylic acids.
2 Determine the common and I.U.P.A.C. names of carboxylic acids.
3 Describe the biological, medical, or environmental significance of several carboxylic acids.
4 Write equations representing acid-base reactions of carboxylic acids.
5 Write structures and describe the physical properties of esters.
6 Determine the common and I.U.P.A.C. names of esters.
7 Write equations representing the synthesis and hydrolysis of an ester.
8 Define the term saponification and describe how soap works in the emulsification of grease and oils.
9 Classify amines as primary, secondary, or tertiary.
10 Describe the physical properties of amines.
11 Draw and name simple amines using common and systematic nomenclature.
12 Write equations showing the basicity and neutralization of amines.
13 Describe the structure of quaternary ammonium salts and discuss their use as antiseptics and disinfectants.
14 Describe the physical properties of amides.
15 Draw the structure and write the common and I.U.P.A.C. names of amides.
16 Write equations representing the hydrolysis of amides.

Introduction

Carboxylic acids are characterized by the **carboxyl group,** and have the following general structure:

$$
\underset{\text{Aromatic carboxylic acid}}{Ar - \overset{\displaystyle O}{\overset{\displaystyle \|}{C}} - OH}
\qquad\qquad
\underset{\text{Aliphatic carboxylic acid}}{R - \overset{\displaystyle O}{\overset{\displaystyle \|}{C}} - OH}
$$

The term *carboxylic acid* tells us that the carboxyl group is derived from a carbonyl group and hydroxyl group. It further tells us that these molecules are acids.

Esters are produced in the reaction between a carboxylic acid and an alcohol and have the following general structure:

177

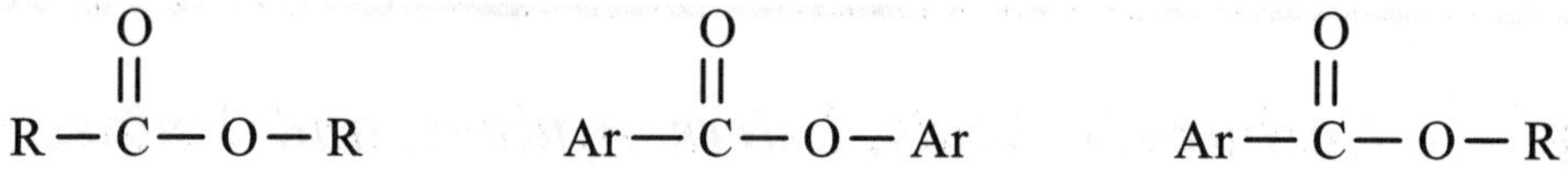

$$\underset{\overset{\|}{R-C-O-R}}{O} \qquad \underset{\overset{\|}{Ar-C-O-Ar}}{O} \qquad \underset{\overset{\|}{Ar-C-O-R}}{O}$$

Examples of aliphatic and aromatic esters

The following structure is called the **acyl group:**

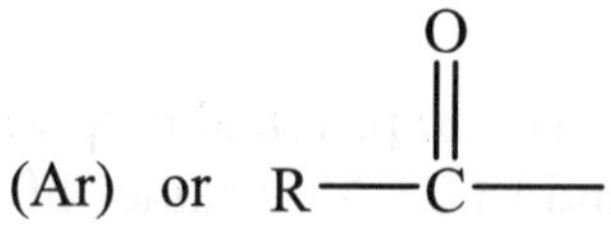

$$(Ar) \quad or \quad R-\overset{\overset{\displaystyle O}{\|}}{C}-$$

The acyl group is the functional group of the carboxylic acid derivatives, including the esters and amides.

Amines are organic molecules that contain the amino group, $-NH_2$, or substituted amino group. They may be aromatic or aliphatic and have the general formula (Ar–) or $R-NH_2$. You can think of the amines as substituted ammonia molecules in which one or more of the hydrogens has been substituted by an organic group. Amides are carboxylic acid derivatives. The amide group is made up of two portions: one portion from a carboxyl group and the other from an amine. The amino group is found in many important biological molecules, particularly the proteins and nucleic acids (DNA and RNA), and is important in understanding the properties of these large, complex molecules.

Example 1

The following general formulas are used to represent different classes of organic compounds. Name each of the appropriate families.

1. ROH   4. ROR
2. RCHO   5. RCOOH
3. RCOR   6. RCOOR

Answer
1. alcohol   4. ether
2. aldehyde   5. carboxylic acid
3. ketone   6. ester

13.1 Carboxylic Acids

Structure and Physical Properties

Learning Goal 1:   Write structures and describe the physical properties of carboxylic acids.

The carboxyl group consists of two very polar functional groups, the carbonyl group and the hydroxyl group. Thus, carboxylic acids are very polar compounds.

In addition, carboxylic acids can hydrogen bond to one another. As a result, they boil at higher temperatures than aldehydes, ketones, or even alcohols of comparable molecular weight.

Carboxylic acids can form intermolecular hydrogen bonds with water molecules. Thus, small carboxylic acids are water-soluble. However, solubility falls off dramatically as the carbon chain length increases.

The lower molecular weight carboxylic acids have sharp, sour tastes and unpleasant aromas. The longer chain carboxylic acids are called fatty acids and are components of many biologically important lipids.

Example 2

Which member of each of the following sets of compounds has the highest boiling point?

1. ethane or ethanol or ethanoic acid
2. propanal or propanone or 1-propanol
3. methanol or methanal or methanoic acid

Work

The boiling points of most alkanes, alcohols, aldehydes, and carboxylic acids that have similar molecular weights obey the following relationship:

alkane < aldehyde or ketone < alcohol < carboxylic acid
lowest   highest

Answer

1. ethanoic acid   2. 1-propanol   3. methanoic acid

Nomenclature

Learning Goal 2:	Determine the common and I.U.P.A.C. names of carboxylic acids.

In the I.U.P.A.C. System, carboxylic acids are named by replacing the -e ending of the parent alkane with the suffix -oic acid. The suffix –dioic acid is used if there are two carboxyl groups. The parent chain is numbered so that the carboxyl carbon is carbon-1. Other groups are named and numbered in the usual way.

To name the acyl group of a carboxylic acid, replace the -oic acid suffix with –yl. For instance, the acyl group of acetic acid is the acetyl group.

The carboxylic acid derivatives of cycloalkanes are named by adding the suffix carboxylic acid to the name of the cycloalkane or substituted cycloalkane. The carboxyl group is defined to be on carbon-1.

Common names of carboxylic acids are frequently used. Table 13.1 in the text shows the I.U.P.A.C. and common names of several carboxylic acids, as well as their sources and the Latin or Greek words that gave rise to the common names.

Aromatic carboxylic acids are usually named as derivatives of benzoic acid. The *-oic acid* or *-ic acid* suffix is attached to the appropriate prefix. However, common names of substituted benzoic acids are frequently used.

Example 3

Give the I.U.P.A.C. names for the following carboxylic acids:

1. CH_3COOH 2. CH_3CH_2COOH 3. $CH_3CHBrCH_2COOH$

Work

1. Determine the longest continuous chain (parent) of carbon atoms that includes the carboxyl group. The carbon atom of the carboxyl group is defined as 1 and no number is included in the name.
2. Then drop the *-e* of the parent alkane name and add *-oic acid*.

Answer

1. ethanoic acid   2. propanoic acid   3. 3-bromobutanoic acid

---

Example 4

Give the common name of each of the following carboxylic acids:

1. CH_3COOH
2. $CH_3CH_2CH_2COOH$
3. $HCOOH$
4. $C_{17}H_{35}COOH$

Answer

1. acetic acid
2. butyric acid
3. formic acid
4. stearic acid

Some Important Carboxylic Acids

Learning Goal 3:	Describe the biological, medical, or environmental significance of several carboxylic acids.

Many carboxylic acids found in nature are listed in Table 13.1 of the text. Fatty acids can be isolated from a variety of fats and oils. More complex carboxylic acids are also found in a variety of foodstuffs. Citric acid is found in citrus fruits and is added to foods to give a sharp taste (sour candies) or as a preservative and antioxidant. Adipic acid adds tartness to soft drinks and helps to retard spoilage. Bacteria in milk produce lactic acid as a product of fermentation of sugars. It contributes a tangy flavor to yogurt and buttermilk and acts as a food preservative. Lactic acid is also produced by muscles during strenuous exercise.

Reactions Involving Carboxylic Acids

Preparation of Carboxylic Acids

Many of the small carboxylic acids are prepared by the oxidation of the corresponding primary alcohol or aldehyde. A variety of oxidizing agents can be used in these reactions and a catalyst is often required. The general reaction is:

$$RCH_2OH \xrightarrow{[O]} R-\overset{\overset{\displaystyle O}{\|}}{C}-H \xrightarrow{[O]} R-\overset{\overset{\displaystyle O}{\|}}{C}-OH$$

Primary alcohol   Aldehyde   Carboxylic acid

Acid-Base Reactions

> **Learning Goal 4:** Write equations representing acid-base reactions of carboxylic acids.

The carboxylic acids behave as weak acids because they are proton donors. They are weak acids (typically less than 5% dissociation) that produce a carboxylate ion and a hydrogen ion in water, as seen in the following example:

$$R-\overset{\overset{\displaystyle O}{\|}}{C}-OH \ + \ H_2O \ \rightleftharpoons \ R-\overset{\overset{\displaystyle O}{\|}}{C}-O^- \ + \ H_3O^+$$

Carboxylic acid   Water   Carboxylate anion   Hydronium ion

When strong bases are added to a carboxylic acid, neutralization occurs:

$$R-\overset{\overset{\displaystyle O}{\|}}{C}-OH \ + \ NaOH \ \longrightarrow \ R-\overset{\overset{\displaystyle O}{\|}}{C}-O^-Na^+ \ + \ H_2O$$

Carboxylic acid   Strong base   Carboxylic acid salt   Water

The salt of a carboxylic acid is named by replacing the *-ic acid* suffix with *-ate*. This name is preceded by the name of the appropriate cation, for instance, sodium.

The salts are ionic substances and, hence, are quite soluble in water. The long-chain carboxylic acid salts (fatty acid salts) are good **soaps.** Soaps are made from water, strong base, and natural fats and oils obtained from animals or plants. Soaps form micelles around grease and oil. The hydrophobic alkane tail of the soap dissolves the oils and grease, and the hydrophilic carboxylate end of the molecule remains associated with water. This produces a micelle, a tiny sphere with the alkyl group tails of the soap molecules, grease, and oils in the center and the carboxylate on the outside of the sphere, dissolved in the water.

Example 5

Show the reactions of the following carboxylic acids with water:

1. CH_3COOH
2. HCOOH
3. $CH_3CH_2CH_2COOH$

Answer

1. Acetic acid dissociates to produce the acetate ion and hydronium ion.
   $$CH_3COOH + H_2O \rightleftharpoons CH_3COO^- + H_3O^+$$

2. Formic acid dissociates to produce the formate ion and hydronium ion.
   $$HCOOH + H_2O \rightleftharpoons HCOO^- + H_3O^+$$

3. Butyric acid dissociates to produce the butyrate ion and hydronium ion.
   $$CH_3CH_2CH_2COOH + H_2O \rightleftharpoons CH_3CH_2CH_2COO^- + H_3O^+$$

13.2 Esters

Structure and Physical Properties

Learning Goal 5:   Write structures and describe the physical properties of esters.

Esters are mildly polar and have pleasant aromas. Many esters are found in natural foodstuffs. Esters boil at approximately the same temperature as aldehydes or ketones of comparable molecular weight. The simpler ones are reasonably soluble and nonreactive in water.

Nomenclature

Learning Goal 6:   Determine the common and I.U.P.A.C. names of esters.

Esters are formed in the reaction of a carboxylic acid with an alcohol and both of these families are reflected in the naming of the ester. The *alkyl* or *aryl* portion of the alcohol name is used as the prefix and the *-ic acid* ending of the name of the carboxylic acid is replaced with *-ate*.

Example 6

Show the general reaction of an alcohol (R′OH) with a carboxylic acid (RCOOH) to produce an ester.

Answer

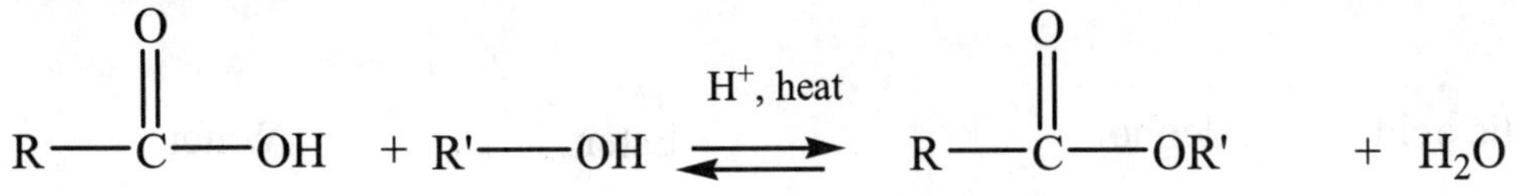

$$R-\overset{\overset{\displaystyle O}{\|}}{C}-OH \; + \; R'-OH \underset{\xleftarrow{\hspace{1cm}}}{\overset{H^+,\ heat}{\xrightarrow{\hspace{1cm}}}} \; R-\overset{\overset{\displaystyle O}{\|}}{C}-OR' \; + \; H_2O$$

Carboxylic acid   Alcohol   Ester   Water

Example 7

Give the common names of the following esters:

1. 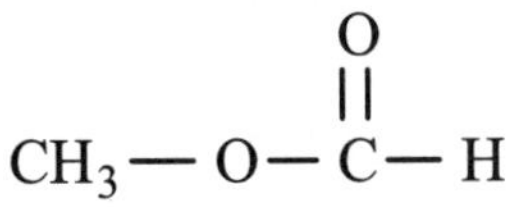 $CH_3-O-\overset{\overset{\displaystyle O}{\|}}{C}-H$

3. phenyl $-\overset{\overset{\displaystyle O}{\|}}{C}-O-CH_3$

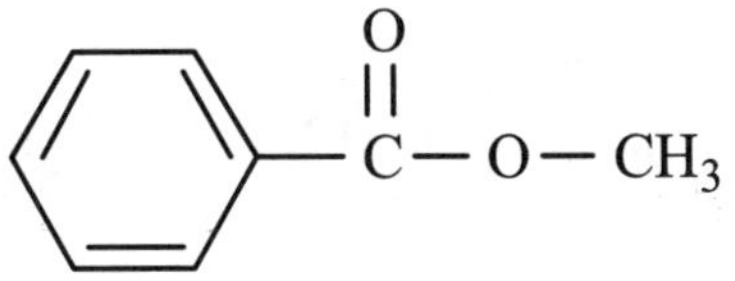

2. 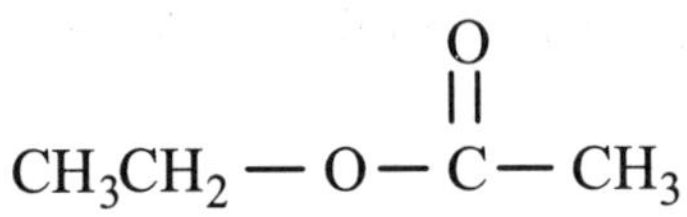 $CH_3CH_2-O-\overset{\overset{\displaystyle O}{\|}}{C}-CH_3$

4. 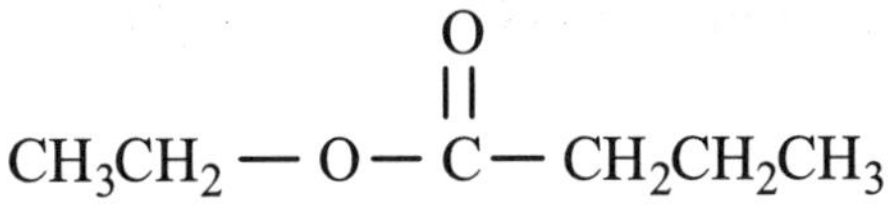 $CH_3CH_2-O-\overset{\overset{\displaystyle O}{\|}}{C}-CH_2CH_2CH_3$

Work

Esters are formed from the reaction of a carboxylic acid with an alcohol. The alkyl or aryl portion of the alcohol name is used as a prefix and is followed by the name of the carboxylic acid in which the *-ic acid* (common or I.U.P.A.C.) is replaced by *-ate*.

Answer
1. methyl formate
2. ethyl acetate
3. methyl benzoate
4. ethyl butyrate

Reactions Involving Esters

Learning Goal 7: Write equations representing the synthesis and hydrolysis of an ester.

Preparation of Esters

Carboxylic acids react with alcohols to form esters and water according to the following general reaction:

$$R-\overset{\overset{\displaystyle O}{\|}}{C}-OH \;+\; R'-OH \;\underset{\text{}}{\overset{H^+,\ \text{heat}}{\rightleftharpoons}}\; R-\overset{\overset{\displaystyle O}{\|}}{C}-OR' \;+\; H_2O$$

Carboxylic acid   Alcohol   Ester   Water

The conversion of a carboxylic acid to an ester requires heat and is catalyzed by a trace of acid (H^+).

Hydrolysis of Esters

Esters undergo **hydrolysis** reactions in water. This reaction requires heat and a small amount of acid (H^+) or base (OH^-) to catalyze the reaction, as seen in the following general equations:

$$R-\overset{\overset{\displaystyle O}{\|}}{C}-OR' \;+\; H_2O \;\underset{\text{Heat}}{\overset{H^+}{\rightleftharpoons}}\; R-\overset{\overset{\displaystyle O}{\|}}{C}-OH \;+\; R'-OH$$

Ester   Water   Carboxylic acid   Alcohol

$$R-\overset{\overset{\displaystyle O}{\|}}{C}-OR' \;+\; NaOH \;\longrightarrow\; R-\overset{\overset{\displaystyle O}{\|}}{C}-O^-Na^+ \;+\; R'-OH$$

Ester   Strong base   Carboxylic acid salt   Alcohol

Learning Goal 8:	**Define the term saponification and describe how soap works in the emulsification of grease and oil.**

The base-catalyzed hydrolysis of an ester is called **saponification.** The product is a carboxylic acid salt. The carboxylic acid is formed when the reaction mixture is neutralized with an acid. Saponification is used to hydrolyze fats and oils, which are esters, to the salts of long chain fatty acids—*soaps*.

184

13.3 Amines

Structure and Physical Properties

Learning Goal 9:   Classify amines as primary, secondary, or tertiary.

Like ammonia, amines are pyramidal molecules. The nitrogen atom is attached to three groups and has a nonbonding pair of electrons. They are classified by the number of hydrocarbon groups attached to the nitrogen. **Primary (1°) amines** have one R group; **secondary (2°) amines** have two R groups; and **tertiary (3°) amines** have three R groups.

Example 8
Classify the following amines as either primary, secondary, or tertiary:

1. CH_3NH_2
2. CH_3NHCH_3
3. $CH_3N(CH_3)_2$

Work
Amines are classified according to the number of alkyl or aryl groups that are directly attached to the nitrogen atom.

Answer

1.
$$H_3C - \underset{\underset{H}{|}}{\overset{\overset{H}{|}}{N}} - H$$

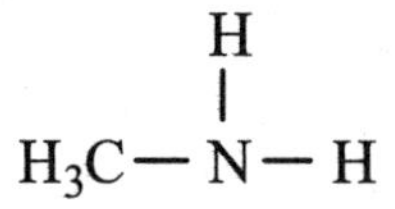

primary: one directly attached carbon atom

2.
$$H_3C - \underset{\underset{H}{|}}{N} - CH_3$$

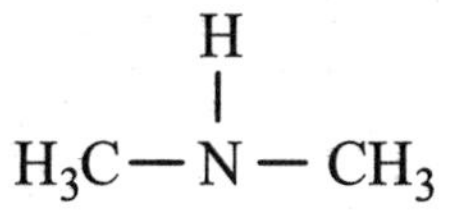

secondary: two attached carbon atoms

3.
$$H_3C - \underset{\underset{CH_3}{|}}{N} - CH_3$$

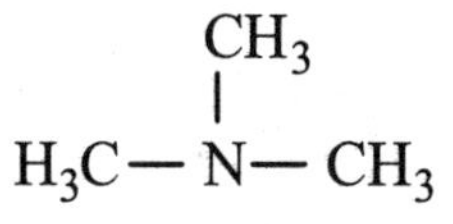

tertiary: three attached carbon atoms

Example 9
Write the general formulas for the following:
1. primary amine
2. secondary amine
3. tertiary amine

Answer
1. RNH_2
2. R_2NH
3. R_3N

Learning Goal 10:   Describe the physical properties of amines.

The N–H bond is polar, and therefore hydrogen bonding occurs between amine molecules. This feature determines the physical properties of the amines, such as boiling point and water solubility. The –NH group is less polar than the –OH group. As a result, the boiling points of amines are lower than comparable alcohols, but higher than comparable ethers or alkanes. The smaller amines are readily soluble in water, but as the size of the hydrocarbon groups increases, their solubility in water decreases.

Example 10
Which member of each of the following sets of compounds with similar formula weights has the highest boiling point?
　　　　1. propane or dimethyl ether or ethanamine
　　　　2. ethanol or dimethyl ether or ethanamine

Work
The normal boiling points of alkanes, ethers, primary amines, and primary alcohols obey the following relationship:

$$alkane < ether < primary\ amine < primary\ alcohol$$

Answer
1. ethanamine
2. ethanol

Example 11
Show the hydrogen bonding of methylamine with water.

Work
Since nitrogen is more electronegative than hydrogen, the N–H group is polar.

Example 11 continued

Answer

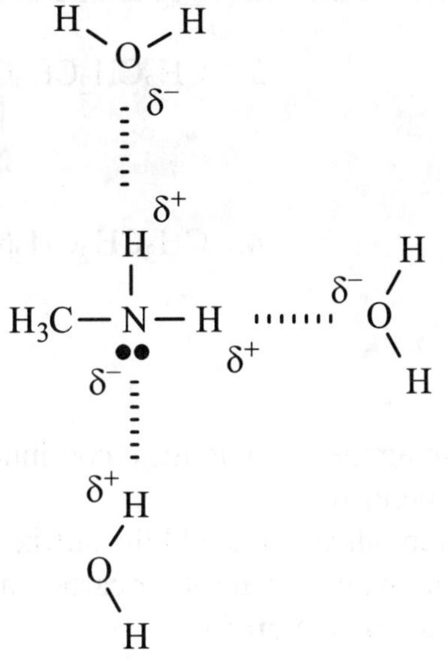

Example 12
The boiling points of methylamine, dimethylamine, and trimethylamine are –6.3°C, 7°C, and 3.5°C, respectively. What causes these differences?

Answer
There is a formula weight difference between each of these molecules (15 grams/mole) due to differences in the number of –CH_3 units. The boiling point of trimethylamine is lower than that of dimethylamine because trimethylamine is less polar owing to its symmetrical structure.

Nomenclature

Learning Goal 11:   **Draw and name simple amines using common and systematic nomenclature.**

In the systematic nomenclature system, the final -*e* of the name of the parent compound is dropped and the suffix -*amine* is added. For secondary or tertiary amines, the prefix *N*-alkyl is added to the name of the parent compound. Many aromatic amines have special names. An example of this is aniline, a benzene molecule with a substituent amino group.

Common names are also used, especially for the simple amines. The common names of the alkyl groups bonded to the amine nitrogen are followed by the suffix -*amine*. Each alkyl group is listed alphabetically as one continuous word followed by the ending -*amine*.

Example 13

Determine the systematic names of the following amine compounds:

1. CH_3NH_2

3. $CH_3CH_2CH_2CHCH_3$
   $\overset{|}{NH_2}$

2. $CH_3 \overset{\overset{H}{|}}{\underset{\underset{NH_2}{|}}{C}} CH_3$

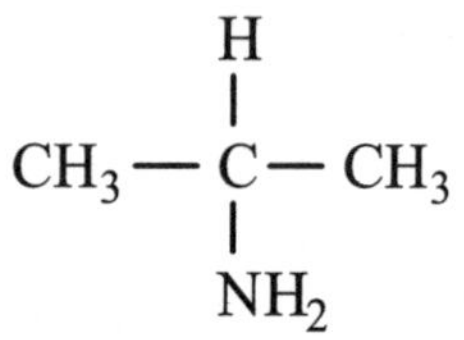

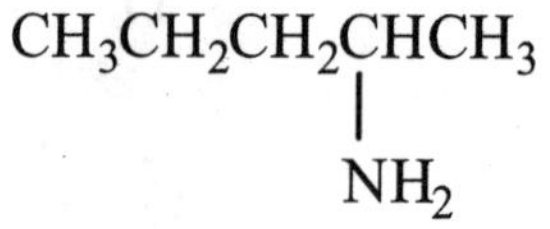

4. $CH_3CH_2CH_2NHCH_3$

Work

1. The parent compound of an amine is the longest continuous chain of carbon atoms to which the amino group is bonded.
2. Drop the final -e of the parent alkane and add the suffix -*amine*.
3. If a substituent is present on the nitrogen (other carbon atoms are directly bonded to the nitrogen), it is designated by the prefix *N-*.

Answer

1. methanamine
2. 2- propanamine
3. 2- pentanamine
4. *N*-methyl-1-propanamine

Example 14

Determine the common names for each of the following amines.

1. CH_3NH_2

2. $(CH_3)_2NH$

3. $(CH_3CH_2)_2NH$

4. 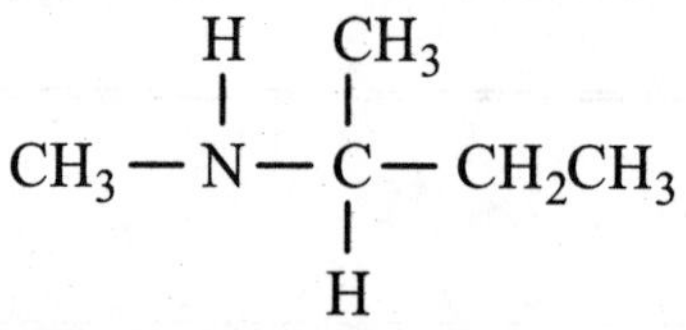 $CH_3 \overset{\overset{H \quad CH_3}{| \quad \; |}}{\underset{\underset{H}{|}}{N - C}} CH_2CH_3$

Work

The common names of the amines are derived from the various alkyl groups bonded to the amine nitrogen atom.

Answer

1. methylamine
2. dimethylamine
3. diethylamine
4. methyl-*sec*-butylamine

Medically Important Amines

Medically important amines include decongestants such as pseudoephedrine, antimicrobials such as the sulfa drugs, and anesthetics such as novocaine.

Reactions Involving Amines

Basicity

Amines have a nonbonding pair of electrons that can be shared with an electron-deficient group to form a new bond. For instance, an amine may react with a proton (H^+), producing a new N–H bond. The original unshared pair of electrons on the nitrogen atom is shared with the electron-deficient proton. The product is an **alkylammonium ion.**

Neutralization

Since amines are moderately strong bases, they react with most acids to form alkylammonium salts. The generalized reaction is represented here:

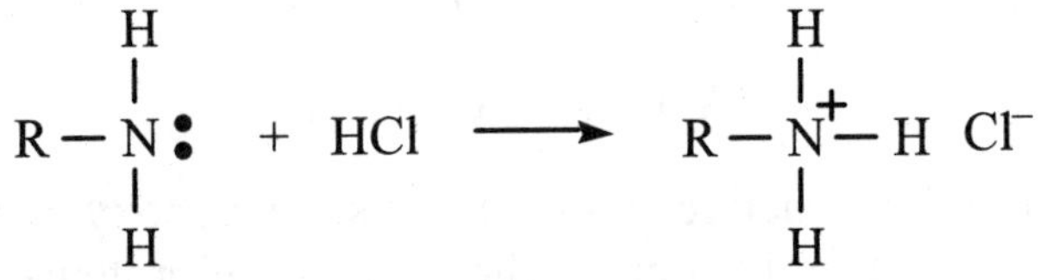

$$\underset{\underset{H}{|}}{\overset{\overset{H}{|}}{R-N:}} \;+\; HCl \;\longrightarrow\; \underset{\underset{H}{|}}{\overset{\overset{H}{|}}{R-\overset{+}{N}-H}} \; Cl^-$$

The product, an *alkylammonium salt*, is named by replacing the term *amine* with the term *ammonium* followed by the name of the anion. The salts are quite soluble in water.

Example 15

Complete the following reactions and name the products:

1. $CH_3CH_2NH_2 + HCl \rightarrow$?
2. $(CH_3)_2NH + HI \rightarrow$?
3. $CH_3NH_2 + H_2O \;\rightleftharpoons$

Work

Amines are weak to moderately strong bases and will react with most acids to form salts.

Answer

1. $CH_3CH_2NH_3^+ \; Cl^-$ ethylammonium chloride
2. $(CH_3)_2NH_2^+ \; I^-$ dimethylammonium iodide
3. $CH_3NH_3^+ \; OH^-$ methylammonium hydroxide

Because amines act as bases (proton acceptors) and amine salts act as acids (proton donors), they are useful as *buffers*. Many naturally occurring amines serve as biological buffers. For instance, the protein hemoglobin helps maintain the acid/base (pH) balance of the blood.

Several important drugs are amines. They are generally administered as alkylammonium salts because the salts are more soluble in water and body fluids.

Quaternary Ammonium Salts

Learning Goal 13:   **Describe the structure of quaternary ammonium salts and discuss their use as antiseptics and disinfectants.**

Quaternary ammonium salts are ammonium salts having four organic groups bonded to the nitrogen. They have the general structure R_4N^+ A^-.

Quaternary ammonium salts with very long carbon chains are often used as disinfectants and antiseptics because they have detergent activity.

13.4 Amides

Amides are an important class of nitrogen-containing organic compounds with the functional group shown here:

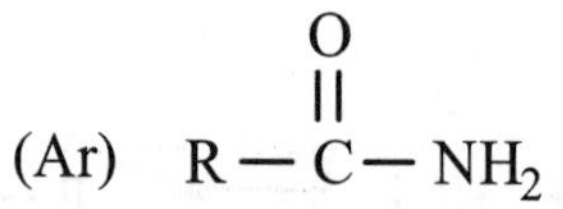

$$\text{(Ar)}\quad R - \overset{\overset{\displaystyle O}{\|}}{C} - NH_2$$

As careful inspection of this structure shows, amides are carboxylic acid derivatives.

The amide bond is the central feature in the structure of proteins. The amide bond between two amino acids is called a peptide bond.

Structure and Physical Properties

Learning Goal 14:   **Describe the physical properties of amides.**

Most amides are solids at room temperature, with boiling points even higher than the corresponding carboxylic acid. The simpler amides are quite soluble in water. Both of these properties reflect the strong intermolecular hydrogen bonding. The water solubility decreases as the molecular weight increases.

Nomenclature

Learning Goal 15:   **Draw the structure and write the common and I.U.P.A.C. names of amides.**

Amides are named by removing the *-ic acid* ending of the common name or the *-oic acid* ending of the I.U.P.A.C. name of the carboxylic acid and replacing it with *-amide*. If

190

there are substituents on the nitrogen, they are placed as prefixes and are indicated by *N-*, followed by the name of the substituent. There are no spaces between the prefix and the amide name.

Example 16

Determine the I.U.P.A.C. names of the following amides:

1. 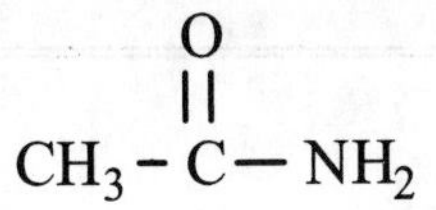 $CH_3-\overset{\overset{\displaystyle O}{\|}}{C}-NH_2$

2. $CH_3CH_2-\overset{\overset{\displaystyle O}{\|}}{C}-\overset{\overset{\displaystyle H}{|}}{N}-CH_3$

3. $CH_3CH_2CH_2CH_2-\overset{\overset{\displaystyle O}{\|}}{C}-\overset{\overset{\displaystyle H}{|}}{N}-CH_2CH_3$

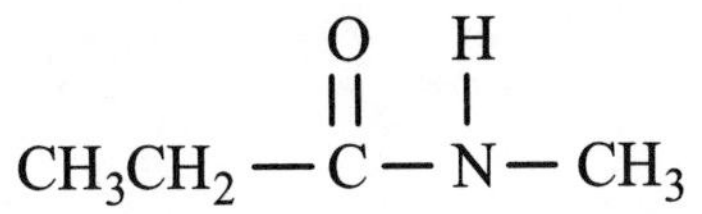

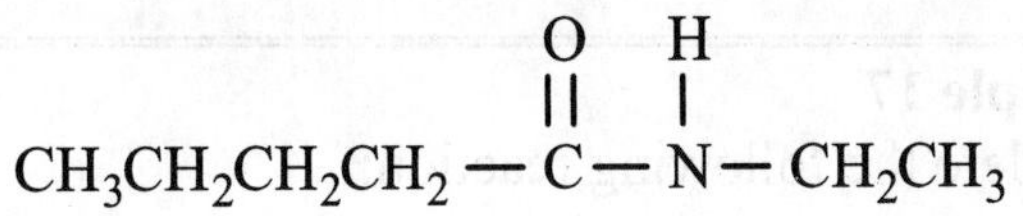

4. $CH_3CH_2CONH_2$

Work

1. The common and I.U.P.A.C. names of the amides are derived from the common and I.U.P.A.C. names of the corresponding carboxylic acids.

2. Amides are named by removing the *-ic acid* ending of the common name or the *-oic acid* ending of the I.U.P.A.C. name of the carboxylic acid and replacing it with *-amide*.

3. Substituents on the nitrogen atom are placed as prefixes and are indicated by *N-* followed by the name of the substituent.

Answer

1. ethanamide
2. *N*-methylpropanamide
3. *N*-ethylpentanamide
4. propanamide

Medically Important Amides

Medically important amides include barbiturates and acetaminophen, the active ingredient in Tylenol.

Reactions Involving Amides

Hydrolysis of Amides

Learning Goal 16:   Write equations showing the hydrolysis of amides.

Hydrolysis of an amide results in breaking the amide bond to produce the carboxylic acid and ammonia or an amine. This reaction requires heat and the presence of a strong acid or base.

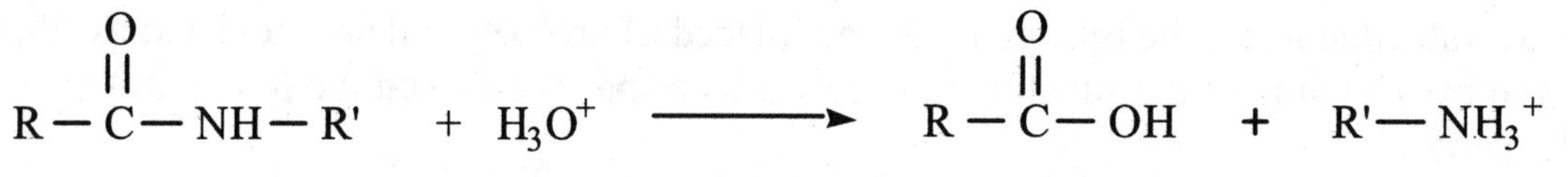

$$R-\overset{\overset{O}{\|}}{C}-NH-R' \;\;+\;\; H_3O^+ \;\;\longrightarrow\;\; R-\overset{\overset{O}{\|}}{C}-OH \;\;+\;\; R'-NH_3^+$$

| Amide | Strong acid | Carboxylic acid | Amine |

Example 17

Complete the following reaction:

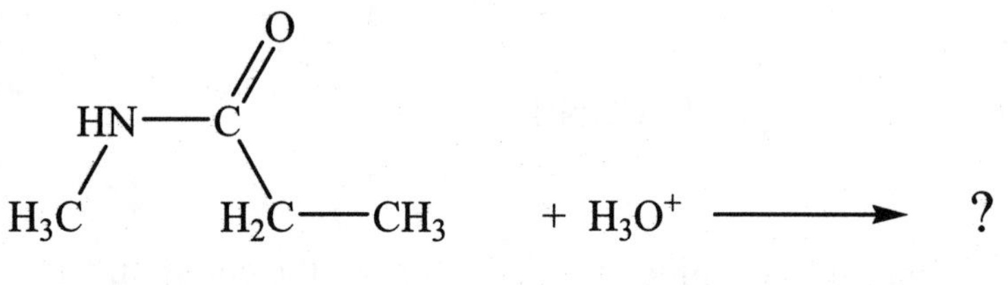

N-methylpropanamide

Answer

The products of this reaction are:

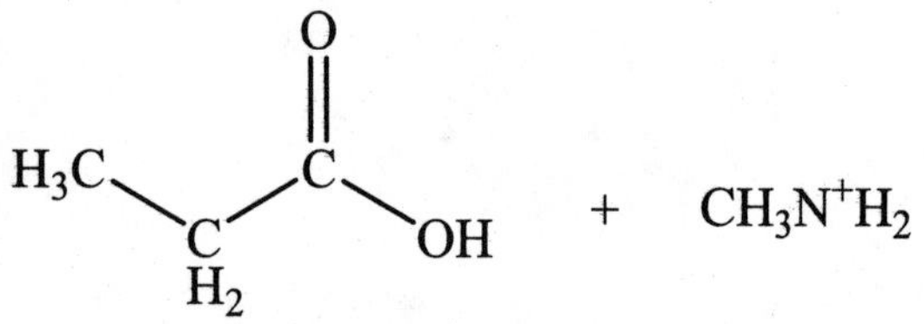

Propanoic acid   Methylamine

Glossary of Key Terms in Chapter Thirteen

acyl group (13.2) the functional group that contains the carbonyl group attached to one alkyl or aryl group.

alkylammonium ion (13.3) the ion formed when the lone pair of electrons of the nitrogen atom of an amine is shared with a proton from a water molecule.

amide (13.4) the family of organic compounds formed by the reaction between a carboxylic acid derivative and an amine.

amide bond (13.4) the bond between the carbonyl carbon of a carboxylic acid and the amino nitrogen of an amine.

amine (13.3) the family of organic molecules with the general formula $R\text{-}NH_2$, R_2NH, or R_3N (R- can equal R- or Ar-); they may be viewed as substituted ammonia molecules in which one or more of the ammonia hydrogens has been substituted by a more complex organic group.

carboxyl group (13.1) the -COOH functional group; the functional group found in carboxylic acids.

carboxylic acid (13.1) the family of organic compounds that contains the -COOH functional group.

condensation polymer (13.2) a polymer, which is a large molecule formed by combination of many small molecules (monomers), that results from joining monomers in a reaction that forms a small molecule, such as water.

ester (13.2) a carboxylic acid derivative formed by the reaction of a carboxylic acid and an alcohol.

fatty acid (13.1) any member of the family of continuous-chain carboxylic acids that generally contain 4 to 20 carbon atoms; the naturally occurring members of this class contain an even number of carbon atoms.

hydrolysis (13.2) a chemical reaction that involves the reaction of a molecule with water; results in the splitting off of the water molecule as new bonds are formed.

primary (1°) amine (13.3) an amine with the general formula RNH_2.

quaternary ammonium salt (13.3) an amine salt with the general formula $R_4N^+ A^-$ (in which R– can be an alkyl or aryl group or a hydrogen atom and A^- can be any anion.

saponification (13.2) a reaction in which a soap is produced; generally, the base-catalyzed hydrolysis of an ester.

secondary (2°) amine (13.3) an amine with the general formula R_2-NH.

soap (13.2) any of a variety of the alkali metal salts of fatty acids.

tertiary (3°) amine (13.3) an amine with the general formula R_3-N.

Self Test for Chapter Thirteen

1. Give the general formula for each of the following classes of organic compounds:

   a. alcohol   d. aldehyde

   b. ether   e. carboxylic acid

   c. ester   f. ketone

2. Draw the general structure of an acyl group.

3. Give the I.U.P.A.C. names of the following:

   a. HCOOH

   b. $CH_3(CH_2)_4COOH$

   c. CH_3COOH

   d. $CH_3CH(CH_3)CH_2COOH$

4. Draw the following carboxylic acids:

   a. acetic acid

   b. formic acid

   c. benzoic acid

   d. capric acid

5. Provide the common names of each of the following carboxylic acids:

   a. $CH_3(CH_2)_2\overset{\displaystyle O}{\overset{\displaystyle \|}{C}}-OH$ b. CH_3CH_2COOH c. $CH_3\underset{\underset{\displaystyle CH_3}{\displaystyle |}}{CH}-\overset{\displaystyle O}{\overset{\displaystyle \|}{C}}-OH$

193

6. Give the common name of the following esters:

a. 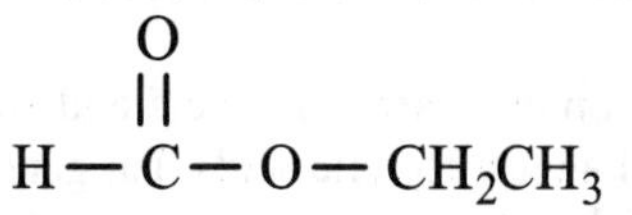
$$H-\overset{\overset{\displaystyle O}{\|}}{C}-O-CH_2CH_3$$

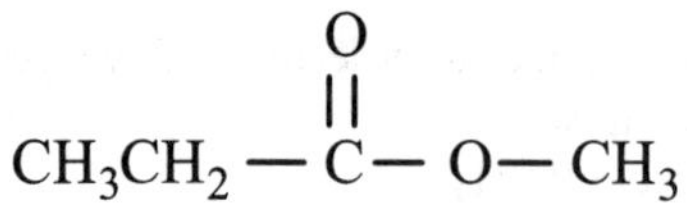

b.
$$CH_3CH_2-\overset{\overset{\displaystyle O}{\|}}{C}-O-CH_3$$

c.
$$CH_3CH_2CH_2CH_2CH_2-O-\overset{\overset{\displaystyle O}{\|}}{C}-CH_3$$

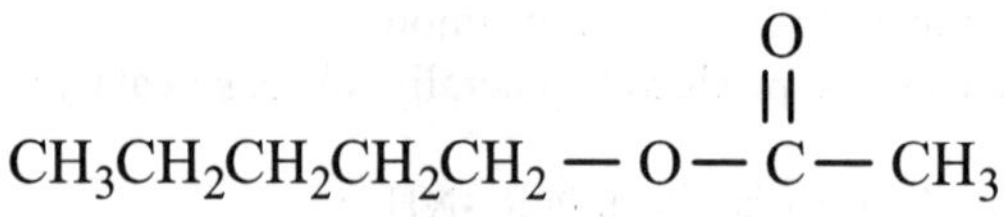

7. Give the I.U.P.A.C. names of the following esters:

a. methyl acetate

b. 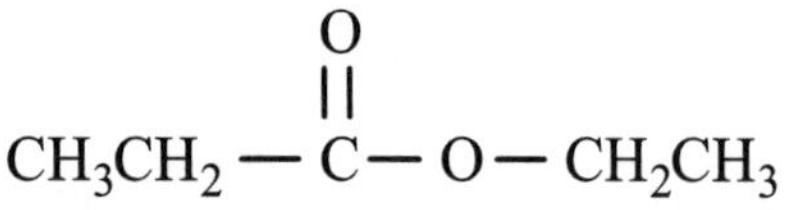
$$CH_3CH_2-\overset{\overset{\displaystyle O}{\|}}{C}-O-CH_2CH_3$$

8. Which of the following compounds that have nearly the same formula weight has the highest boiling point?

a. acetic acid

b. propanal

c. propanone

d. 1-propanol

9. Complete the following reactions:

a. $CH_3COOH + H_2O \rightleftharpoons$?

b. $CH_3COOH + NaOH$?

c. $CH_3COOH + CH_3OH$ (acid as a catalyst) ?

10. What are the hydrolysis products of an ester?

11. What two ions are formed when a soluble carboxylic acid reacts with a molecule of water?

12. For the following reactions, name the products that are formed:

a. butyric acid + NaOH

b. acetic acid + $Ca(OH)_2$

13. The long-chain carboxylic acid salts (fatty acid salts) are good _____.

14. What is the term for the large, nonpolar hydrocarbon end of a soap?

15. What two compounds must be reacted to form methyl propanoate?

16. Name the functional group found in all carboxylic acids.

17. Provide the I.U.P.A.C. name for the product of the reaction between ethanoic acid and methanol in the presence of heat and a trace of acid.

18. A primary alcohol is mixed with potassium permanganate solution. What reaction occurs and what is the function of the $KMnO_4$?

19. Is the carboxylate part of soap hydrophobic or hydrophilic?

20. What is the term for the "particles" formed when molelcules of oil or grease are surrounded by soap molecules?

21. Provide the missing reactant in the following equation:

$$CH_3CH_2CH_2COOH \ + \ ? \ \longrightarrow \ CH_3CH_2CH_2 - \overset{\overset{\displaystyle O}{\|}}{C} - O - CH_3$$

22. Identify the following as either a primary, secondary, or tertiary amine or as a quaternary ammonium salt.

   a. $(CH_3)_3N$

   b. $(CH_3)_3N^+CH_2CH_3 \ Cl^-$

   c. $CH_3CH_2NH_2$

   d. $CH_3CH_2NH\text{-}CH_2CH_3$

23. Provide the systematic names of the following amine compounds:

   a. $CH_3CH_2NH_2$

   b. $(CH_3)_2CHNH_2$

   c. $(CH_3)_3N$

24. Complete each of the following reactions:

   a. $CH_3CONH_2 + H_3O^+$

   b. $CH_3CH_2\text{-}CONH\text{-}CH_3 + H_3O^+$

25. Complete the following reactions:

   a. $(CH_3)_2NH + HI \rightarrow \ ?$

   b. $(CH_3)_3N + HCl \rightarrow \ ?$

26. Write the general formulas for the following:

   a. primary amine

   b. tertiary amine

27. Determine the I.U.P.A.C. names of the following amides:

   a. acetamide

   b. $CH_3(CH_2)_4CONH\text{-}CH_2CH_2CH_3$

   c. $CH_3CH_2CH_2CONH\text{-}CH_3$

28. Amines may be viewed as compounds that are substituted products of which inorganic compound?

29. Amines are classified according to the number of alkyl or aryl groups directly attached to what atom?

30. What product is formed in the reaction of an amine with an acid?

31. Name the product that results when methylamine reacts with HCl.

32. What is the physical state of most amides at room temperature?

33. What two products are released by the hydrolysis of an amide?

34. Why are quaternary ammonium salts effective as disinfectants and antiseptics?

Vocabulary Quiz for Chapter Thirteen

1. A(n) _____ is a carboxylic acid derivative formed by the reaction of a carboxylic acid and an alcohol.

2. _____ is the hydrolysis of an ester by an aqueous base.

3. The conversion of an alcohol to a carboxylic acid is an example of a(n) _____ reaction.

4. The characteristic functional group of the carboxylic acids is the _____.

5. An amine having the general formula RNH_2 is a(n) _____.

6. The functional group that is produced by a reaction between a carboxylic acid and an amine is the _____.

7. A(n) _____ is an amine with three alkyl groups attached to the nitrogen atom.

8. A(n) _____ is the cation formed when the nonbonding electron pair of an amine nitrogen is shared with a proton.

9. A(n) _____ has the general formula $R_4N^+A^-$.

10. The collection of organic molecules with the general formulas RNH_2, R_2NH, or R_3N is the _____ family.

14 *Carbohydrates*

Learning Goals

1 Explain the difference between complex and simple carbohydrates and know the amounts of each recommended in the daily diet.
2 Apply the systems of classifying and naming monosaccharides according to the functional group and number of carbons in the chain.
3 Explain stereoisomerism.
4 Determine whether a molecule has a chiral center.
5 Identify monosaccharides as either D- or L-.
6 Draw and name the common monosaccharides using structural formulas.
7 Given the linear structure of a monosaccharide, draw the Haworth projection of its α- and β-cyclic forms and vice versa.
8 By inspection of the structure, predict whether a sugar is reducing or non-reducing.
9 Discuss the use of Benedict's reagent to measure the level of glucose in urine.
10 Draw and name the common disaccharides and discuss their significance in biological systems.
11 Describe the difference between galactosemia and lactose intolerance.
12 Discuss the structural, chemical, and biochemical properties of starch, glycogen, and cellulose.

Introduction

Learning Goal 1:	Explain the difference between complex and simple carbohydrates and know the amounts of each recommended in the daily diet.

Carbohydrates are produced in plants by photosynthesis. They are the main source of energy for both plants and animals. They are found in many natural sources, such as grains and cereals, breads, fruits, sugar cane, and sugar beets. A healthy diet should include both complex and simple carbohydrates. It is recommended that 58% of the diet should be carbohydrate and that no more than 10% of the daily caloric intake should be sucrose.

14.1 Types of Carbohydrates

Carbohydrates may be categorized by size. **Monosaccharides** are composed of a single (mono-) sugar (saccharide) unit. A **disaccharide** consists of two monosaccharides. Intermediate in size are the **oligosaccharides,** polymers consisting of three to ten monosaccharide units. The largest and most complex carbohydrates are the **polysaccharides;** polymers consisting of greater than ten monosaccharide units. The largest and most complex carbohydrates are the polysaccharides. Oligosaccharides and polysaccharides are chains of monosaccharides held together by **glycosidic bonds** through "bridging" oxygen atoms.

Example 1

List each of the following as either a monosaccharide, disaccharide, or polysaccharide:

1. starch   5. glycogen
2. glucose   6. sucrose
3. ribose   7. cellulose
4. maltose   8. glyceraldehyde

Work

Monosaccharides are those carbohydrates that cannot be broken down into any simpler substance by hydrolysis. *Disaccharides* are composed of two monosaccharides joined together by a glycosidic bond. *Polysaccharides* are composed of many monosaccharides joined together by glycosidic bonds between each pair.

Answer

1. polysaccharide   5. polysaccharide
2. monosaccharide   6. disaccharide
3. monosaccharide   7. polysaccharide
4. disaccharide   8. monosaccharide

14.2 Monosaccharides

Learning Goal 2:   **Apply the systems of classifying and naming monosaccharides according to the functional group and number of carbons in the chain.**

If a monosaccharide is a ketone, it is called a **ketose.** If it is an aldehyde, it is called an **aldose.** All carbohydrates contain a large number of hydroxyl groups and are, therefore, *polyhydroxyaldehydes* or *polyhydroxyketones*.

A second system of nomenclature is based on the number of carbon atoms in the main skeleton. A **triose** has three carbons, a **tetrose** has four carbons, and so on. By combining the two systems, a name is derived that provides information about both the structure and composition of a sugar. For instance, a sugar may be an aldotriose, aldohexose, ketotriose, ketotetrose, etc.

In addition, each carbohydrate also has a specific unique name, such as glucose and fructose. Finally, it is important to indicate the isomeric form of the sugar; thus, a D- or L-designation is placed in front of the name.

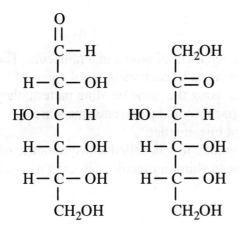

Aldose   Ketose
Hexose   Hexose
Aldohexose   Ketohexose
D-Glucose   **D-Fructose**

Example 2

Classify each of the following monosaccharides as an aldose or a ketose, and also give the number of carbon atoms found in each:

1. ribose   3. galactose
2. glyceraldehyde   4. fructose

Work

A more detailed system of nomenclature may be used for monosaccharides. They may be classified as an aldose (which means they have an aldehyde structure) or as a ketose (ketone structure). Also, by using suffixes, we can indicate the number of carbon atoms found in each. For example, glucose is classified as an aldohexose, since it has six carbon atoms, and it also contains an aldehyde structure.

Answer

1. aldopentose   3. aldohexose
2. aldotriose   4. ketohexose

14.3 Stereoisomers and Stereochemistry

Learning Goal 3:   **Explain stereoisomerism.**

Learning Goal 4:   **Determine whether a molecule has a chiral center.**

Learning Goal 5:   **Identify monosaccharides as either D- or L-.**

Stereoisomers

Stereochemistry is the study of the spatial arrangement of atoms in a molecule. The prefixes D- and L- are used to distinguish **stereoisomers.** In each member of a pair of stereoisomers, all of the atoms are bonded together using the same bonding pattern; they differ only in the arrangements of their atoms in space. D- and L-stereoisomers are **enantiomers,** nonsuperimposable mirror images of one another.

A carbon atom that has *four different* groups bonded to it is called an asymmetric or **chiral carbon.** A molecule that has a chiral carbon (a chiral molecule) can exist as a pair of enantiomers.

Rotation of Plane-Polarized Light

Each member of a pair of stereoisomers will rotate a plane of polarized light in different directions. A polarimeter is used to measure the direction of rotation of plane-polarized light. Compounds that rotate light in a clockwise direction are termed **dextrorotatory** and are designated by a plus sign (+). Compounds that rotate light in a counterclockwise direction are called **levorotatory** and are indicated by a minus sign (–). So, (+)-glyceraldehyde will rotate plane-polarized light to the right, and (–)-glyceraldehyde will rotate plane-polarized light by the same amount to the left.

The Relationship Between Molecular Structure and Optical Activity

Louis Pasteur first recognized the relationship between the structure of a compound and the effect of that compound on plane-polarized light. He noticed mirror-image crystals of tartaric acid were formed in wine. He then found that these compounds that were physical mirror images of one another also rotated plane-polarized light in opposite directions. His work, and that of others, resulted in the recognition that enantiomers rotate plane-polarized light to the same degree as one another, but in opposite directions.

Fischer Projection Formulas

The **Fischer Projection** is a two-dimensional drawing of a molecule that shows a chiral carbon at the intersection of two lines. Horizontal lines represent bonds projecting out of the page and vertical lines represent bonds that project into the page. The most oxidized carbon is always represented at the "top" of the structure.

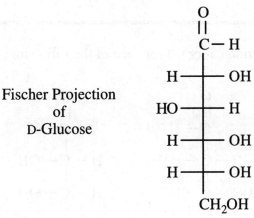

Fischer Projection
of
D-Glucose

The D- and L- System of Nomenclature

In Figure 14.4 of the text, the stereoisomers of glyceraldehyde are presented. When the hydroxyl group is drawn to the right of the chiral carbon, the isomer is the D-enantiomer. When the hydroxyl group is drawn to the left of the chiral carbon, the molecule is the L-enantiomer.

Example 3
Draw and label the two different stereoisomers of glyceraldehyde.

Answer

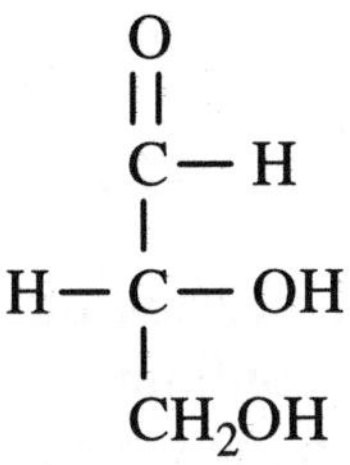

$$
\begin{array}{cc}
\begin{array}{c}
O \\
\| \\
C-H \\
| \\
H-C-OH \\
| \\
CH_2OH
\end{array}
&
\begin{array}{c}
O \\
\| \\
C-H \\
| \\
HO-C-H \\
| \\
CH_2OH
\end{array}
\end{array}
$$

D-Glyceraldehyde 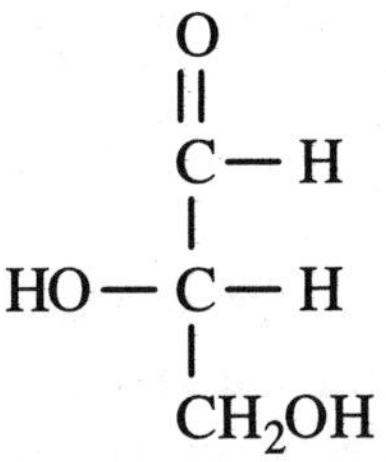   L-Glyceraldehyde

---

For sugars that have more than three carbons, the D- or L- configuration is determined by the chiral carbon farthest from the aldehyde or ketone group. If the –OH on this carbon is to the right, the stereoisomer is of the D- family. If the –OH is on to the left, the stereoisomer is of the L- family.

201

Example 4

Determine the configuration (D- or L-) for each of the following:

i.

$$
\begin{array}{c}
O \\
\parallel \\
C-H \\
\mid \\
HO-C-H \\
\mid \\
CH_2OH
\end{array}
$$

2.

$$
\begin{array}{c}
O \\
\parallel \\
C-H \\
\mid \\
H-C-OH \\
\mid \\
HO-C-H \\
\mid \\
HO-C-H \\
\mid \\
H-C-OH \\
\mid \\
CH_2OH
\end{array}
$$

3.

$$
\begin{array}{c}
O \\
\parallel \\
C-H \\
\mid \\
H-C-OH \\
\mid \\
H-C-OH \\
\mid \\
H-C-OH \\
\mid \\
CH_2OH
\end{array}
$$

4.

$$
\begin{array}{c}
O \\
\parallel \\
C-H \\
\mid \\
H-C-H \\
\mid \\
H-C-OH \\
\mid \\
H-C-OH \\
\mid \\
CH_2OH
\end{array}
$$

Work

By convention, the D- isomer is the one in which the –OH on the chiral carbon farthest from the carbonyl group is on the right-hand side of the structure. The L- form is just the opposite.

Answer

1. L-glyceraldehyde
2. D-galactose
3. D-ribose
4. D-2-deoxyribose

Note that the (+) and (–) designations are a result of a physical measurement with a polarimeter, but the D- and L- designations are based solely on the configuration of the chiral carbon furthest from the aldehyde or ketone group. To determine whether a specific D-sugar is (+) or (–), it must be analyzed using a polarimeter. To determine whether a specific (+) sugar is D- or L- (if the chemical structure is unknown), further experimentation will need to be done.

As an example, given a sample of D-galactose, it is impossible to predict whether it will rotate plane-polarized light to the right or to the left until the sample is actually measured in a polarimeter. It is certain, however, that L-galactose will rotate plane-polarized light by the same amount as D-galactose, but *in the opposite direction.*

14.4 Biologically Important Monosaccharides

Learning Goal 6:   **Draw and name the common monosaccharides using structural formulas.**

Glucose

Glucose (dextrose), found in fruits and honey, is the preferred energy source of many tissues, especially the brain. Thus, the blood glucose concentration must be carefully controlled to allow the body to function optimally. The enantiomer of glucose found in nature is D-glucose.

Learning Goal 7:	Given the linear structure of a monosaccharide, draw the Haworth projection of its α- and β- cyclic forms and vice versa.

Generally sugars of five or more carbons exist in cyclic form under physiological conditions. This results from the reaction of the carbonyl group at C-1 of glucose with the hydroxyl group at C-5 to produce a six-membered ring.

Two isomers are formed in this reaction because cyclization creates a new asymmetric carbon, in this case C-1. These isomers are designated as either α- or β-forms. In the α-isomer, the C-1 hydroxyl group is below the ring. In the β-isomer, it is above the ring.

Haworth projections are used to depict the three-dimensional configuration of cyclic monosaccharide molecules. To draw a Haworth projection, begin with the structural formula of the linear form of the sugar (refer to Example 14.3 in the text). Chemical groups to the left of the carbon chain are placed above the ring in the Haworth projection. Chemical groups to the right of the carbon chain are drawn beneath the carbon ring in the Haworth projection.

The Haworth projections for α- and β-D-glucose are shown below:

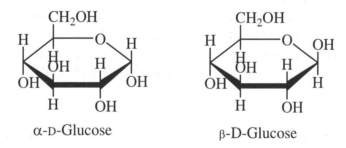

α-D-Glucose   β-D-Glucose

Fructose

Fructose, the sweetest of all sugars, is found in honey and fruits. It is also called levulose and fruit sugar. Fructose is a ketohexose.

Cyclization of D-fructose produces α- and β-D-fructose. The most common ring structure formed by D-fructose consists of a five-carbon skeleton.

Galactose

Galactose is a hexose (six-carbon sugar) found most commonly as a component of the disaccharide lactose. Lactose, or milk sugar, is the most abundant sugar found in milk. β-D-galactose and a modified form, *N*-acetyl-β-D-galactosamine, are found in the oligosaccharides on the surface of red blood cells. These oligosaccharides are referred to as the *blood group antigens*. They determine whether an individual has Type A, B, AB, or O blood.

Ribose and Deoxyribose, Five-Carbon Sugars

Ribose, a pentose, is a component of ribonucleic acids (RNA) and of several coenzymes, compounds required for the action of some enzymes. The molecule that carries the genetic information in the cell is deoxyribonucleic acid (DNA). DNA contains a modified form of D-ribose in which the –OH group at C-2 has been replaced by a hydrogen. This sugar is called D-*2-deoxyribose.*

Reducing Sugars

Learning Goal 8:	By inspection of the structure, predict whether a sugar is reducing or nonreducing.

Learning Goal 9:	Discuss the use of Benedict's reagent to measure the level of glucose in urine.

The aldehyde group of aldoses is easily oxidized. Thus, **Benedict's reagent** can be used to detect the presence of aldoses, or **reducing sugars.** The aldehyde group is oxidized to a carboxylate group, and Cu^{2+} is reduced to Cu^+.

Benedict's reagent can be used to detect the presence of glucose in the urine. Use of Benedict's reagent to test urine glucose levels has largely been replaced by chemical tests that provide more accurate results. The most common technology is a strip test in which a strip is impregnated with the enzyme glucose oxidase and other reagents to produce a color change when the enzyme catalyzes the oxidation of glucose.

Frequently, physicians recommend that diabetics monitor *blood* glucose levels several times a day. This provides a more accurate indication of how well the diabetes is being controlled.

Example 5

Write an equation representing the oxidation of L-glyceraldehyde by Benedict's reagent.

Answer

204

Example 6

Write an equation representing the oxidation of D-glucose by Benedict's solution.

Answer

$$
\begin{array}{c}
\text{O} \\
\| \\
\text{C}-\text{H} \\
| \\
\text{H}-\text{C}-\text{OH} \\
| \\
\text{HO}-\text{C}-\text{H} \\
| \\
\text{H}-\text{C}-\text{OH} \\
| \\
\text{H}-\text{C}-\text{OH} \\
| \\
\text{CH}_2\text{OH}
\end{array}
\;+\; 2\,Cu^{2+} \;+\; 5\,OH^- \longrightarrow
\begin{array}{c}
\text{O} \\
\| \\
\text{C}-\text{O}^- \\
| \\
\text{H}-\text{C}-\text{OH} \\
| \\
\text{HO}-\text{C}-\text{H} \\
| \\
\text{H}-\text{C}-\text{OH} \\
| \\
\text{H}-\text{C}-\text{OH} \\
| \\
\text{CH}_2\text{OH}
\end{array}
\;+\; Cu_2O \;+\; 3\,H_2O
$$

14.5 Biologically Important Disaccharides

Learning Goal 10: **Draw and name the common disaccharides and discuss their significance in biological systems.**

The α- or β-hydroxyl group of a cyclic monosaccharide can react with a hydroxyl group of another sugar to form an "oxygen bridge" (ether bond) and produce a disaccharide. This reaction is illustrated in Figure 14.8 in the text for two molecules of glucose.

If the disaccharide has an unreacted –OH group at C-1, it is a reducing sugar, because the cyclic structure can break open, forming a free aldehyde. When drawing the structure of a disaccharide, the nonreducing end is written to the left. Most disaccharides are reducing sugars. The single exception is sucrose, which is a **nonreducing sugar.**

Maltose

Maltose is a disaccharide composed of two glucose molecules. Since the C-1 hydroxyl group of one α-D-glucose molecule is attached to C-4 of another glucose molecule (which could be α or β), the bond between the two monosaccharides is called an α(1→4) glycosidic bond. When the reducing end of maltose has a β-hydroxyl group, this disaccharide is called β-maltose.

Lactose

Learning Goal 11: **Describe the difference between galactosemia and lactose intolerance.**

Milk sugar, or **lactose,** is a dimer of β-D-galactose and either α- or β-D-glucose. In lactose the C-1 hydroxyl group of β-D-galactose is bonded to the C-4 hydroxyl group of glucose. Thus the bond between these two monosaccharides is called a β(1→4) glycosidic bond.

Lactose, the principal sugar in milk, must be broken down to glucose and galactose before it can be used by the body as an energy source. Glucose can be used directly in energy-harvesting metabolic reactions. However, galactose must be converted into a phosphorylated form of glucose before it can enter the energy harvesting glycolysis pathway. **Galactosemia** is a genetic disease caused by the absence of an enzyme needed for this conversion. A toxic compound formed from galactose accumulates to toxic levels, causing severe mental retardation, cataracts, and early death. Galactosemic infants must be provided a diet that does not contain any milk or milk products.

The inability to digest lactose, called **lactose intolerance,** is caused by the absence of the enzyme lactase. The symptoms include intestinal cramping, diarrhea, and dehydration. Because some intestinal bacteria metabolize lactose and release organic acids and CO_2 gas, the individual suffers further discomfort. Avoiding milk and milk products will eliminate these unpleasant symptoms. Lactase is available in tablet form and can be ingested along with dairy products to avoid the symptoms of lactose intolerance.

Sucrose

Sucrose, common table sugar, is a disaccharide of α-D-glucose joined to β-D-fructose. Examine the structure of sucrose in Figure 14.11 in the text. The glycosidic linkage between α-D-glucose and β-D-fructose is quite different from those in lactose and maltose. This bond is called an (α1→β2)-glycosidic linkage, because it involves the C-1 carbon of glucose and the C-2 carbon of fructose. Sucrose will not react with Benedict's reagent and is, therefore, not a reducing sugar.

14.6 Polysaccharides

Learning Goal 12:	Discuss the structural, chemical, and biochemical properties of starch, glycogen, and cellulose.

Most carbohydrates found in nature are **polysaccharides,** high molecular weight polymers of glucose.

Starch

Plants use the energy of sunlight to produce glucose from CO_2 and H_2O. Most plants store glucose in the form of the polysaccharide **starch.**

Starch is composed of the glucose polymers **amylose** (20%) and **amylopectin** (80%). Amylose is a linear polymer of up to 4,000 α-D-glucose molecules connected by α(1→4) glycosidic bonds. Amylose exists as a helix that coils at every sixth glucose unit.

During digestion, amylose is degraded by two enzymes. The enzyme α-amylase cleaves the glycosidic bonds of amylose chains at random along the chain, and β-amylase sequentially removes maltose molecules from the reducing end of the amylose chain.

Amylopectin is a highly branched form of amylose. The main chain consists of $\alpha(1\rightarrow4)$ glycosidic bonds, and the branches of 20–25 glucose molecules are attached to the C-6 hydroxyl groups by $\alpha(1\rightarrow6)$ glycosidic bonds.

Glycogen

Glycogen is the principal glucose storage form of animals. It is stored in granules in liver and muscle cells. The structure of glycogen is like that of amylopectin, except that glycogen has more branches, and they are shorter.

The liver regulates the level of blood glucose by the formation and degradation of glycogen. When blood glucose levels are high, liver cells take up glucose from the blood and convert it to glycogen. When the blood glucose levels are too low, the liver breaks down glycogen and releases the glucose into the blood stream.

Cellulose

Cellulose, a polymer of about 3,000 β-D-glucose units linked by $\beta(1\rightarrow4)$ glycosidic bonds, is the most abundant organic molecule in the world. Cellulose is an unbranched polymer that forms long fibrils. These long, straight, parallel chains of cellulose form a rigid cage that serves as a structural component of plant cell walls.

We cannot digest cellulose because we lack the enzyme *cellulase*. Cellulose serves as a source of dietary fiber, a necessary component of a healthful diet.

Example 7

Write the "word-reactions" for the hydrolysis of the following carbohydrates:

1. starch   2. maltose   3. sucrose   4. glycogen   5. lactose

Answer
1. starch + many H_2O molecules → many glucose molecules
2. maltose + H_2O → 2 glucose molecules
3. sucrose + H_2O → glucose + fructose
4. glycogen + many H_2O molecules → many glucose molecules
5. lactose + H_2O → glucose + galactose

For each of the above reactions, acid conditions are required.

> **Example 8**
>
> Why are termites, cows, and goats able to digest cellulose, while humans are unable to do so?
>
> **Answer**
>
> The termites, cows, and goats have microorganisms within their digestive systems that produce the enzyme *cellulase*, which allows them to break down cellulose. Humans do not have the cellulase enzyme needed for the hydrolysis of cellulose.

Glossary of Key Terms in Chapter Fourteen

aldose (14.2) a sugar that contains an aldehyde (carbonyl) group.

amylopectin (14.6) a highly branched form of amylose; the branches are attached to the C-6 hydroxyl by $\alpha(1\rightarrow6)$ glycosidic linkage; a component of starch.

amylose (14.6) a linear polymer of α-D-glucose in $\alpha(1\rightarrow4)$ glycosidic linkage that is a component of starch; a polysaccharide storage form.

Benedict's reagent (14.4) a buffered solution of Cu^{2+} ions that can be used to test for reducing sugars.

carbohydrate (14.1) generally sugars and polymers of sugars; the primary source of energy for the cell.

cellulose (14.6) the most abundant organic compound in the world; a polymer of β-D-glucose linked by $\beta(1\rightarrow4)$ glycosidic bonds.

chiral carbon (14.3) ina carbon bonded to four different groups.

chiral molecule (14.3) molecules capable of existing in mirror image forms.

disaccharide (14.1) a sugar composed of two monosaccharides joined through an oxygen atom bridge.

enantiomers (14.3) stereoisomers that are nonsuperimposable mirror images of one another.

Fischer Projection (14.3) a two-dimensional drawing of a molecule, which shows a chiral carbon at the intersection of two lines with the horizontal lines representing bonds projecting out of the page and the vertical lines representing bonds that project into the page.

fructose (14.4) a ketohexose that is also called levulose and fruit sugar; the sweetest of all sugars, abundant in honey and fruit.

galactose (14.4) an aldohexose that is a component of lactose, milk sugar.

galactosemia (14.5) a human genetic disease caused by the inability to convert galactose to a phosphorylated form of glucose (glucose-1-phosphate) that can be used in cellular metabolic reactions..

glucose (14.4) an aldohexose, the most abundant monosaccharide; it is a component of many disaccharides, such as lactose and sucrose, and polysaccharides, such as cellulose, starch, and glycogen.

glyceraldehyde (14.3) an aldotriose that is the simplest carbohydrate; phosphorylated forms of glyceraldehyde are important intermediates in cellular metabolic reactions.

glycogen (14.6) the glucose storage form of animals; a linear backbone of α-D-glucose in $\alpha(1\rightarrow4)$ linkage, with numerous short branches attached to the C-6 hydroxyl group by $\alpha(1\rightarrow6)$ linkage.

glycosidic bond (14.1) the bond between the hydroxyl group of C-1 carbon of one sugar and a hydroxyl group of another sugar.

Haworth projections (14.4) a means of representing the orientation of substituent groups around a cyclic sugar molecule.

hexose (14.2) a six carbon monosaccharide.

ketose (14.2) a sugar that contains a ketone (carbonyl) group.

lactose (14.5) a disaccharide composed of β-D-galactose and either α- or β-D-glucose in $\beta(1\rightarrow4)$ glycosidic linkage; milk sugar.

lactose intolerance (14.5) the inability to produce the enzyme lactase, which degrades lactose to galactose and glucose.

maltose (14.5) a disaccharide composed of α-D-glucose and either α- or β-D-glucose in α(1→4) glycosidic linkage.

monosaccharide (14.1) the simplest type of carbohydrate consisting of a single saccharide unit.

nonreducing sugar (14.5) a sugar that cannot be oxidized by Benedict's reagent.

oligosaccharide (14.1) an intermediate-sized carbohydrate composed of from three to ten monosaccharides.

pentose (14.2) a five carbon monosaccharide.

polysaccharide (14.1) a large, complex carbohydrate composed of long chains of monosaccharides.

reducing sugar (14.4) a sugar that can be oxidized by Benedict's reagent. This includes all monosaccharides and most disaccharides.

ribose (14.4) a five carbon monosaccharide that is a component of RNA and many coenzymes.

saccharide (14.1) a sugar molecule.

stereochemistry (14.3) the study of the spatial arrangement of atoms in a molecule.

stereoisomers (14.3) a pair of molecules having the same structural formulas and bonding patterns, but differing in the arrangement of the atoms in space.

sucrose (14.5) a disaccharide composed of α-D-glucose and β-D-fructose in (α1→β2) glycosidic linkage; table sugar.

tetrose (14.2) a four carbon monosaccharide.

triose (14.2) a three carbon monosaccharide.

Self Test for Chapter Fourteen

1. Classify each of the following as a monosaccharide, disaccharide, or polysaccharide:

   a. deoxyribose   d. starch

   b. galactose   e. sucrose

   c. glucose   f. fructose

2. Classify each of the following as an aldose or a ketose. Also, include in the classification the number of carbon atoms in each.

   a. glucose   d. ribose

   b. fructose   e. deoxyribose

   c. galactose   f. glyceraldehyde

3. What is the term for a pair of stereoisomers that are nonsuperimposable mirror images of one another?

4. Which polysaccharide serves as a structural element in plant cell walls?

5. Which enantiomeric form of glucose is found in humans?

6. Name the simplest carbohydrate.

7. What two enzymes are involved in the degradation of amylose?

8. Name the bond that holds two monosaccharides together in a disaccharide.

9. What is the structural difference between an aldose and a ketose?

10. What is the difference between a triose and a pentose?

11. Determine the D- or L- configurations for each of the following:

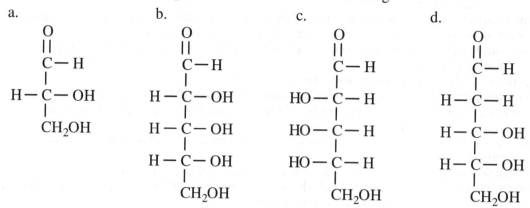

12. List three common names for glucose.

13. Which monosaccharide is the sweetest of all carbohydrates?

14. Which polysaccharide is the major glucose storage form of animals?

15. Give word-reactions for the hydrolysis of the following carbohydrates:

    a. glucose

    b. lactose

    c. sucrose

    d. starch

    e. cellulose

16. Which of the disaccharides will not react with Benedict's solution?

17. What substances are described as either polyhydroxyaldehydes or polyhydroxyketones?

18. Why are humans unable to digest cellulose?

19. What is the term for the condition that results from an inability to digest lactose?

20. Which carbohydrate is found in blood in concentrations up to 0.1% (W/V)?

21. Which form of cyclic D-glucose has the –OH at carbon-1 below the ring?

22. What test was once used to test for the amount of glucose in the urine?

23. What monosaccharide is found in RNA?

24. Give the molecular formula of ribose.

25. What is the name of the human genetic disease in which galactose cannot be converted into a form that can be used in cellular metabolic reactions?

26. What is the term for carbohydrates that can reduce metal ions?

27. What is the term for a sugar composed of two monosaccharide units?

28. What are the major sites of glycogen synthesis in the human body?

29. What test might be used to distinguish between a monosaccharide and polysaccharide?

30. What test might be used to distinguish between a reducing sugar and a nonreducing sugar?

Vocabulary Quiz for Chapter Fourteen

1. _____ is a human genetic disease caused by the inability to convert galactose into a phosphorylated form of glucose (glucose-1-phosphate) that can be used in cellular metabolic reactions.

2. _____ is a two-dimensional drawing of a molecule, which shows a chiral carbon at the intersection of two lines with the horizontal lines representing bonds projecting out of the page and the vertical lines representing bonds that project into the page.

3. A carbon atom bonded to four different chemical groups is called a(an) _____ carbon.

4. Two stereoisomers that are nonsuperimposable mirror images are called _____.

5. _____ is the glucose storage polymer found in the liver and muscle of animals.

6. The most abundant organic compound in the world, composed of a polymer of β-D-glucose linked by β(1→4) glycosidic bonds is _____.

7. The bond between the hydroxyl group of one sugar and a hydroxyl group of another sugar is called a _____.

8. A sugar that can be oxidized by Benedict's reagent is a _____.

9. The study of the spatial arrangement of atoms in a molecule is _____.

10. _____ is the disaccharide composed of β-D-galactose and either α- or β-D-glucose in β(1→4) glycosidic linkage. It is also called milk sugar.

15 *Lipids and Their Functions in Biochemical Systems*

Learning Goals

1 Discuss the physical and chemical properties and biological function of each of the families of lipids.
2 Write the structures of saturated and unsaturated fatty acids.
3 Compare and contrast the structure and properties of saturated and unsaturated fatty acids.
4 Write equations representing the reactions that fatty acids undergo.
5 Describe the functions of prostaglandins.
6 Discuss the mechanism by which aspirin reduces pain.
7 Draw the structure of a phospholipid and discuss its amphipathic nature.
8 Discuss the general classes of sphingolipids and their functions.
9 Draw the structure of the steroid nucleus and discuss the functions of steroid hormones.
10 Describe the function of lipoproteins in triglyceride and cholesterol transport in the body.
11 Draw the structure of the cell membrane and discuss its functions.
12 Discuss passive and facilitated diffusion of materials through a cell membrane.
13 Describe the mechanism of action of a Na^+-K^+ ATPase.

Introduction

Lipids are a diverse collection of organic molecules grouped together on the basis of their solubility in nonpolar solvents. The four groups of lipids that will be considered in this chapter are fatty acids, glycerides, nonglyceride lipids, and complex lipids.

Example 1

List two properties that all lipids have in common.

Answer

Most are insoluble in water but very soluble in nonpolar solvents.

15.1 Biological Functions of Lipids

Lipids are involved in a variety of biological processes. They are structural components of cell membranes. They are an energy storage form in the body, stored in adipocytes (fat cells). Some lipids, like the steroids, are hormones. Others are vitamins, required for processes such as blood clotting, proper bone development, and vision.

Example 2

List four important functions of the class of compounds called the lipids.

Answer

1. Cell membranes are composed of lipids, including phospholipids and steroids. The cell membrane creates a barrier between the cell and its environment and provides a means for the controlled passage of materials into and out of the cell.
2. Most of the available stored energy in animals is in the form of lipids known as triglycerides.
3. Several of the lipids have hormonelike properties.
4. Others are lipid-soluble vitamins.

15.2 Fatty Acids

Structure and Properties

Fatty acids are long-chain (12–24 carbons) monocarboxylic acids. The general formula for a **saturated fatty acid** is

$$CH_3(CH_2)_n COOH$$

where n is an even integer between 10 and 22. In a saturated fatty acid, each carbon in the chain is bonded to the maximum number of hydrogen atoms.

In the case of **unsaturated fatty acids,** there is at least one carbon-carbon double bond, as seen in this example:

$$CH_3(CH_2)_7 CH = CH(CH_2)_7 COOH$$

As we observed with alkanes, the melting points of saturated fatty acids increase with increasing carbon number. This general trend is also seen for unsaturated fatty acids, except that the melting point also decreases markedly as the number of carbon-to-carbon double bonds increases.

Example 3

List the properties that most fatty acids have in common.

Answer

1. They are long-chain (12–24) monocarboxylic acids.

2. They often contain an even number of carbon atoms.

3. They form continuous chains of carbon atoms with no branches.

4. They are either saturated, with only single bonds in the R group, or unsaturated, with some double bonds in the R group.

Example 4

What physical property clearly distinguishes a saturated and unsaturated fatty acid with the same number of carbon atoms?

Answer

The melting points of saturated fatty acids are much higher. The greater the number of double bonds in the unsaturated fatty acid, the lower its melting point.

Example 5

Is the fatty acid $C_{13}H_{25}COOH$ saturated or unsaturated?

Answer

It is unsaturated, since all saturated fatty acids (similar to alkanes) will fit the general formula of $C_nH_{2n-1}COOH$. If it were saturated, its formula would have to be $C_{13}H_{27}COOH$; thus, the $C_{13}H_{25}COOH$ has two fewer hydrogen atoms than the saturated one. This means that the $C_{13}H_{25}COOH$ contains one double bond.

Chemical Reactions of Fatty Acids

Learning Goal 4:   Write equations representing the reactions that fatty acids undergo.

The reactions of fatty acids are similar to those of short-chain carboxylic acids.

Esterification

Esterification is the reaction of a fatty acid with an alcohol to form an ester and water.

214

Example 6

Complete the following esterification reaction:

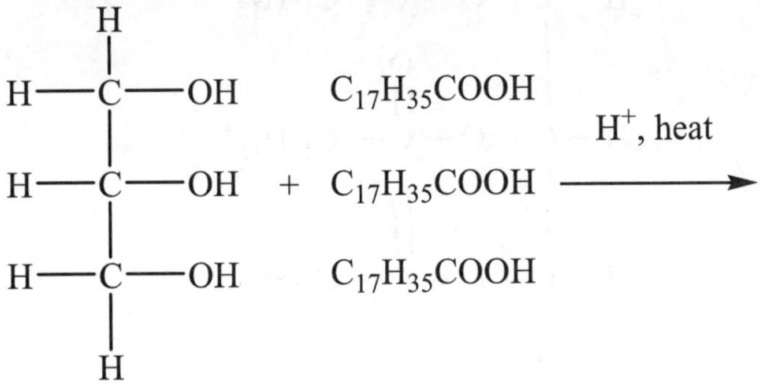

$$
\begin{array}{ccc}
& H & \\
& | & \\
H-C-OH & & C_{17}H_{35}COOH \\
& | & \\
& & \xrightarrow{\ H^{+},\ heat\ } \\
H-C-OH & + & C_{17}H_{35}COOH \\
& | & \\
H-C-OH & & C_{17}H_{35}COOH \\
& | & \\
& H &
\end{array}
$$

Work

A triglyceride is produced by the reaction of three fatty acids and glycerol. Line up the carboxyl group of each fatty acid with the hydroxyl groups (–OH) of the glycerol molecule. Then remove a water molecule between each to form an ester structure at each of the three positions. The resulting triglyceride will always have the following general structure:

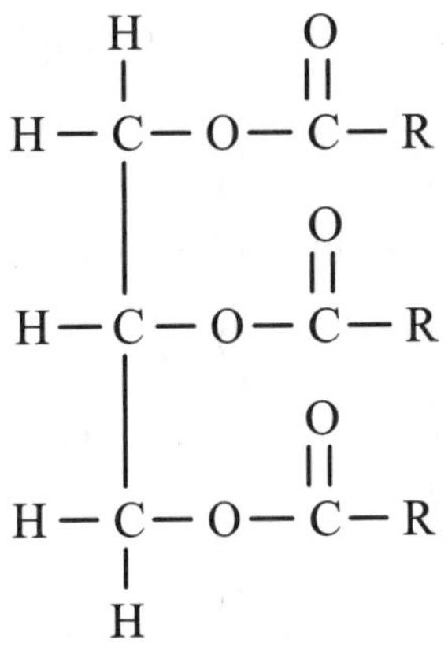

$$
\begin{array}{cccc}
H & & O & \\
| & & || & \\
H-C-O- & & C-R & \\
| & & O & \\
& & || & \\
H-C-O- & & C-R & \\
| & & O & \\
& & || & \\
H-C-O- & & C-R & \\
| & & & \\
H & & &
\end{array}
$$

Answer

The R groups will always be the R groups that were found in each specific fatty acid. Notice that there are no hydroxyl or carboxyl groups present; only three ester groups are now holding the glycerol to the three fatty acids.

215

Example 6 continued

The following triglyceride would be formed, along with three molecules of water:

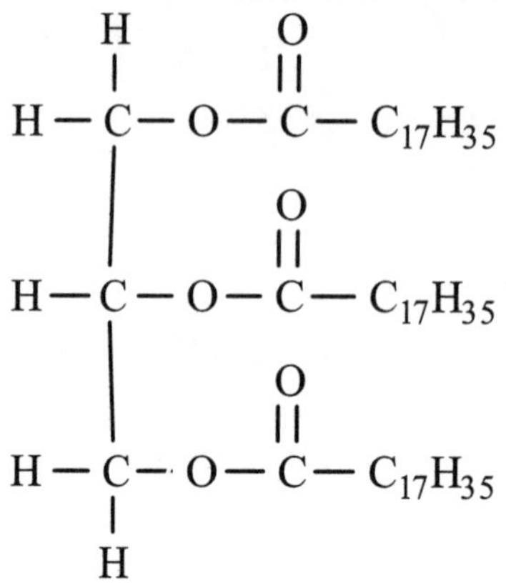

$$
\begin{array}{c}
\text{H} \qquad\quad \text{O} \\
| \qquad\qquad || \\
\text{H} - \text{C} - \text{O} - \text{C} - \text{C}_{17}\text{H}_{35} \\
| \\
\qquad\qquad \text{O} \\
\qquad\qquad || \\
\text{H} - \text{C} - \text{O} - \text{C} - \text{C}_{17}\text{H}_{35} \\
| \\
\qquad\qquad \text{O} \\
\qquad\qquad || \\
\text{H} - \text{C} - \text{O} - \text{C} - \text{C}_{17}\text{H}_{35} \\
| \\
\text{H}
\end{array}
$$

Acid Hydrolysis

Hydrolysis is the addition of water to a fatty acid ester to produce a fatty acid and an alcohol.

Example 7

Give the complete acid hydrolysis of the following triglyceride:

$$
\begin{array}{c}
\text{H} \qquad\quad \text{O} \\
| \qquad\qquad || \\
\text{H} - \text{C} - \text{O} - \text{C} - (\text{CH}_2)_{12}\text{CH}_3 \\
| \\
\qquad\qquad \text{O} \\
\qquad\qquad || \\
\text{H} - \text{C} - \text{O} - \text{C} - (\text{CH}_2)_{16}\text{CH}_3 \\
| \\
\qquad\qquad \text{O} \\
\qquad\qquad || \\
\text{H} - \text{C} - \text{O} - \text{C} - (\text{CH}_2)_{14}\text{CH}_3 \\
| \\
\text{H}
\end{array}
$$

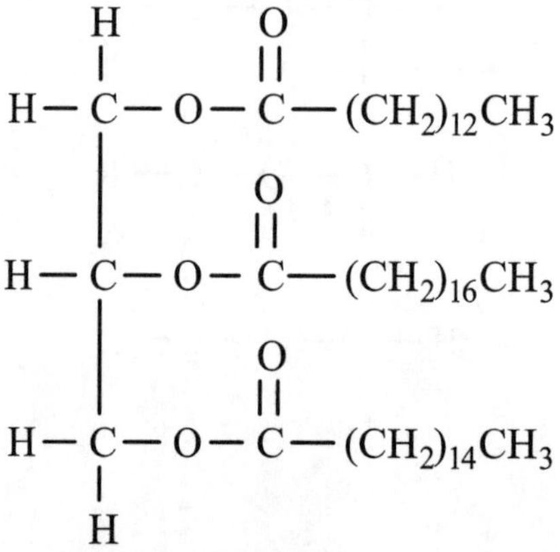

1-Myristoyl-2-stearoyl-3-palmitoylglycerol

Work

The hydrolysis of a triglyceride requires three water molecules and a catalyst (generally acid). The products formed are glycerol and one molecule of each of the fatty acids used to prepare the triglyceride. The name of the above triglyceride tells us that glycerol plus myristic acid, stearic acid, and palmitic acid were used to make it.

216

Example 7 continued

Answer

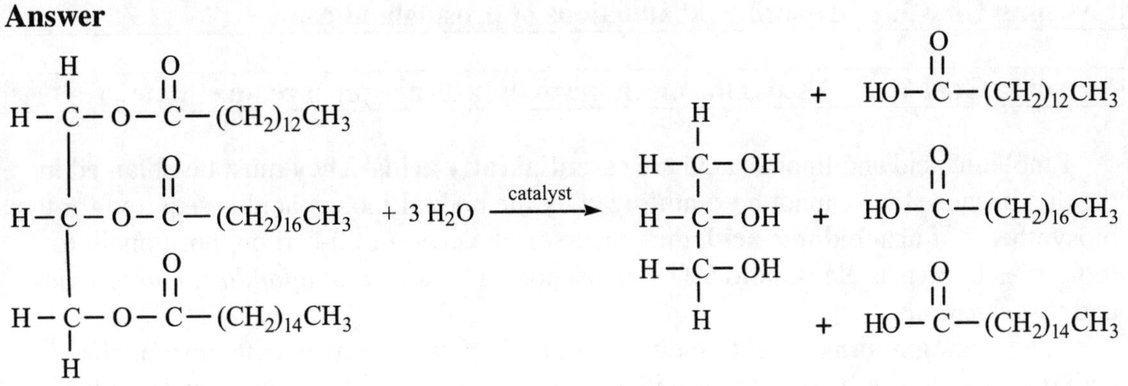

$$H-\underset{\underset{H}{\overset{\overset{H}{|}}{|}}{C}}-O-\overset{\overset{O}{||}}{C}-(CH_2)_{12}CH_3$$

(reaction showing triacylglycerol hydrolysis)

$$H-\overset{\overset{O}{||}}{C}-O-\overset{\overset{O}{||}}{C}-(CH_2)_{16}CH_3 \quad + \ 3\ H_2O \xrightarrow{\text{catalyst}}$$

$$H-\overset{\overset{O}{||}}{C}-O-\overset{\overset{O}{||}}{C}-(CH_2)_{14}CH_3$$

products:

$$H-\overset{|}{C}-OH$$
$$H-\overset{|}{C}-OH$$
$$H-\overset{|}{C}-OH$$

$$+ \quad HO-\overset{\overset{O}{||}}{C}-(CH_2)_{12}CH_3$$
$$+ \quad HO-\overset{\overset{O}{||}}{C}-(CH_2)_{16}CH_3$$
$$+ \quad HO-\overset{\overset{O}{||}}{C}-(CH_2)_{14}CH_3$$

Saponification

  Saponification is the base-catalyzed hydrolysis of an ester. The product of this reaction, an ionized salt, is a soap.

Reaction at the Double Bond (Unsaturated Fatty Acids)

  Hydrogenation of an unsaturated fatty acid is the addition of hydrogen to a double bond. This is an example of an addition reaction. Hydrogenation is used in the food industry to convert polyunsaturated vegetable oils into solid fats.

Example 8

Show how linoleic acid can be converted in the laboratory into stearic acid.

Answer

$$CH_3(CH_2)_4CH=CHCH_2CH=CH(CH_2)_7COOH \ + \ 2\ H_2$$

$$\downarrow \begin{array}{l}\text{heat}\\\text{pressure}\\\text{Ni}\end{array}$$

$$CH_3(CH_2)_{16}COOH$$

Stearic acid

217

Eicosanoids: Prostaglandins, Leukotrienes, and Thromboxanes

Learning Goal 5:   Describe the functions of prostaglandins.

Learning Goal 6:   Discuss the mechanism by which aspirin reduces pain.

Linolenic acid and linoleic acid are **essential fatty acids.** They must be obtained in the diet because they cannot be synthesized by the body. Linoleic acid is required for the biosynthesis of **arachidonic acid,** the precursor of a class of 20-carbon, hormonelike molecules known as **eicosanoids.** The eicosanoids include *prostaglandins, leukotrienes,* and *thromboxanes.*

The **prostaglandins** are extremely potent biological molecules with hormonelike activity. They are carboxylic acids with a five-carbon ring. All are composed of a basic twenty-carbon skeleton and are grouped under the designations of A, B, E, F, and so on, which indicate the basic structural variations. Each prostaglandin also has a number designation that indicates the number of carbon-carbon double bonds in the compound. Figure 15.3 in the text provides examples of prostaglandin structure and nomenclature.

Prostaglandins are made in all tissues and exert biological effects on the cells that produce them and on neighboring cells. Their functions are briefly summarized below.

1. *Blood clotting:* Thromboxane A_2, produced by blood platelets, enhances blood clotting. PGI_2 (prostacyclin), produced by the cells lining the blood vessels, inhibits the clotting process. Working together, these molecules ensure that blood clots form only when necessary.
2. *The inflammatory response:* Prostaglandins promote the pain and fever associated with the inflammatory response.
3. *Reproductive system:* PGE_2 stimulates smooth muscle contraction, especially uterine contractions. Dysmenorrhea (painful menstruation) is the result of an excess of two prostaglandins. Drugs that inhibit prostaglandin synthesis provide relief from symptoms.
4. *Gastrointestinal tract:* Prostaglandins inhibit the secretion of stomach acid and increase the secretion of a protective mucous layer into the stomach. Aspirin inhibits prostaglandin synthesis; thus, prolonged use may contribute to stomach ulcers.
5. *Kidneys:* Prostaglandins increase the excretion of water and electrolytes.
6. *Respiratory tract:* Leukotrienes promote the constriction of the bronchi associated with asthma. Other prostaglandins have the opposite effect, bronchodilation.

Aspirin relieves pain by inhibiting prostaglandin synthesis. It works by inhibiting the enzyme cyclooxygenase, which catalyzes the first step in prostaglandin synthesis. The acetyl group from aspirin is covalently bound to the enzyme, inhibiting its activity.

15.3 Glycerides

Glycerides are lipids that contain the alcohol glycerol. They may be subdivided into two classes: **neutral glycerides** and **phosphoglycerides.**

Neutral Glycerides

The esterification of glycerol with one, two, or three fatty acids produces a **mono-, di-,** or **triglyceride.** These are also referred to as *mono-, di-,* or *triacylglycerols.* Although mono- and diglycerides are present in nature, the most common is the triglyceride, the major storage form of lipids found in fat cells.

Neutral glycerides do not dissociate into charged species, because bonding throughout the molecule is covalent and nonpolar. Consequently, they readily stack with one another and are easily stored in adipocytes. In fact, their major function in the body is energy storage. If more nutrients are consumed than are needed for daily metabolic processes, the excess is converted to triglycerides and stored in adipocytes, which form *adipose tissue.* If energy is needed, the triglycerides are broken down, and their stored energy is released for use by the body.

Phosphoglycerides

Learning Goal 7:	Draw the structure of a phospholipid and discuss its amphipathic nature.

Phospholipids contain a phosphoryl group (PO_4^{3-}). The presence of the phosphoryl group produces an *amphipathic* molecule, which means that the molecule has regions with very different properties. Specifically, phospholipids have a polar head (the phosphoryl group) that is hydrophilic and a nonpolar tail (the hydrocarbon chain of the fatty acid) that is hydrophobic.

Phosphoglycerides contain acyl groups derived from long-chain fatty acids esterified at C-1 and C-2 of glycerol-3-phosphate. The simplest, phosphatidate, contains a free phosphoryl group. More complex phosphoglycerides are formed when the phosphoryl group is bonded to another hydrophilic molecule.

Phosphatidycholine (lecithin), phosphatidylserine, and *phosphatidylethanolamine* (*cephalin*) are commonly found in cell membranes. The structures of these three phospholipids are shown in Figure 15.7 of the text. Each possesses a polar "head" and a nonpolar "tail." The ionic "head" is hydrophilic and interacts with water molecules, and the nonpolar "tail" is hydrophobic and interacts with nonpolar molecules. This bipolar, or amphipathic, nature is essential to the structure and properties of biological membranes.

Example 9

Draw a phosphatidylcholine that contains the R groups from linolenic acid (position 1) and linoleic acid (position 2), and then label the hydrophobic and hydrophilic ends.

Answer

The general structure of a phosphatidylcholine is shown on the left. The R groups of linolenic and linoleic acids are esterified at positions 1 and 2 on the glycerol molecule, respectively, in the structure on the right.

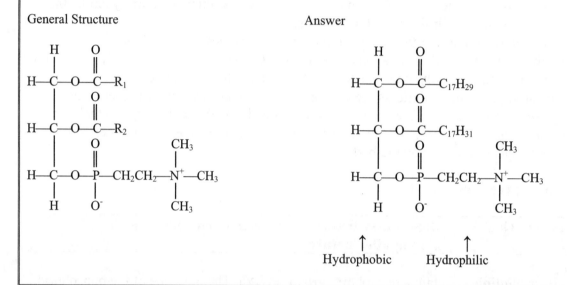

15.4 Nonglyceride Lipids

Sphingolipids

| Learning Goal 8: | Discuss the general classes of sphingolipids and their functions. |

Sphingolipids are lipids that are not derived from glycerol. They are phospholipids derived from the amino alcohol sphingosine.

Sphingomyelin, a sphingolipid, is abundant in the myelin sheath that surrounds and insulates cells of the central nervous system and is essential to proper nerve transmission.

Glycosphingolipids, or *glycolipids,* include the cerebrosides, sulfatides, and gangliosides. They are built on a ceramide backbone structure, a fatty acid amide of sphingosine. Two common cerebrosides are glucocerebroside, found primarily in the membranes of macrophages, and galactocerebroside, found almost exclusively in brain cells. Sulfatides contain a sulfate group and, thus, have a negative charge at physiological pH. The gangliosides are found in most tissues of the body, although they were first isolated from nervous tissue.

Steroids

Steroids are an important family of lipids derived from cholesterol. The steroids include the bile salts that aid in the emulsification and digestion of lipids, and the sex hormones testosterone and estrone. All steroids are structured around the steroid nucleus (steroid carbon skeleton), which consists of four fused rings. Two fused rings share one or more common bonds as part of their ring backbones.

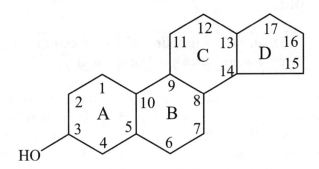

Cholesterol is found in most cell membranes, where it functions to regulate the fluidity of the membrane. A high serum cholesterol concentration is associated with heart disease, especially **atherosclerosis** or hardening of the arteries. Cholesterol and other substances coat the inside of the arteries. This causes the arteries to become narrower, and more pressure is needed to cause blood to flow through them. This results in elevated blood pressure (hypertension), which is also linked to heart disease.

Many steroids play roles in the reproductive cycle. *Progesterone*, the most important hormone associated with pregnancy, is synthesized from cholesterol. Produced in the ovaries and placenta, it prepares the uterine lining to accept the fertilized egg. Progesterone is also involved in fetal development and suppression of further ovulation during pregnancy.

Progesterone is the precursor of *testosterone*, a male sex hormone, and estrone, a female sex hormone. Both these hormones are involved in the development of secondary sexual characteristics.

Development of birth control agents has involved application of steroid chemistry. One of the first effective synthetic birth control agents was 19-norprogesterone. Unfortunately, it had to be taken by injection. Norlutin (17-ethynyl-19-nortestosterone) is equally effective and can be taken orally. These compounds all act by inducing a false pregnancy which prevents ovulation. Currently "combination" oral contraceptives are prescribed most often. These include a progesterone and an estrogen. There are at least thirty such combination pills available today.

Cortisone is involved in carbohydrate metabolism and is an important drug used to treat rheumatoid arthritis, asthma, gastrointestinal, and skin disorders. Care must be taken in the use of cortisone because of the possible side effects.

Waxes

The chemical composition of **waxes** is highly variable. All are insoluble in water and are solid at room temperature. Paraffin wax is a mixture of solid straight-chain hydrocarbons. Carbowax is a synthetic polyether. The natural waxes are composed of a long-chain fatty acid esterified to a long-chain alcohol.

Examples of naturally occurring waxes include beeswax; lanolin, used in skin creams; carnauba wax, used in automobile polish; and whale oil, once used as a fuel and for candles.

15.5 Complex Lipids

Learning Goal 10:	Describe the function of lipoproteins in triglyceride and cholesterol transport in the body.

Complex lipids are lipids that are bonded to other types of molecules. **Plasma lipoproteins** are complex lipids that transport other lipids through the blood stream. Because lipids are only slightly soluble in water, their movement from organ to organ requires such a transport system. Lipoprotein particles consist of a core of hydrophobic lipids surrounded by a shell of polar lipids and proteins.

There are four major classes of human plasma lipoproteins. **Chylomicrons** carry triglycerides from the intestine to other tissues. **Very low-density lipoproteins (VLDL)** carry triglycerides synthesized in the liver to other tissues for storage. **Low-density lipoproteins (LDL)** carry cholesterol to peripheral tissues and help regulate cholesterol levels. **High-density lipoproteins (HDL)** transport cholesterol from peripheral tissues to the liver.

The path of lipid transport begins with dietary fat, which is emulsified in the small intestine. Triglycerides are hydrolyzed by lipase, releasing fatty acids and monoglycerides that are absorbed by intestinal cells and reassembled into triglycerides. Chylomicrons are produced, which eventually enter the bloodstream for transport to cells throughout the body. Triglycerides and cholesterol synthesized in the liver are transported in lipoproteins. Triglycerides are carried in VLDL particles, and cholesterol is transported in LDL particles.

High concentrations of LDL in the blood are associated with hardening of the arteries; high levels of HDL in the blood appear to reduce the incidence of atherosclerosis.

15.6 The Structure of Biological Membranes

Learning Goal 11:	Draw the structure of the cell membrane and discuss its functions.

Biological membranes are *lipid bilayers* consisting of phospholipids and cholesterol. The hydrophobic, hydrocarbon tails stack in the center of the bilayer, and the ionic head groups are exposed on the surfaces.

Fluid Mosaic Structure of Biological Membranes

Membranes are fluid, having the consistency of olive oil. The degree of fluidity is determined by the amounts of saturated and unsaturated fatty acids and by the length of the fatty acid tails. Shorter, more unsaturated fatty acid tails produce a more fluid membrane. Floating in the sea of phospholipids are many proteins that are critical to normal cellular function. When viewed by electron microscopy, these proteins look like a mosaic. Because of the fluid consistency of the membrane and the presence of numerous proteins, our concept of membrane structure is called the **fluid mosaic theory.**

Peripheral proteins are found only on the surfaces of the membrane. **Transmembrane proteins** are embedded within the membrane and extend completely through it.

Membrane Transport

The cell membrane is responsible for the controlled passage of molecules into and out of the cell. Transport of some molecules across the membrane is regulated by transmembrane transport proteins, while other molecules pass through the membranes unassisted, by **passive transport.**

Passive Diffusion: The Simplest Form of Membrane Transport

Learning Goal 12:	Discuss passive and facilitated diffusion of materials through a cell membrane.

Small uncharged molecules, such as O_2 and CO_2, pass through the membrane by **simple diffusion,** the net movement of a substance from an area of high concentration to an area of low concentration. In this case, diffusion involves the movement of solutes across a cellular membrane. Diffusion is a form of passive transport because it requires no energy expenditure on the part of the cell.

Large or highly charged molecules cannot pass through the lipid bilayer directly because the membrane is selectively permeable. That is, the membrane allows the diffusion of some molecules, but not others.

Facilitated Diffusion: Specificity of Molecular Transport

Most molecules diffuse across membranes through specific protein carriers called **permeases.** This process, called **facilitated diffusion,** is also a means of passive transport because the direction of movement will depend on the concentration of the substance on each side of the membrane. Facilitated diffusion occurs through pores within the permease. Each pore has a shape complementary to the shape of the molecule to be transported. Only molecules having the right shape can pass through the pore and the rate of diffusion is limited by the number of permease molecules in the membrane as well as the concentration of solute on either side of the membrane.

Energy Requirements for Transport

Learning Goal 13:	Describe the mechanism of action of a Na+-K+ ATPase.

To survive, it is often necessary for cells to accumulate high concentrations of nutrients that are in low concentration in the environment. In this circumstance, the cell must expend energy to move nutrients from an area of low concentration to an area of higher concentration. This process is called **active transport.** Most ions and food molecules are imported by active transport through the cell membrane. An example of active transport is the Na+-K+ pump that maintains a high concentration of Na+ outside the cell and a high concentration of K+ inside the cell.

Glossary of Key Terms in Chapter Fifteen

active transport (15.6) the movement of molecules across a membrane against a concentration gradient.

arachidonic acid (15.2) a fatty acid that is derived from linolenic acid; the precursor of the prostaglandins.

atherosclerosis (15.4) deposition of excess plasma cholesterol and other lipids and proteins on the walls of arteries, resulting in a decreased artery diameter and increased blood pressure.

cholesterol (15.4) a 27-carbon steroid ring structure that serves as the precursor of the steroid hormones.

chylomicron (15.5) a plasma lipoprotein (aggregate of protein and triglycerides) that carries triglycerides from the intestine to all body tissues via the bloodstream.

complex lipid (15.5) a lipid that is bonded to other types of molecules.

diglyceride (15.3) the product of the esterification of glycerol with fatty acids at two positions.

eicosanoid (15.2) any of the derivatives of 20-carbon fatty acids, including the prostaglandins, leukotrienes, and thromboxanes.

emulsifying agent (15.3) a bipolar molecule that aids in the suspension of fats in water.

essential fatty acid (15.2) the fatty acids linoleic acid and linolenic acid, which must be supplied in the diet because they cannot be synthesized by the body.

esterification (15.2) the reaction between a fatty acid and an alcohol to form an ester and water.

facilitated diffusion (15.6) movement of a solute across a membrane from an area of high concentration to an area of low concentration through a transmembrane protein or permease.

fatty acid (15.2) any member of the family of continuous-chain carboxylic acids that generally contain 12–20 carbon atoms; the most concentrated source of energy used by the cell.

fluid mosaic model (15.6) the model of membrane structure that describes the fluid nature of the phospholipid bilayer and the presence of numerous proteins embedded within the membrane.

glyceride (15.3) a lipid that contains glycerol.

high–density lipoprotein (15.5) a plasma lipoprotein that transports cholesterol from peripheral tissue to the liver.

hydrogenation (15.2) a reaction in which hydrogen (H_2) is added to a double or triple bond.

lipid (15.1) a member of the group of organic molecules of varying composition that are classified together on the basis of their solubility in nonpolar solvents.

low–density lipoprotein (15.5) a plasma lipoprotein that carries cholesterol to peripheral tissues and helps to regulate cholesterol levels in those tissues.

monoglyceride (15.2) the product of the esterification of glycerol with one fatty acid.

neutral glyceride (15.3) the product of the esterification of glycerol with fatty acids at one, two, or three positions.

passive transport (15.6) the net movement of a solute from an area of high concentration to an area of low concentration.

peripheral membrane protein (15.6) a protein that is bound to either the inner or outer surface of a membrane.

phosphatidate (15.3) the simplest phosphoglyceride; a molecule with fatty acids esterified to C-1 and C-2 of glycerol and a free phosphoryl group esterified at C-3.

phosphoglyceride (15.3) a molecule with fatty acids esterified at the C-1 and C-2 positions of glycerol and a phosphoryl group esterified at C-3. There may be an additional hydrophilic group bonded to the phosphoryl group.

phospholipid (15.3) a lipid containing a phosphoryl group.

plasma lipoprotein (15.5) a complex composed of lipid and protein that is responsible for the transport of lipids throughout the body.

prostaglandin (15.2) a family of hormonelike substances derived from the 20-carbon fatty acid arachidonic acid; produced by many cells of the body, they regulate many body functions.

saponification (15.2) a reaction in which a soap is produced; more generally, the hydrolysis of an ester by an aqueous base.

saturated fatty acid (15.2) a long-chain monocarboxylic acid in which each carbon of the chain is bonded to the maximum number of hydrogen atoms.

sphingolipid (15.4) a phospholipid that is derived from the amino alcohol sphingosine, rather than from glycerol.

sphingomyelin (15.4) a sphingolipid that is found in abundance in the myelin sheath that surrounds and insulates cells of the central nervous system.

steroid (15.4) a lipid that is derived from cholesterol and composed of one five-sided ring and three six-sided rings; the steroids include sex hormones and anti-inflammatory compounds.

transmembrane protein (15.6) a protein that is embedded within a membrane and crosses the lipid bilayer, protruding from the membrane both inside and outside the cell.

triglyceride (15.3) triacylglycerol; a molecule composed of glycerol esterified to three fatty acids.

unsaturated fatty acid (15.2) a long-chain monocarboxylic acid having at least one carbon-to-carbon double bond.

very low–density lipoprotein (VLDL) (15.5) a plasma lipoprotein that binds triglycerides synthesized by the liver and carries them to adipose tissue for storage.

wax (15.4) a collection of lipids that are generally considered to be esters of long-chain alcohols.

Self Test for Chapter Fifteen

1. What is the simplest type of lipid?

2. Why are most lipids insoluble in water?

3. What is the main energy storage compound in animals?

4. List two properties of all fatty acids.

5. What products are formed by the hydrolysis of phosphatidate?

6. What products are formed by hydrolysis of a natural wax?

7. List the four major classes of human plasma lipoproteins.

8. What products are released by the complete hydrolysis of a triglyceride?

9. Evidence indicates that high levels of which plasma lipoprotein help reduce the incidence of atherosclerosis?

10. Patients whose diets are high in saturated fat tend to have high levels of what substance in their blood?

11. In the lipid bilayer of cell membranes, which part of the phospholipid is found in the center of the bilayer?

12. What substance dissolved in the hydrophobic region of a biological membrane helps to regulate membrane fluidity?

13. What features of the fatty acid tails of membrane phospholipids determine the degree of fluidity of a biological membrane?

14. Write an equation for the reaction of lauric acid with sodium hydroxide.

15. What structure is found in all steroids?

16. Name foods in your diet that have a high concentration of cholesterol.

17. Which steroid is responsible for both the successful initiation and completion of pregnancy?

18. Which class of biological compounds has members of varying chemical composition, all of which are grouped together on the basis of their solubility?

19. Saturated fatty acids containing ten or more carbon atoms are found in what physical state at room temperature?

20. What products are formed in the reaction of a fatty acid with an alcohol?

21. List two ions found in hard water.

22. Hydrogenation is used in the food industry to convert polyunsaturated vegetable oils into what?

23. List the three different kinds of eicosanoids.

24. Prostaglandins produced in the kidneys cause what effect on renal blood vessels?

25. What are the properties of neutral glycerides?

26. What two regions characterize phosphoglycerides?

27. Which lipid is found in egg yolks and soybeans and is used as an emulsifying agent in ice cream?

28. Which type of lipoprotein is bound to plasma cholesterol and transports it from the peripheral tissues to the liver?

29. Which steroid is used to suppress the inflammatory response in the treatment of rheumatoid arthritis?

30. What is the purpose of the orally ingested synthetic steroid hormone called norlutin?

Vocabulary Quiz for Chapter Fifteen

1. The 27-carbon steroid ring molecule that is the precursor for sex hormones, is required for the regulation of membrane fluidity, and has been implicated in atherosclerosis and high blood pressure is _____.

2. The current model of the structure of biological membranes is called the _____.

3. The most common lipids found in biological membranes that consist of two fatty acids esterified at the C-1 and C-2 positions of glycerol, a phosphoryl group esterified at the C-3 position, and having another hydrophilic molecule attached to the phosphoryl group are called _____.

4. The most common lipid storage form in the body, which consists of three fatty acids esterified to a glycerol molecule, is called a(n) _____.

5. _____ are long-chain monocarboxylic acids.

6. The sphingolipid found in abundance in the sheath that surrounds and insulates cells of the central nervous system is _____.

7. The plasma lipoprotein that carries triglycerides from the intestine to various tissues throughout the body is the _____.

8. A fatty acid that cannot be synthesized by the body and must be obtained in the diet is called a(n) _____.

9. _____ are lipids that are generally considered to be esters of long-chain alcohols.

10. A(n) _____ is a bipolar molecule that aids in the suspension of lipids in water.

16 Protein Structure and Enzymes

Learning Goals

1 List the functions of proteins.
2 Draw the general structure of an amino acid and classify amino acids based on their R groups.
3 Describe the primary structure of proteins and draw the structure of the peptide bond.
4 Describe the types of secondary structure of a protein and discuss the forces that maintain secondary structure.
5 Describe the structure and function of fibrous proteins.
6 Describe the tertiary and quaternary structure of a protein and list the R group interactions that maintain protein shape.
7 Describe the roles of hemoglobin and myoglobin.
8 Describe how extremes of pH and temperature cause denaturation of proteins.
9 Classify enzymes according to the type of reaction catalyzed.
10 Describe the effect that enzymes have on the activation energy of a reaction.
11 Discuss the role of the active site and the importance of enzyme specificity and describe the difference between the lock-and-key model and the induced fit model of enzyme-substrate complex formation.
12 Discuss the roles of cofactors and coenzymes in enzyme activity.
13 Explain how pH and temperature affect the rate of an enzyme-catalyzed reaction.
14 Discuss the mechanisms by which certain chemicals inhibit enzyme activity.
15 Provide examples of medical uses of enzymes.

Introduction

Proteins are polymers made up of α-amino acids. Most proteins are enormous molecules, composed of many amino acids, and therefore have very high molecular weights. Proteins carry out all the enzymatic reactions of the cell and are essential structural elements of the body.

16.1 Cellular Functions of Proteins

Learning Goal 1:   List the functions of proteins.

Enzymes are proteins that serve as biological catalysts. **Immunoglobulins,** or **antibodies,** are specific protein molecules produced in the immune system in response to foreign **antigens. Transport proteins** carry materials from one place to another in the body. They include permeases and **receptors** that move molecules across cell membranes and soluble proteins, such as hemoglobin, which transports oxygen from the lungs to the tissues. **Regulatory proteins** control cell function and include hormones such as insulin and glucagon. **Structural proteins** provide mechanical support and outer covering to animals. **Movement proteins** are necessary for all forms of movement. Our muscles contract and expand through the interaction of actin and myosin proteins. **Storage proteins** are sources of amino acids for embryos and infants.

Example 1

Provide a summary of the cellular functions of proteins.

Answer

1. Most *enzymes* are proteins. Enzymes catalyze almost all the chemical reactions that occur in living cells. Reactions that would take days in the laboratory occur in a matter of seconds or minutes in the presence of enzymes.

2. *Antibodies* are proteins that help stop infections by binding specifically to an antigen and then causing its destruction or removal from the body.

3. *Transport proteins* carry materials from one place to another in the body.

4. *Regulatory proteins* control many aspects of cell function, including metabolism and reproduction.

5. *Structural proteins* provide mechanical support to large animals and provide them with their outer coverings.

6. *Movement proteins* allow a single-celled organism or higher organism to have different types of motility.

7. *Storage proteins* are sources of amino acids for infants and embryos.

16.2 The α-Amino Acids

Structures of Amino Acids

Learning Goal 2:	Draw the general structure of an amino acid and classify amino acids based on their R groups.

The general structure of an **α-amino acid** is shown below:

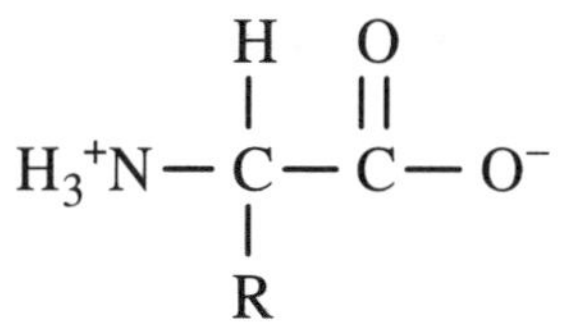

$$\text{H}_3{}^+\text{N} - \overset{\overset{\text{H}}{|}}{\underset{\underset{\text{R}}{|}}{\text{C}}} - \overset{\overset{\text{O}}{||}}{\text{C}} - \text{O}^-$$

The α-carbon is attached to a carboxylate group ($-CO_2{}^-$), a protonated amino group ($-NH_3{}^+$), a hydrogen, and an R group. The carboxylate group and protonated amino group are necessary for the covalent binding of amino acids to one another to form a protein. The R groups cause the proteins to fold into precise, three-dimensional configurations that determine their ultimate function.

Stereoisomers of Amino Acids

The α-carbon of all the α-amino acids, except glycine, is attached to four different groups and is therefore chiral. The configuration of all the naturally occurring α-amino acids isolated from proteins is L-. See Figure 16.2 for a comparison of D- and L-glyceraldehyde with D- and L-alanine.

Example 2
Draw and compare the structures of glyceraldehyde and alanine isolated from natural sources.

Answer
In living organisms, we find that glyceraldehyde is in the D-isomer form, but the amino acids exist in the L-isomer form.

D-Glyceraldehyde   L-Alanine

Classes of Amino Acids

The amino acids are grouped according to the polarity of their side chains. The side chains of some amino acids are nonpolar. These amino acids are **hydrophobic** (water-fearing), and they are generally found buried in the interior of proteins. The nonpolar amino acids include alanine, valine, leucine, isoleucine, proline, glycine, methionine, phenylalanine, and tryptophan.

Example 3
List the amino acids that contain R groups (side chains) that are hydrophobic. Where are these usually found in a protein molecule?

Answer
Alanine, valine, leucine, isoleucine, proline, glycine, methionine, phenylalanine and tryptophan contain hydrophobic R groups. These amino acids are generally found buried in the interior of proteins, where they can associate with one another and remain isolated from interaction with water molecules.

The side chains of the remaining amino acids are polar and therefore are **hydrophilic** (water-loving); they are often found on the surfaces of proteins. These polar amino acids can be subdivided into three classes:

1. Polar, neutral amino acids

2. Negatively charged (acidic) amino acids

3. Positively charged (basic) amino acids

The amino acids are often referred to using an abbreviation of three letters. Most of these abbreviations are the first three letters of their names. (For example, the abbreviation for glutamate is Glu, but glutamine is Gln.)

Example 4
List and categorize the amino acids that have hydrophilic side chains.

Answer
1. Arginine, aspartate, glutamate, histidine, lysine, asparagine, glutamine, serine, cysteine, threonine, and tyrosine have side chains that are hydrophilic.
2. These polar amino acids are divided into three classes:
   a. Polar, neutral amino acids: These have side chains that have a high affinity for water, but they are not ionic. Serine, threonine, tyrosine, cysteine, asparagine, and glutamine fall into this category.
   b. Acidic amino acids are those that have R groups that can ionize to form the negatively charged carboxylate ion ($-CO_2^-$ ion). Aspartate and glutamate are in this category.
   c. Basic amino acids are those with R groups that can ionize to form the positively charged amine ion ($-NH_3^+$). These amino acids act like bases in water, since the side chains react with water, picking up a proton and releasing a hydroxide ion. Lysine, arginine, and histidine are in this category.

16.3 The Peptide Bond

Learning Goal 3: Describe the primary structure of proteins and draw the structure of the peptide bond.

Proteins are polymers of L–amino acids joined by **peptide bonds.** This is the covalent bond formed between the α-carboxylate group of one amino acid and the α-amino group of another amino acid. The amino acid with a free $\alpha-NH_3^+$ group is known as the amino terminal, or simply the **N-terminal amino acid,** and the amino acid with a free $\alpha-CO_2^-$ group is known as the carboxyl, or **C-terminal amino acid.** The N-terminal amino acid is always drawn on the left, and the C-terminal amino acid on the right when depicting a series of covalently linked amino acids.

Example 5

Draw the tripeptide made from alanine + glycine + serine.

Work

1. Begin by drawing the backbone of the tripeptide. Make sure that the N-terminal amino acid is on the left and the C-terminal amino acid is on the right, and that the amino acids are joined correctly by amide bonds.

$$H_3{}^+N-\underset{\underset{R}{|}}{\overset{\overset{H}{|}}{C}}-\underset{}{\overset{\overset{O}{||}}{C}}-\underset{\underset{H}{|}}{\overset{\overset{H}{|}}{N}}-\underset{\underset{R}{|}}{\overset{\overset{H}{|}}{C}}-\underset{}{\overset{\overset{O}{||}}{C}}-\underset{\underset{H}{|}}{\overset{\overset{H}{|}}{N}}-\underset{\underset{R}{|}}{\overset{\overset{H}{|}}{C}}-\underset{}{\overset{\overset{O}{||}}{C}}-O^-$$

N-terminal amino acid

C-terminal amino acid

2. Next, fill in the correct R– groups to correspond with the correct amino acids.

Answer

$$H_3{}^+N-\underset{\underset{CH_3}{|}}{\overset{\overset{H}{|}}{C}}-\underset{}{\overset{\overset{O}{||}}{C}}-\underset{\underset{H}{|}}{\overset{\overset{H}{|}}{N}}-\underset{\underset{H}{|}}{\overset{\overset{H}{|}}{C}}-\underset{}{\overset{\overset{O}{||}}{C}}-\underset{\underset{H}{|}}{\overset{\overset{H}{|}}{N}}-\underset{\underset{CH_2}{|}}{\overset{\overset{H}{|}}{C}}-\underset{}{\overset{\overset{O}{||}}{C}}-O^-$$

OH

The names of peptides are derived from the C-terminal amino acid, which receives its entire name. For all other amino acids, the ending *-ine* is changed to *-yl*. Thus, the dipeptide tryptophanyl-leucine has leucine as the C-terminal amino acid and tryptophan as the N-terminal amino acid. Example 16.1 in the text describes the procedure for drawing the structure of a peptide chain.

Example 6
Show the hydrolysis reaction for alanyl-glycyl-cysteine.

Answer

$$H_3^+N-\underset{\underset{CH_3}{|}}{\overset{\overset{H}{|}}{C}}-\overset{\overset{O}{||}}{C}-\underset{\underset{H}{|}}{\overset{\overset{H}{|}}{N}}-\underset{\underset{H}{|}}{\overset{\overset{H}{|}}{C}}-\overset{\overset{O}{||}}{C}-\underset{\underset{H}{|}}{\overset{\overset{H}{|}}{N}}-\underset{\underset{CH_2}{|}}{\overset{\overset{H}{|}}{C}}-\overset{\overset{O}{||}}{C}-O^- \quad + 2\,H_2O$$

SH

↓ acid

$$H_3^+N-\underset{\underset{CH_3}{|}}{\overset{\overset{H}{|}}{C}}-COO^- \quad + \quad H_3^+N-\underset{\underset{H}{|}}{\overset{\overset{H}{|}}{C}}-COO^- \quad + \quad H_3^+N-\underset{\underset{CH_2}{|}}{\overset{\overset{H}{|}}{C}}-COO^-$$

Alanine   Glycine   Cysteine   SH

16.4 The Primary Structure of Proteins

Learning Goal 3:   Describe the primary structure of proteins and draw the structure of the peptide bond.

    The **primary protein structure** is the linear sequence of amino acids. It results from the covalent bonding of amino acids to one another via peptide bonds. The primary structure of the protein is dictated by the genetic information in the DNA. Since the sequence of amino acids determines where the R groups will be located, it also determines how the protein will fold into its final three-dimensional shape. This three-dimensional shape ultimately determines the biological function of the protein.

16.5 The Secondary Structure of Proteins

Learning Goal 4:   Describe the types of secondary structure of a protein and discuss the forces that maintain secondary structure.

    Secondary protein structure results from folding of the chain of covalently linked amino acids into regularly repeating structures. The folding pattern is maintained by numerous hydrogen bonds between the amide hydrogen and the carbonyl oxygen of the peptide chain background.

α-Helix

The most common type of secondary structure is a right-handed helical conformation known as the **α-helix.** Each carbonyl oxygen in the helix is hydrogen bonded to an amide hydrogen four amino acids away in the chain, producing an array of hydrogen bonds that are parallel to the long axis of the helix.

The **α-keratins** are α-helical proteins. These **fibrous proteins** form the covering (hair, wool, and fur) of most land animals. The individual α-helices of the keratins coil together in a bundle, producing a three-stranded protofibril that is part of an array known as a microfibril. These molecular pigtails possess great mechanical strength.

β-Pleated Sheet

The second common secondary structure in proteins resembles the pleated folds of drapery, and is known as **β-pleated sheet.** In this secondary structure, the polypeptide chain is nearly completely extended, with all the carbonyl oxygens and amide hydrogens involved in hydrogen bonds. Silk fibroin is an example of a protein whose structure is a β-pleated sheet.

16.6 The Tertiary Structure of Proteins

Most cellular proteins are **globular.** This globular, three-dimensional structure is called the **tertiary protein structure.** The peptide chain, with its regions of secondary structure, further folds to form the tertiary structure.

The forces that maintain the tertiary structure of a protein include the following:

1. *van der Waals attractions* between the R groups of nonpolar amino acids
2. *hydrogen bonds* between the polar R groups of the polar amino acids
3. *ionic bonds* between the R groups of oppositely charged amino acids
4. *covalent bonds* between the thiol-containing amino acids; two cysteines can be oxidized to a dimeric amino acid, cystine, that can be a cross-link between different proteins or hold two segments within a protein together.

The tertiary structure of the protein determines its biological function; therefore, the weak interactions that hold the protein in its correct three-dimensional shape are extremely important.

16.7 The Quaternary Structure of Proteins

Learning Goal 6:	Describe the tertiary and quaternary structure of a protein and list the R group interactions that maintain protein shape.

The active form of some proteins is an aggregate of two or more smaller globular proteins. This is the **quaternary protein structure.** The attractions that hold two or more peptides together are the same as those that maintain tertiary structure: hydrogen bonds, ionic bridges, disulfide bridges, and van der Waals forces. Some proteins must be bound to a nonprotein **prosthetic group** in order to be functional. Hemoglobin, composed of four protein subunits, is an example of a protein with quaternary structure. It also has four prosthetic groups, called heme groups. **Glycoproteins** are proteins with covalently bonded sugar groups.

Example 7

Summarize the four types of protein structure and their relationship to one another.

Answer

1. The *primary structure of a protein* is the linear sequence of amino acids bonded to one another through amide (peptide) bonds.

$$H_3{}^+N - \underset{R}{\overset{\overset{\displaystyle H}{|}}{C}} - \underset{}{\overset{\overset{\displaystyle O}{||}}{C}} - \underset{H}{\overset{\overset{\displaystyle H}{|}}{N}} - \underset{R'}{\overset{\overset{\displaystyle H}{|}}{C}} - \underset{}{\overset{\overset{\displaystyle O}{||}}{C}} - \underset{H}{\overset{\overset{\displaystyle H}{|}}{N}} - \underset{R''}{\overset{\overset{\displaystyle H}{|}}{C}} - \underset{}{\overset{\overset{\displaystyle O}{||}}{C}} - O^-$$

amino acid 1   amino acid 2   amino acid 3

2. The *secondary structure* involves the hydrogen bonding that can occur between the carbonyl oxygen and the amide hydrogen of the peptide bonds. α-Helix and the β-pleated sheet structures are the most common kinds.

3. The *tertiary structure* involves the overall folding of the entire protein molecule. Both noncovalent interactions between the R groups and covalent –S–S– bridges play a role in determining the overall tertiary structure.

4. The *quaternary structure* involves the aggregation of two or more peptide chains with respect to one another. The quaternary structure is also maintained by R group interactions.

16.8 Myoglobin and Hemoglobin

| Learning Goal 7: | Describe the roles of hemoglobin and myoglobin. |

Myoglobin and Oxygen Storage

Myoglobin is the oxygen storage protein of skeletal muscle. It is bound to a heme group that serves as the site for oxygen binding. Myoglobin has a greater affinity for oxygen than hemoglobin does, and therefore it serves as an efficient molecule to receive and store oxygen in the muscle.

Hemoglobin and Oxygen Transport

Hemoglobin (Hb) is the blood protein responsible for oxygen transport from the lungs to other tissues. It is composed of four peptide subunits, two α subunits and two β subunits, each of which contains a heme group.

Example 8
Explain the general structure and function of hemoglobin.

Answer
Hemoglobin is the oxygen transport protein of the blood. It is composed of four separate peptide subunits. There are two identical α chains and two identical β chains. In addition, each subunit of hemoglobin contains a heme group. A hemoglobin molecule therefore has the ability to bind and carry four molecules of oxygen, as shown in the following equation:

$$Hb \ + \ 4O_2 \ \rightarrow \ Hb(O_2)_4$$

$$\text{Deoxyhemoglobin} \qquad\qquad \text{Oxyhemoglobin}$$

Oxygenation of hemoglobin in the lungs is favored by the high partial pressure of oxygen in the air we breathe and the low partial pressure of oxygen in the blood. Thus, oxygen diffuses from the region of high partial pressure to the region of low partial pressure. When the blood reaches actively metabolizing tissues, this situation is reversed and oxygen diffuses into tissues from the blood.

Sickle Cell Anemia

Sickle cell anemia is a human genetic disease caused by a mutation in the gene that encodes the β-subunit of hemoglobin. This mutation results in the synthesis of sickle cell hemoglobin (Hb S). There is only a single amino acid difference between normal and sickle cell hemoglobin. That single change causes deoxyhemoglobin to polymerize, causing the red blood cells to sickle. This can result in damage to many organs and death at an early age.

In individuals with sickle cell anemia, both copies of the β-globin gene produce the mutant protein. Individuals with *sickle cell trait* have one mutant and one normal β-globin gene. Typically, they do not suffer the symptoms of the disease; however, they may pass the mutant gene on to their children.

Example 9
Explain what causes sickle cell anemia.

Answer
Sickle cell anemia is a human genetic disease that afflicts about 0.4% of African Americans. These individuals produce a mutant hemoglobin known as sickle cell hemoglobin (Hb S). Hb S differs from normal hemoglobin by a single amino acid in the β-chains; a valine has replaced glutamic acid. When Hb S molecules release their oxygen, they bind to one another as long fibers that alter the shape of the red blood cell, causing sickling. The sickled cells are unable to pass through the small capillaries, thereby causing severe medical problems.

16.9 Denaturation of Proteins

Learning Goal 8: **Describe how extremes of pH and temperature cause denaturation of proteins.**

Denaturation occurs when the weak interactions that maintain the three-dimensional structure of the protein are disrupted. Environmental factors that can cause proteins to denature include elevated temperature and a rise or drop in pH. Other factors that cause denaturation include organic solvents, detergents, heavy metals, and mechanical stress, such as whipping egg whites.

16.10 Enzymes

Enzymes are proteins that function as catalysts in cell processes. They are essential for the thousands of metabolic reactions that allow life to exist. Enzymes are remarkably specific, usually recognizing and binding to only a single type of **substrate,** or reactant, and facilitating its conversion to a **product.**

Example 10
Describe some characteristics of enzymes.

Answer
1. Life depends upon the simultaneous occurrence of hundreds of chemical reactions that must take place rapidly under mild conditions. Enzymes allow these critical reactions to occur under the mild conditions required for life.

2. The enzymes are proteins that speed up biochemical reactions by lowering the energy of activation and increasing the rate of the reaction.

Example 10 continued

3. An enzyme is very specific. It generally recognizes only one, or occasionally a few, molecules (substrates) upon which it will work its magic.

4. Enzyme-catalyzed reactions often occur from 1 to 100 million times faster than the uncatalyzed reactions.

Classification of Enzymes

Learning Goal 9:   Classify enzymes according to the type of reaction catalyzed.

Enzymes may be classified according to the type of reaction that they catalyze. These six classes are as follows: (1) **oxidoreductases,** (2) **transferases,** (3) **hydrolases,** (4) **lyases,** (5) **isomerases,** and (6) **ligases.**

Example 11
Explain the functions of oxidoreductases.

Answer
These enzymes catalyze oxidation-reduction reactions. They are responsible for the removal of protons and electrons from a substrate to cause its oxidation, or the addition of protons and electrons to a substrate to cause its reduction. Lactate dehydrogenase is a good example. This enzyme transiently binds the coenzyme NAD^+, which accepts a hydride anion ($H:^-$) from the substrate lactate. The product is pyruvate, as shown below:

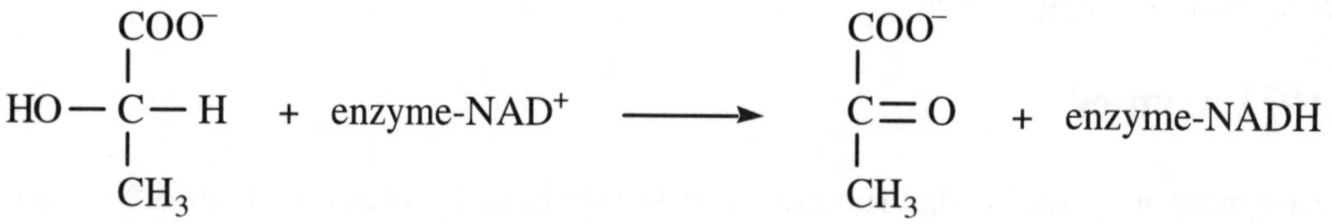

$$COO^- \qquad\qquad\qquad\qquad\qquad COO^-$$
$$HO-\overset{|}{\underset{|}{C}}-H \;+\; \text{enzyme-NAD}^+ \longrightarrow \overset{|}{\underset{|}{C}}=O \;+\; \text{enzyme-NADH}$$
$$CH_3 \qquad\qquad\qquad\qquad\qquad CH_3$$

lactate	lactate		pyruvate	lactate
(reduced form)	dehydrogenase		(oxidized form)	dehydrogenase
	(oxidized form)			(reduced form)

Example 12
List the six types of enzymes classified according to the type of reaction that they catalyze. Briefly describe the function of each class.

Answer
1. *Oxidoreductases:* these enzymes catalyze electron transfers from one molecule to another.

Example 12 continued

2. *Transferases:* these enzymes catalyze the transfer of functional groups from one molecule to another.
3. *Isomerases:* these enzymes catalyze the rearrangement of functional groups within a molecule to convert the substrate into a different isomeric form.
4. *Hydrolases:* these enzymes catalyze hydrolysis reactions.
5. *Lyases:* these enzymes catalyze the addition of a group to a double bond or the removal of a group to form a double bond.
6. *Ligases:* these enzymes catalyze the condensation or joining of two molecules.

Example 13
To which class does each of the following enzymes belong?

1. pyruvate kinase
2. lipase (hydrolysis of triglycerides)

3. triose isomerase
4. lactate dehydrogenase

Answer
1. transferase
2. hydrolase

3. isomerase
4. oxidoreductase

Naming of Enzymes

An enzyme's name often tells us the substrate of the reaction and the nature of the reaction. For instance, the enzyme sucrase hydrolyzes the disaccharide sucrose and the enzyme lactate dehydrogenase removes hydrogen ($H:^-$) from lactate ions. Some enzymes have historical names that do not reveal the nature of the substrate or of the reaction. The substrate and reaction catalyzed by such enzymes as catalase and trypsin must simply be memorized.

Example 14
Explain how the common names of enzymes are derived and give four examples.

Answer
1. The common name of an enzyme is often derived from the name of the substrate with which the enzyme interacts, and/or the type of reaction that it catalyzes. Most enzyme names end in *-ase*.
2. The following are specific examples of common names for enzymes:
   a. Urea is the substrate acted on by the enzyme urease.
   b. Lactose is the substrate of lactase.
   c. Dehydrogenase is an enzyme that removes hydrogen from a substrate.
   d. Decarboxylase is an enzyme that removes a carboxyl group from a substrate.

The Effect of Enzymes on the Activation Energy of a Reaction

Learning Goal 10:	Describe the effect that enzymes have on the activation energy of a reaction.

An enzyme speeds up a reaction by changing the path of the reaction, providing a lower energy route for the conversion of the substrate into the product. Thus, enzymes increase the rate of a chemical reaction by lowering the activation energy.

The Enzyme-Substrate Complex

Learning Goal 11:	Discuss the role of the active site and the importance of enzyme specificity and describe the difference between the lock-and-key model and the induced fit model of enzyme-substrate complex formation.

The first step in an enzyme-catalyzed reaction is the binding of the substrate by the enzyme to form the **enzyme-substrate complex.** The groove or pocket in the enzyme that binds to the substrate is the **active site.** The active site of an enzyme is small compared to the overall size of the enzyme. The substrate is held within the active site by weak, noncovalent interactions. The conformation (shape and charge distribution) of the active site is complementary to the conformation of the substrate. *Thus, the conformation of the active site determines the specificity of the enzyme.* Only those substrates that fit into the active site can bind the enzyme.

Example 15
Describe the formation of the enzyme-substrate complex.

Answer
The first step in an enzyme-catalyzed reaction involves the enzyme binding to the substrate to form the enzyme-substrate complex. The portion of the enzyme that is in contact with the substrate is called the active site.

The **lock-and-key model** of enzyme activity describes a rigid enzyme active site into which the substrate snaps, just as two pieces of a jigsaw puzzle might snap together. The **induced fit model** of enzyme activity describes a flexible active site that approximates the shape of the substrate. When the substrate enters the pocket, the active site "molds" itself around the substrate, producing a perfect fit.

The Transition State and Product Formation

The overall process of an enzymatic reaction can be summarized by the following set of reversible reactions:

$$\text{Step I} \qquad \text{Step II} \qquad \text{Step III} \qquad \text{Step IV}$$

$$E + S \rightleftharpoons ES \rightleftharpoons ES^* \rightleftharpoons EP \rightleftharpoons E + P$$

| Enzyme + Substrate | Enzyme substrate complex | Transition state | Enzyme + Product | Enzyme product complex |

In the **transition state,** the shape of the substrate is altered, due to its interaction with the enzyme, into an intermediate form having features of both the substrate and the final product. This favors the conversion of the substrate into the product, which is subsequently released. There are several ways in which an enzyme could cause a reaction to proceed more quickly. In some cases, the enzyme exerts "pressure" on a bond, thereby facilitating bond breakage. An enzyme may simply bring reactants into close proximity and in the proper orientation for reaction to occur. Alternatively, the active site of an enzyme may modify the pH of the microenvironment surrounding the substrate by serving as a donor or acceptor of H^+. Remember that the enzyme is unchanged by the chemical reaction. If it transiently loses or gains a proton, it will be converted back to its original form by either regaining or losing a proton, respectively.

Example 16

Summarize the three ways in which an enzyme may lower the activation energy of a reaction.

Answer

1. The enzyme might put stress on a bond in a substrate and therefore facilitate bond breakage.

2. An enzyme may facilitate a reaction by bringing two reactant molecules close together and in the proper orientation, so that a reaction easily occurs.

3. The active site of an enzyme may so modify the pH in the microenvironment of the substrate that a reaction will occur quickly.

Cofactors and Coenzymes

Learning Goal 12: **Discuss the roles of cofactors and coenzymes in enzyme activity.**

Some enzymes require an additional nonprotein **prosthetic group** in order to function. The protein portion of such an enzyme is called the **apoenzyme,** and the nonprotein prosthetic group is called the **cofactor.** Usually, cofactors are metal ions that bind to the enzyme and help maintain the correct shape of the active site. Cofactors may also be organic compounds or organometallic compounds.

Other enzymes require the transient binding of a **coenzyme.** Coenzymes are organic molecules that generally serve as carriers of electrons or chemical groups. They take part in a reaction by either donating or accepting chemical groups. Most coenzymes contain modified vitamins as part of their structure.

Nicotinamide adenine dinucleotide (NAD^+) is a coenzyme that accepts hydride ions (a hydrogen atom with two electrons) from the substrate that is oxidized. The portion of NAD^+ derived from the vitamin niacin is reduced to produce NADH.

Environmental Effects

Learning Goal 13:	Explain how pH and temperature affect the rate of an enzyme-catalyzed reaction.

Effect of pH

If the pH of a solution becomes too acidic or too basic, a protein is denatured. The pH at which an enzyme functions best is called the **pH optimum.** Enzyme function decreases as the pH rises above or falls below the pH optimum, and at extremes of pH, they are denatured and cease to function.

Most cellular enzymes function optimally at a pH near 7. However, **pepsin,** a proteolytic enzyme found in the stomach where the pH is very low, has a pH optimum of 2. Another proteolytic enzyme, **trypsin,** functions under the conditions of higher pH found in the intestine and has a pH optimum around 8.5.

Effect of Temperature

Enzymes are rapidly denatured if the temperature of the solution rises much above 37°C, but they remain stable at much lower temperatures. For this reason, enzymes used for clinical assays are stored in refrigerators or freezers. Since heating enzymes destroys their activity, cells can't survive extremes of temperatures. Thus, heat is an effective means of sterilizing medical instruments and solutions. Instruments may be sterilized by dry heat of 160°C for at least two hours. Alternatively, use of an autoclave is much quicker and more reliable. The pressure of steam in the chamber allows the temperature to rise to 121°C. Twenty minutes at this temperature will destroy all bacteria and viruses, as well as the heat-resistant spores produced by organisms such as the anthrax bacillus.

Some bacteria and yeast can survive very high temperatures, living in active volcanoes or in hot springs where the temperature is near the boiling point of water. The proteins of these bacteria have a structure that is stable at these extraordinary temperatures.

Inhibition of Enzyme Activity

Learning Goal 14:	Discuss the mechanisms by which certain chemicals inhibit enzyme activity.

Enzyme inhibitors are chemicals that bind to enzymes and either eliminate or drastically reduce their catalytic ability. They are classified on the basis of whether the inhibition is reversible or irreversible, competitive or noncompetitive.

Irreversible Inhibition

Irreversible inhibitors, such as arsenic, bind very tightly, sometimes even covalently, to the enzyme. This binding irreversibly blocks substrate binding or enzyme catalysis.

Reversible, Competitive Inhibitors

Generally, these inhibitors are **structural analogs,** molecules that "look like" the structure of the natural substrate for an enzyme because of similarities in shape and charge distribution. Because of this resemblance, the inhibitor can occupy the enzyme active site, but no reaction can occur. Since it is only bound by weak interactions, the inhibitor is easily removed from the active site, providing an opportunity for the substrate to bind. This is competitive inhibition because the degree of inhibition depends on the relative concentrations of substrate and inhibitor and their relative affinities for the active site.

Use of Enzymes in Medicine

Learning Goal 15:   Provide examples of medical uses of enzymes.

Enzymes are used as diagnostic tools in medicine. They are also used to treat a variety of human genetic disorders.

Liver disease is indicated by elevated levels of one of the isoenzymes of lactate dehydrogenase (LDH$_5$), as well as elevated levels of alanine aminotransferase/serum glutamate-pyruvate transaminase (ALT/SGPT), and aspartate aminotransferase/serum glutamate-oxaloacetate transaminase (AST/SGOT) in blood serum. Elevated blood serum concentrations of amylase and lipase are indications of pancreatitis, an inflammation of the pancreas.

In the clinical laboratory, enzymes are valuable analytical reagents. For example, enzymes are used in the BUN test (Blood Urea Nitrogen test). Direct measurement of urea levels is difficult due to the complexity of blood. So the enzyme urease is added, which converts each molecule of urea into two molecules of ammonia (the indicator of urea). Ammonia concentration is easily measured.

The enzyme *glucocerebrosidase* is used to treat Gaucher's disease. Those who suffer this genetic disorder are incapable of producing active glucocerebrosidase and are thus unable to carry out the degradation of the glycolipid called glucocerebroside. Glucocerebroside accumulates in the macrophages of the liver, spleen, and bone marrow. If untreated, these abnormal macrophages will displace normal bone marrow cells, causing severe anemia. Glucocerebrosidase, administered intravenously, degrades glucocerebroside to glucose and ceramide, which are further metabolized.

Glossary of Key Terms in Chapter Sixteen

active site (16.10) the cleft in the surface of an enzyme that is the site of substrate binding.
α-amino acids (16.2) the basic subunits of proteins; each is composed of an α-carbon bonded to a carboxylate group, a protonated amino group, a hydrogen atom, and a variable "R" group.

antibody (16.1) one of the specific proteins produced by cells of the immune system in response to invasion by infectious agents.

antigen (16.1) any substance able to stimulate the immune system; they are usually proteins or large carbohydrates.

apoenzyme (16.10) the protein portion of an enzyme that requires a cofactor in order to function in catalysis.

C-terminal amino acid (16.3) the amino acid at the end of the protein chain that has a free α–CO_2^- group; this is the last amino acid in a peptide.

coenzyme (16.10) an organic group required by some enzymes; generally a donor or acceptor of electrons or functional groups in a reaction.

cofactor (16.10) metal ions, organic compounds, or organometallic compounds that must be bound to an apoenzyme to maintain the correct configuration of the active site.

competitive inhibitor (16.10) a structural analog; a molecule that has a structure very similar to the natural substrate of an enzyme, competes with the natural substrate for binding to the enzyme active site, and inhibits the reaction.

defense proteins (16.1) proteins that defend the body against infectious diseases; antibodies are defense proteins.

denaturation (16.9) the process by which the organized structure of a protein is disrupted resulting in a completely disorganized, nonfunctional form of the protein.

enzyme (16.10) a protein that serves as a biological catalyst.

enzyme-substrate complex (16.10) a molecular aggregate formed when the substrate binds to the active site of the enzyme.

fibrous protein (16.5) a protein composed of peptides arranged in long sheets or fibers.

globular protein (16.6) a protein composed of polypeptide chains that are tightly folded into a compact spherical shape.

glycoprotein (16.7) a protein with sugars as prosthetic groups; often these are receptors on the cell surface.

α-helix (16.5) a right-handed coiled secondary structure maintained by hydrogen bonds between the amide hydrogen of one amino acid and the carbonyl oxygen of an amino acid four residues away.

hemoglobin (16.8) the major protein component of red blood cells; the function of this red, iron-containing protein is transport of oxygen.

holoenzyme (16.10) an active enzyme consisting of an apoenzyme bound to a cofactor.

hydrolase (16.10) an enzyme that catalyzes a hydrolysis reaction.

hydrophilic amino acid (16.2) "water loving," one of the polar or ionic amino acids that has a high affinity for water.

hydrophobic amino acid (16.2) "water fearing," a nonpolar amino acid that prefers contact with other nonpolar amino acids over contact with water.

induced fit model (16.10) the theory of enzyme-substrate binding that assumes that the enzyme is a flexible molecule and that both the substrate and enzyme change their shapes to accommodate one another as the enzyme-substrate complex forms.

irreversible enzyme inhibitor (16.10) a chemical that binds strongly to the R groups of an amino acid in the active site and eliminates enzyme activity.

isomerase (16.10) an enzyme that catalyzes the conversion of one isomer to another.

α-keratin (16.5) fibrous proteins that form the covering of most land animals; they are major components of fur, skin, beaks, and nails.

ligase (16.10) an enzyme that catalyzes the joining of two molecules.

lock-and-key model (16.10) the theory of enzyme-substrate binding that describes enzymes as inflexible molecules and a substrate that fits into the rigid active site in the same way a key fits into a lock.

lyase (16.10) an enzyme that catalyzes a reaction involving double bonds.

movement protein (16.1) a protein involved in any aspect of movement in an organism, for instance, actin and myosin in muscle tissue and flagellin that composes bacterial flagella.

myoglobin (16.8) the oxygen storage protein found in muscle.

N-terminal amino acid (16.3) the amino acid at the end of a protein chain that has a free α-N^+H_3 group; this is the first amino acid of a peptide.

oxidoreductase (16.10) an enzyme that catalyzes oxidation-reduction reactions.

peptide bond (16.3) the covalent linkage between two amino acids in a peptide chain; an amide formed by a condensation reaction between two amino acids.

pH optimum (16.10) the pH at which an enzyme catalyzes the reaction at maximum efficiency.

β-pleated sheet (16.5) a common secondary structure that resembles the pleats of an oriental fan.

primary structure (of a protein) (16.4) the sequence of amino acids in a protein; this is determined by the genetic information of the gene for each protein.

product (16.10) the chemical species that results from an enzyme-catalyzed reaction.

prosthetic group (16.7) a nonprotein group that is attached to a protein and is essential to the biological activity of that protein; often a complex organic compound.

protein (16.1) a macromolecule whose primary structure is a linear sequence of α-amino acids and whose final structure results from folding of the chain into a specific three-dimensional structure; they have many functions in the organism, including serving as catalysts, structural components, and nutritional elements.

quaternary structure (of a protein) (16.7) aggregation of two or more globular proteins to form a functional protein.

regulatory protein (16.1) a protein that controls cell functions such as metabolism and reproduction.

reversible, competitive enzyme inhibitor (16.10) a chemical that resembles the structure and charge distribution of the natural substrate and competes with it for the active site.

secondary structure (of a protein) (16.5) folding of the primary structure of a protein into an α-helix or a β-pleated sheet; the forces that maintain the folding are hydrogen bonds between the amide hydrogen and the carbonyl oxygen of the peptide bond.

sickle cell anemia (16.8) a human genetic disease resulting from inheriting mutant hemoglobin genes from both parents.

storage protein (16.1) a protein that serves as a source of amino acids for embryos or fetuses.

structural analog (16.10) a chemical having a structure and charge distribution very similar to a natural enzyme substrate.

structural protein (16.1) a protein that provides mechanical support for large plants and animals.

substrate (16.10) the reactant in a chemical reaction that binds to the enzyme active site and is converted to the product.

temperature optimum (16.10) the temperature at which an enzyme functions optimally and the rate of reaction is maximal.

tertiary structure (of a protein) (16.6) the globular, three-dimensional structure of a protein that results from folding the regions of secondary structure; this folding occurs spontaneously as a result of interactions of the side chains or R groups of the amino acids.

transferase (16.10) an enzyme that catalyzes the transfer of a functional group from one molecule to another.

transition state (16.10) the unstable intermediate in catalysis in which the enzyme has altered the form of the substrate so that it now shares properties of both the substrate and the product.

transport protein (16.1) a protein that transports materials across the cell membrane or soluble proteins, such as hemoglobin, that carry substances through the body.

vitamin (16.10) an organic substance that is required in the diet in small amounts; water-soluble vitamins are used in the synthesis of coenzymes required for the function of cellular enzymes; lipid-soluble vitamins are involved in calcium metabolism, vision, and blood clotting.

Self Test for Chapter Sixteen

1. Name the simplest amino acid.

2. Which amino acid contains the following structure as its R group?

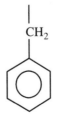

245

3. What are enzymes?

4. Where and when are antibodies produced?

5. What two proteins are responsible for the transport and storage of oxygen in higher organisms?

6. What protein is found in hair and fingernails?

7. What is produced by the controlled hydrolysis of proteins?

8. Draw the structure of serine as it appears in a living cell.

9. Which amino acids have negatively charged side chains at physiological pH?

10. Which amino acids have positively charged side chains at physiological pH?

11. Give the three-letter abbreviations used for the following amino acids:

    a. alanine   c. glycine   e. lysine

    b. cysteine   d. histidine

12. What is the term for the linkage between two amino acids in a peptide?

13. What level of protein structure is defined by the sequence of amino acids in the protein?

14. Which level of protein structure is caused by hydrogen bonding between the carbonyl oxygens and amide hydrogens of the peptide bonds?

15. What level of protein structure is defined by the binding of two or more peptides to produce an active functional protein?

16. What kind of secondary structure characterizes the fibrous proteins of muscles?

17. Which amino acid is a secondary amine?

18. What type of secondary structure characterizes silk fibroin?

19. What type of covalent bond is involved in maintaining the tertiary structure of a protein?

20. What is meant by the term *hydrophobic*?

21. How does an enzyme speed up a biological chemical reaction?

22. What is the term for the reactant in an enzyme-catalyzed reaction?

23. Write a balanced equation for the reaction catalyzed by catalase.

24. What is the substrate for each of the following enzymes:

    a. succinate dehydrogenase

    b. sucrase

    c. glycogen phosphorylase

25. Write an equation representing the specific reaction catalyzed by lactate dehydrogenase.

26. What effect does an enzyme have on the energy of activation of a reaction?

27. What is the first step in an enzyme-catalyzed reaction?

28. What are apoenzymes and cofactors?

29. Name a coenzyme that might be used by an enzyme that is a dehydrogenase?

30. Why are the water-soluble vitamins important?

31. What is the term for the pH at which an enzyme has the greatest activity?

32. At extreme pH values, an enzyme may lose the normal three-dimensional shape of the active site. What is the term used to describe this condition?

33. List two classes of enzyme inhibitors.

34. How do the sulfa drugs work as inhibitors?

35. Where are the catalytic groups of enzymes located?

36. Most enzymes are rapidly destroyed if the temperature is much higher than what value?

Vocabulary Quiz for Chapter Sixteen

1. An amino acid that is nonpolar and prefers contact with other nonpolar amino acids over contact with water is described as _____.

2. The chemical group that is attached to a protein and is essential to the biological activity of that protein is called the _____.

3. _____ are proteins that have sugars as prosthetic groups and often are receptors on the cell surface.

4. The process by which the organized structure of a protein is disrupted resulting in a completely disorganized, nonfunctional form of the protein is called _____.

5. The _____ is the globular, three-dimensional structure of a protein that results from folding the regions of secondary structure. This folding occurs spontaneously as a result of interactions of the side chains or R groups of the amino acids.

6. The _____ is the covalent linkage between two amino acids in a peptide chain. It is formed by a condensation reaction.

7. A common secondary structure that resembles the pleats of an oriental fan is called _____.

8. The _____ is the groove or cleft in the surface of an enzyme that is the site of substrate binding.

9. The _____ is the unstable intermediate in catalysis in which the enzyme has altered the form of the substrate so that it now shares properties of both the substrate and the product.

10. A(n) _____ is a molecule that has a structure very similar to the natural substrate of an enzyme, competes with the natural substrate for binding to the enzyme active site, and inhibits the reaction.

17 *Introduction to Molecular Genetics*

Learning Goals

1 Draw the general structure of DNA and RNA nucleotides.
2 Describe the structure of DNA and compare it with RNA.
3 Explain DNA replication.
4 List three classes of RNA molecules and describe their functions.
5 Explain the process of transcription.
6 List and explain the three types of post-transcriptional modifications of eukaryotic mRNA.
7 Describe the essential elements of the genetic code and develop a "feel" for its elegance.
8 Describe the process of translation.
9 Define mutation and understand how mutations can cause cancer and cell death.

Introduction

Deoxyribonucleic acid (DNA) carries all the genetic information in the cell. The primary structure of all proteins is dictated by the sequence of nucleotides in the DNA. **Ribonucleic acid (RNA)** molecules are the intermediates that carry out the translation of the genetic code into the structure of a protein. Any change in the nucleotide sequence of the DNA is a **mutation.** Mutations may be spontaneous or induced by a mutagen.

17.1 The Structure of the Nucleotide

Learning Goal 1:   Draw the general structure of DNA and RNA nucleotides.

Nucleotide Structure

Nucleotides are made up of a sugar, a nitrogenous base and at least one phosphoryl group. There are two types of nitrogenous bases: the purines and the pyrimidines. In DNA nucleotides, the sugar is 2-D-deoxyribose. In RNA the sugar is D-ribose. Because this large structure contains two cyclic molecules, the sugar and the base, the ring atoms of the sugar are designated with a prime to distinguish them from atoms in the nitrogenous base.

The covalent bond between the sugar and the phosphoryl group is a phosphoester bond. The bond between the base and the sugar is called a β-*N*-glycosidic linkage. The chemical composition of DNA and RNA is summarized below:

Purines in both DNA and RNA

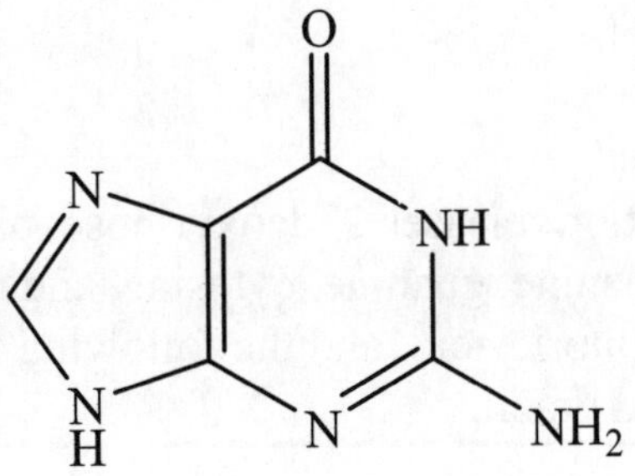

Guanine

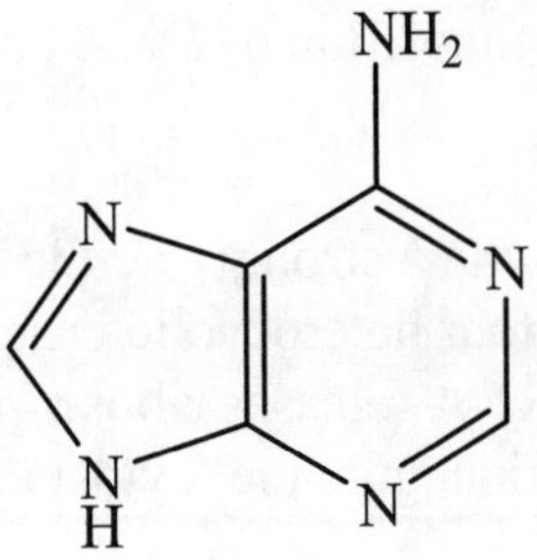

Adenine

Pyrimidines

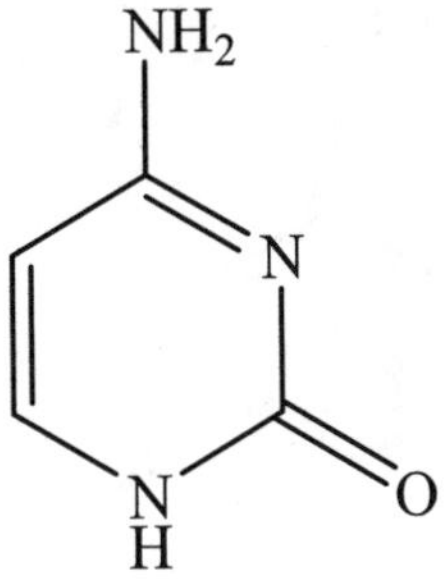

Cytosine
DNA and RNA.

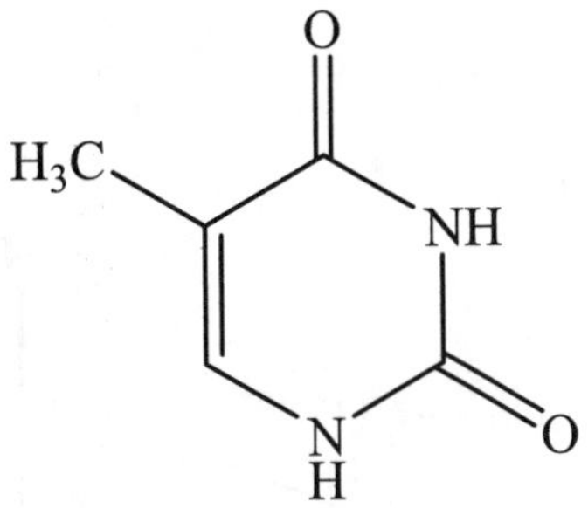

Thymine
DNA only

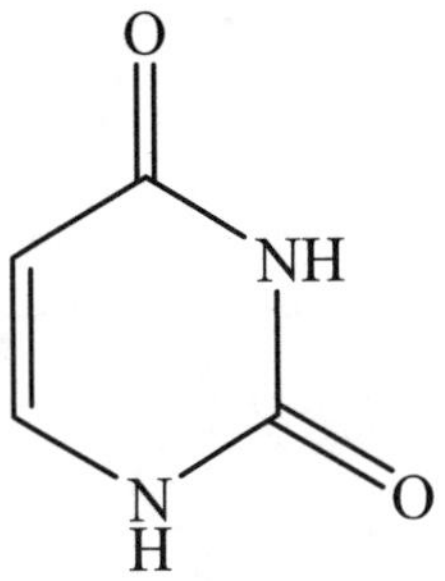

Uracil
RNA only

As noted above, the sugar in DNA is 2'-D-deoxyribose and the sugar in RNA is D-ribose.

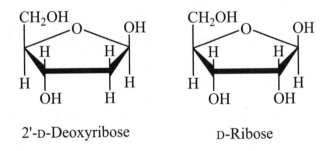

2'-D-Deoxyribose

D-Ribose

In addition, both DNA and RNA contain phosphoryl groups.

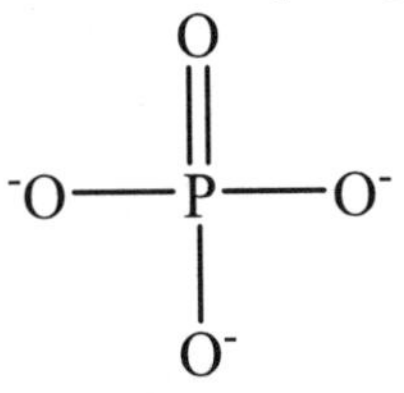

Phosphate group

Example 1
Explain the chemical composition of DNA and RNA.

Answer
The treatment of DNA with a strongly acidic solution releases 2'-deoxyribose, phosphoric acid, and the following four heterocyclic bases: adenine, guanine, cytosine, and thymine. The same treatment of RNA releases ribose, phosphoric acid, and the following four heterocyclic bases: adenine, guanine, cytosine, and uracil.

The structure of a 2'-deoxyribonucleotide composed of the phosphate group, the sugar 2'-deoxyribose, and the pyrimidine base, cytosine.

17.2 The Structure of DNA and RNA

Learning Goal 2:   Describe the structure of DNA and compare it with RNA.

DNA Structure: The Double Helix

A strand of DNA is a polymer of 2'-deoxyribonucleotide units bonded to one another by 3'-5' phosphodiester bonds. The backbone of the polymer is called the sugar-phosphate backbone because it is composed of alternating deoxyribose and phosphoryl groups in phosphodiester linkage. A nitrogenous base is bonded to each sugar by a β-*N*-glycosidic linkage.

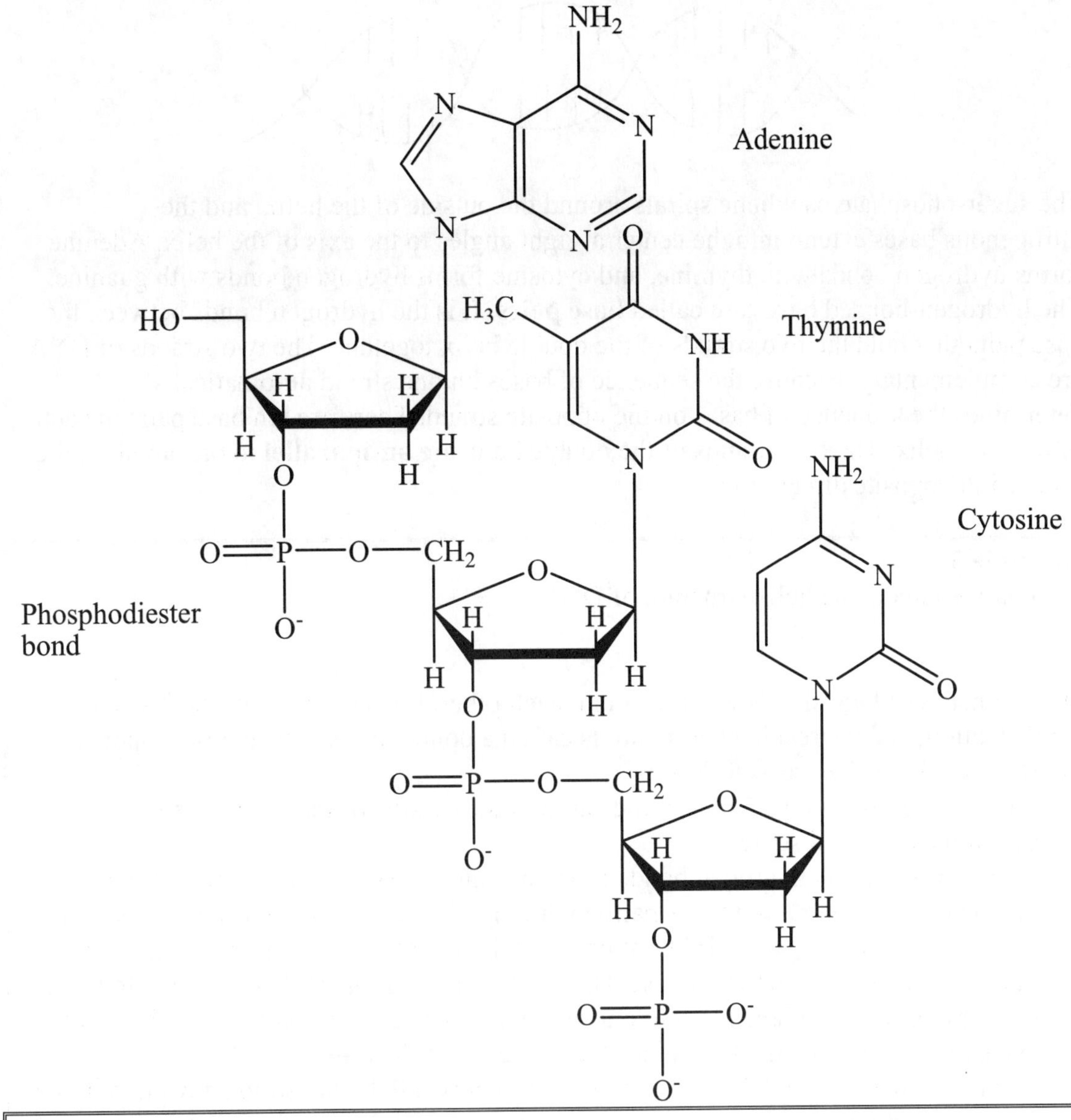

NH₂

Adenine

Thymine

Cytosine

Phosphodiester bond

Example 2
Summarize the primary structure of DNA.

Answer
The primary structure of DNA is the linear sequence of its 2′-deoxyribonucleotides. The nucleotides are linked by 3′-5′ phosphodiester bonds. Each single strand of DNA is characterized by a backbone of alternating deoxyribose and phosphoryl groups in phosphodiester linkages.

DNA is a **double helix** of two strands of DNA held together by hydrogen bonds. The following simple model gives you the idea of the two strands winding around one another and the nitrogenous bases projecting into the center of the helix.

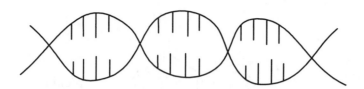

The sugar-phosphate backbone spirals around the outside of the helix, and the nitrogenous bases extend into the center at right angles to the axis of the helix. Adenine forms hydrogen bonds with thymine, and cytosine forms hydrogen bonds with guanine. The hydrogen-bonded bases are called **base pairs.** It is the hydrogen bonds between the base pairs that hold the two strands of the double helix together. The two strands of DNA are complementary, because the sequence of bases on one strand automatically determines the sequence of bases on the opposite strand. There are ten base pairs in each turn of the helix. The two strands of the double helix are **antiparallel** to one another; they proceed in opposite directions.

Example 3
Summarize the double helix structure of DNA.

Answer
DNA consists of two strands wound around each other. Each strand has a helical conformation, and the resultant structure is called a double helix. The major properties of the DNA double helix are as follows:
1. The sugar phosphate backbone winds around the outside of the bases like the handrails of a spiral staircase.
2. Each purine base is hydrogen bonded to a pyrimidine base in the interior of the double helix. Adenine is always paired with thymine, and guanine is always paired with cytosine. Each base pair lies at nearly right angles to the long axis of the helix, like the stairs of the spiral staircase. Due to the base pairing of A to T and G to C, the two strands are complementary to one another. Therefore, the sequence of bases on one strand automatically determines the sequence of the other strand.
3. The two strands of the DNA double helix are antiparallel. One strand advances in the $5' \rightarrow 3'$ direction, and the second strand advances in the $3' \rightarrow 5'$ direction.
4. The double helix of DNA completes one full turn every ten nucleotides.

RNA Structure

RNA molecules are single stranded. The sugar-phosphate backbone of RNA consists of ribonucleotides linked by $3'$-$5'$ phosphodiester bonds. The sugar in RNA is ribose rather than $2'$-deoxyribose, and uracil (U) replaces thymine.

17.3 DNA Replication

Learning Goal 3:   Explain DNA replication.

252

DNA must be replicated before cell division so that each daughter cell inherits a copy of each gene. The mechanism of faithful replication of DNA involves the enzyme DNA polymerase, which "reads" each parental strand, also called the template, and catalyzes the polymerization of a complementary daughter strand. Thus, each daughter DNA molecule consists of one parental strand and one newly synthesized daughter strand. This mode of DNA replication is called **semiconservative replication.** In the diagram below, the dashed lines represent the newly synthesized daughter DNA and the solid lines represent the parental DNA strands.

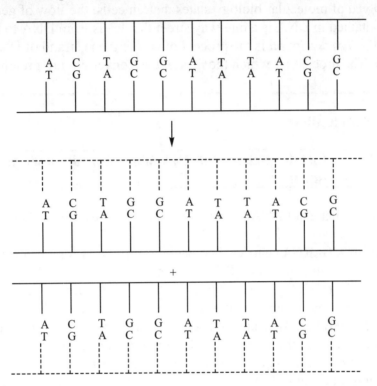

Example 4
Summarize DNA replication.

Answer
DNA replication occurs each time a cell divides; thus, all the genetic information is passed from one generation to the next. As a result of specific base pairing, the sequence of bases along each strand of DNA automatically specifies the sequence of bases in the complementary strand. DNA polymerase makes a copy of each parental strand.

17.4 Information Flow in Biological Systems

The genetic information in the DNA must be expressed to produce the proteins that actually carry out the work of the cell. Gene expression involves two steps. First, DNA is transcribed to produce a variety of RNA molecules. This process is called transcription. Then the RNA molecules participate in translation, a process in which proteins are produced. This unidirectional expression of the genetic information is called the **central dogma** of molecular biology and can be summarized as follows:

$$DNA \rightarrow RNA \rightarrow PROTEIN$$

Example 5

Explain the following terms: central dogma, transcription, and translation.

Answer

The *central dogma* of molecular biology states that, in cells, the flow of genetic information contained in DNA is a one-way street that leads from DNA to RNA to protein synthesis. *Transcription* is the process of copying one strand of DNA into a strand of RNA. The process by which the message is converted into protein is called *translation*.

Classes of RNA Molecules

Learning Goal 4:	**List three classes of RNA molecules and describe their functions.**

There are three classes of RNA found in the cell, and they are classified by their function in gene expression.

Messenger RNA (mRNA) carries the genetic information for a protein from DNA to the ribosomes. It is a complementary RNA copy of a gene on the DNA.

Ribosomal RNA (rRNA) is a structural and functional component of the ribosomes, which are "platforms" on which protein synthesis occurs.

Transfer RNA (tRNA) translates the genetic code of the mRNA into the primary sequence of amino acids in the protein. In addition to the primary structure, tRNA molecules have a cloverleaf-shaped secondary structure resulting from base pair hydrogen bonding (traditional A–U and G–C, as well as other unusual base pairings), and a tertiary structure. The sequence CCA is found at the 3′-end of the tRNA. The 3′-adenosine of this sequence can be covalently attached to an amino acid. Three nucleotides at the base of the cloverleaf structure form the **anticodon.** This triplet of bases forms hydrogen bonds to a **codon** (complementary sequence of bases) in a messenger RNA (mRNA) molecule on the surface of a ribosome during protein synthesis. This interaction assures that the correct amino acid is brought to the site of protein synthesis at the appropriate location in the growing peptide chain.

Example 6

Describe the different classes of RNA molecules.

Answer

An RNA molecule is classified by its cellular location and by its function.

1. *Messenger RNA (mRNA)* carries the genetic information for a protein.
2. *Ribosomal RNA (rRNA)* is a structural and functional component of the ribosomes.
3. *Transfer RNA (tRNA)* is responsible for translating the genetic code of the mRNA.

Transcription

Learning Goal 5:	Explain the process of transcription.

Transcription is catalyzed by the enzyme **RNA polymerase.** RNA polymerase binds to a specific nucleotide sequence, the **promoter,** at the beginning of a gene. It then separates the two strands of DNA so that it can "read" the sequence of the DNA. Chain elongation begins as the RNA polymerase reads the DNA template strand and catalyzes the polymerization of a complementary RNA copy. The final stage of transcription is termination. The RNA polymerase finds a termination sequence at the end of the gene and releases the newly formed RNA molecule.

Post-Transcriptional Processing of RNA

Learning Goal 6:	List and explain the three types of post-transcriptional modifications of eukaryotic mRNA.

In eukaryotes, transcription produces a **primary transcript** that undergoes extensive **post-transcriptional modification** before it is exported from the nucleus for translation.

The first modification is the addition of a **cap structure** to the 5′ end of the RNA. This facilitates efficient translation.

The second modification is the addition of a **poly(A) tail** to the 3′ end of the RNA. This protects the mRNA from enzymatic degradation.

The third modification is **RNA splicing.** Bacterial genes are continuous, and all the nucleotide sequences of the gene are found in the mRNA. Eukaryotic genes are discontinuous; there are extra DNA sequences within the genes. The initial mRNA, or primary transcript, carries both the protein-coding sequences and these extra sequences, which are termed intervening sequences or introns. The **introns** are removed by the process of RNA splicing.

17.5 The Genetic Code

Learning Goal 7:	Describe the essential elements of the genetic code and develop a "feel" for its elegance.

The **genetic code** is a triplet code. Each code word (codon) consists of three nucleotides. The three-letter genetic code contains 64 codons, but there are only 20 amino acids. Thus, there are 44 more codons than are required. Three of the codons (UAA, UAG, and UGA) are translation termination signals, leaving 41 additional codons. The genetic code is said to be highly **degenerate.** Methionine and tryptophan are the only amino acids encoded by only a single codon. All others are encoded by at least two codons, and serine and leucine each have six codons. As a result of this, the genetic code is quite resistant to mutation. For those amino acids that have multiple codons, the first two bases define the amino acid, and the third position is variable. A **point mutation** (the change of a single base) in the third position, therefore, often has no effect upon the amino acid that is incorporated into a protein.

17.6 Protein Synthesis

Protein synthesis, or **translation,** occurs on ribosomes. Ribosomes are complexes of ribosomal RNA and proteins consisting of two subunits, a small and a large ribosomal subunit.

Many ribosomes simultaneously translate each mRNA. These structures are called polyribosomes or **polysomes.**

The Role of Transfer RNA

The molecule that decodes the information on the mRNA molecule into the primary structure of a protein is transfer RNA (tRNA). In order to do this, the tRNA must be covalently linked to one specific amino acid. The enzyme that binds the specific tRNA to its specific amino acid is an **aminoacyl tRNA synthetase.** The product is called an **aminoacyl tRNA.**

The tRNA recognizes the appropriate codon on the mRNA by codon-anticodon hydrogen bonding.

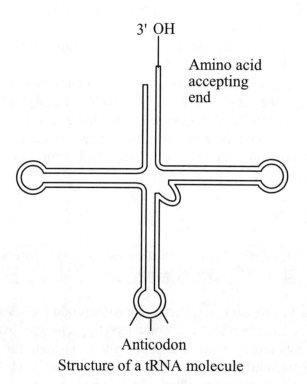

Structure of a tRNA molecule

The Process of Translation

Initiation

Initiation factors mediate the formation of a translation initiation complex composed of an mRNA molecule, the small and large ribosomal subunits, and the initiator tRNA. The initiator tRNA specifically recognizes the **initiation codon,** AUG, on the mRNA. The ribosome has two sites for binding tRNA molecules. The first site is called the **peptidyl tRNA binding site (P-site).** It holds the initiator tRNA in the initiation complex and then carries the tRNA bound to the growing peptide during the remainder of protein synthesis. The second site, called the **aminoacyl tRNA binding site (A-site),** holds the aminoacyl tRNA carrying the next amino acid to be added to the peptide chain.

Chain Elongation

This occurs in three steps that are repeated until protein synthesis is complete. It involves the interaction of several **elongation factors.**

1. An aminoacyl tRNA molecule binds to the empty A-site.
2. Peptidyl transferase catalyzes the formation of the peptide bond, and the peptide chain is shifted to the tRNA that occupies the A-site.
3. The uncharged tRNA molecule is discharged, and the ribosome changes positions (**translocation**), so that the next codon on the mRNA occupies the A-site, shifting the new peptidyl tRNA from the A-site to the P-site.

Termination

When a stop, or **termination codon** (UAA, UAG, or UGA), is encountered, translation is terminated. A **release factor** finds the empty A-site and causes release of the newly formed peptide chain and the ribosomal subunits.

The newly synthesized protein folds into its characteristic three-dimensional shape, and if it has quaternary structure, it may associate with other protein subunits. Some proteins are further modified following protein synthesis by the addition of carbohydrate or lipid molecules.

17.7 Mutation, Ultraviolet Light, and DNA Repair

Learning Goal 9:   Define mutation and understand how mutations can cause cancer and cell death.

The Nature of Mutations

Any change in the nucleotide sequence of a DNA molecule is called a **mutation.** Mutations can arise from errors during DNA replication or may be the result of chemicals, called **mutagens,** which damage the DNA.

Mutations are classified by the kind of change that occurs in the DNA. The substitution of a single nucleotide for another is called a **point mutation.** Loss of one or more nucleotides is a **deletion mutation.** Addition of one or more nucleotides to a DNA sequence is an **insertion mutation.**

The Results of Mutations

Some mutations are **silent;** that is, they cause no change in the organism. Often, however, the result of a mutation has a negative effect on the health of the organism. The effect of a mutation depends on how it alters the genetic code for a protein. There are approximately 4000 human genetic disorders that are known to be caused by mutations, including sickle cell anemia, hemophilia, cystic fibrosis, and color-blindness.

Mutagens and Carcinogens

Often mutagens are also **carcinogens,** cancer-causing chemicals. Most cancers result from mutations in a single normal cell. These mutations result in the loss of normal growth control, causing the abnormal cell to proliferate. If that growth is not controlled or destroyed, it may result in the death of the individual.

Ultraviolet Light Damage and DNA Repair

Ultraviolet (UV) light causes damage to DNA by inducing the formation of **pyrimidine dimers.** Such dimers consist of adjacent pairs of pyrimidine bases covalently bonded to one another. This interferes with normal hydrogen bonding between these pyrimidines and the complementary bases on the opposite strand. As a result, this region of DNA cannot be replicated or transcribed and the cell dies.

Bacteria have four different mechanisms to repair ultraviolet light damage. Mutations occur when the repair system makes an error and causes a change in the nucleotide sequence of the DNA.

Ultraviolet lights (germicidal lamps) are used in hospitals to kill bacteria in the air and on environmental surfaces, such as in a vacant operating room. In addition, excessive exposure to UV light (as during sun tanning) has been correlated with an increased incidence of skin cancer.

Consequences of Defects in DNA Repair

The human repair system for pyrimidine dimers requires at least five enzymes. A mutation in the gene for one of these enzymes results in the genetic skin disorder *xeroderma pigmentosum*. Individuals with this disorder are extremely sensitive to the UV rays of sunlight and usually develop multiple skin cancers before the age of twenty.

Glossary of Key Terms in Chapter Seventeen

aminoacyl tRNA (17.6) the transfer RNA covalently linked to its correct amino acid.

aminoacyl tRNA binding site of ribosome (A-site) (17.6) a pocket on the surface of a ribosome that holds the aminoacyl tRNA during translation.

aminoacyl tRNA synthetase (17.6) an enzyme that recognizes one tRNA and covalently links the appropriate amino acid to it.

anticodon (17.4) a sequence of three ribonucleotides on a tRNA that are complementary to a codon on the mRNA; codon-anticodon binding results in the tRNA bringing the correct amino acid to the site of protein synthesis.

antiparallel strands (17.2) a term describing the polarities of the two strands of the DNA double helix; on one strand, the sugar-phosphate backbone advances in the $5' \rightarrow 3'$ direction while on the opposite, complementary strand, the sugar phosphate backbone advances in the $3' \rightarrow 5'$ direction.

base pair (17.2) a hydrogen-bonded pair of bases within the DNA double helix; the standard base pairs always involve a purine and a pyrimidine; in particular, adenine always base pairs with thymine and cytosine with guanine.

cap structure (17.4) a 7-methylguanosine unit covalently bonded to the 5' end of a mRNA by a 5'-5' triphosphate bridge.

carcinogen (17.7) any chemical or physical agent that causes mutations in the DNA that lead to uncontrolled cell growth or cancer.

central dogma (17.4) a statement of the directional transfer of the genetic information in cells: DNA $\rightarrow$ RNA $\rightarrow$ Protein.

codon (17.4) a group of three ribonucleotides on the mRNA that specifies the addition of a specific amino acid onto the growing peptide chain.

complementary strands (17.2) the opposite strands of the double helix are hydrogen bonded to one another such that adenine and thymine or guanine and cytosine are always paired.

degenerate code (17.5) a term used to describe the fact that different triplet codons may be used to specify a single amino acid.

deletion mutation (17.7) a mutation that results in the loss of one or more nucleotides from a DNA sequence.

deoxyribonucleic acid (DNA) (17.1) the nucleic acid molecule that carries all of the genetic information of an organism; the DNA molecule is a double helix composed of two strands, each of which is composed of phosphate groups, 2'-deoxyribose, and the nitrogenous bases thymine, cytosine, adenine, and guanine.

deoxyribonucleotide (17.1) a nucleotide composed of a nitrogenous base in β-N-glycosidic linkage to the 1' carbon of the sugar 2'-deoxyribose and with one, two, or three phosphoryl groups esterified at the hydroxyl of the 5' carbon.

double helix (17.2) the spiral staircase-like structure of the DNA molecule characterized by two sugar-phosphate backbones wound around the outside and nitrogenous bases extending into the center.

elongation factor (17.6) a protein that facilitates the elongation phase of translation.

exon (17.4) the protein-coding sequences of a gene that are found on the final mature mRNA.

initiation factors (17.6) proteins that are required for formation of the translation initiation complex, which is composed of the large and small ribosomal subunits, the mRNA, and the initiator tRNA.

insertion mutation (17.7) a mutation that results in the addition of one or more nucleotides to a DNA sequence.

intron (17.4) a noncoding sequence within a eukaryotic gene; it must be removed from the primary transcript to produce a functional mRNA.

messenger RNA (mRNA) (17.4) an RNA species produced by transcription, which specifies the amino acid sequence for a protein.

mutagen (17.7) any chemical or physical agent that causes changes in the nucleotide sequence of a gene.

mutation (17.7) any change in the nucleotide sequence of a gene.

nucleotide (17.1) a molecule composed of a nitrogenous base, a five-carbon sugar, and one, two, or three phosphoryl groups.

peptidyl tRNA binding site of ribosome (P-site) (17.6) a pocket on the surface of the ribosome that holds the tRNA bound to the growing peptide chain.

phosphodiester bond (17.2) the bond between two sugars (either ribose or 2'-deoxyribose) through a bridging phosphate group; these linkages form the sugar-phosphate backbone of DNA and RNA.

point mutation (17.7) the substitution of a single base in a codon; this may or may not alter the genetic code of the mRNA resulting in the substitution of one amino acid in the protein.

poly(A) tail (17.4) a tract of 100–200 adenosine monophosphate units covalently attached to the 3' end of a eukaryotic mRNA molecule.

polysome (17.6) complexes of many ribosomes all simultaneously translating a single mRNA.

post-transcriptional modification (17.4) alterations of the primary transcripts produced in eukaryotic cells; these include addition of the poly(A) tail to the 3' end of the transcript, addition of the cap structure to the 5' end of the mRNA, and RNA splicing.

primary transcript (17.4) the RNA product of transcription in eukaryotic cells, before post-transcriptional modifications are carried out.

promoter (17.4) the sequence of nucleotides immediately before a gene that is recognized by the RNA polymerase and signals the start point and direction of transcription.

purine (17.1) a family of nitrogenous bases that are components of DNA and RNA and consist of a six-sided ring fused to a five-sided ring. The common purines in nucleic acids are adenine and guanine.

pyrimidine (17.1) a family of nitrogenous bases that are components of nucleic acids and consist of a single six-sided ring. The common pyrimidines of DNA are cytosine and thymine; the common pyrimidines of RNA are cytosine and uracil.

pyrimidine dimer (17.7) two adjacent pyrimidine bases in a DNA strand become covalently linked to one another; as a result of this ultraviolet light induced damage, there can be no hydrogen bonding to the opposite, complementary strand.

release factor (17.6) a protein that binds to the termination codons in the empty A-site and causes the peptidyl transferase to hydrolyze the bond between the peptide and the peptidyl tRNA.

ribonucleic acid (RNA) (17.1) single-stranded nucleic acid molecules that are composed of phosphate groups, ribose, and the nitrogenous bases uracil, cytosine, adenine, and guanine.

ribonucleotide (17.1) a nucleotide composed of a nitrogenous base in β-*N*-glycosidic linkage to the 1' carbon of the sugar ribose and with one, two, or three phosphoryl groups esterified at the hydroxyl of the 5' carbon of the ribose.

ribosomal RNA (rRNA) (17.4) the RNA species that are a structural and functional component of the small and large ribosomal subunits.

ribosome (17.6) an organelle composed of a large and a small subunit, each of which is made up of ribosomal RNA and proteins; it functions as a platform on which translation can occur and has the enzymatic activity that forms peptide bonds.

RNA polymerase (17.4) the enzyme that catalyzes the synthesis of RNA molecules using DNA as a template.

RNA splicing (17.4) removal of portions of the primary transcript that do not encode protein sequences.

semiconservative replication (17.3) DNA polymerase reads each parental strand of DNA and produces a complementary daughter strand; thus, all newly synthesized DNA molecules consist of one parental and one daughter strand.

silent mutation (17.7) a mutation that changes the sequence of the DNA but does not alter the amino acid sequence of the protein encoded by the DNA.

termination codon (17.6) a triplet of ribonucleotides with no corresponding anticodon on a tRNA. As a result, translation will end, since there is no amino acid to transfer to the peptide chain.

transcription (17.4) the synthesis of RNA from a DNA template.

transfer RNA (tRNA) (17.4) small RNAs that bind to a specific amino acid at the 3' end and mediate its addition at the appropriate site in a growing peptide chain; this is accomplished by recognition of the correct codon on the mRNA by the complementary anticodon on the tRNA.

translation (17.4) the synthesis of a protein from the genetic code carried on the mRNA; the process occurs on ribosomes and the code of the mRNA is decoded by the anticodon of the tRNA.

translocation (17.6) movement of the ribosome along the mRNA during translation; each time an amino acid is added to the chain, the ribosome moves to the next codon on the mRNA.

Self Test for Chapter Seventeen

1. Which molecule in living things is the carrier of genetic information?
2. In which type of nucleic acid is thymine usually found?
3. Give the correct abbreviations for the following compounds:
    a. adenine + deoxyribose + triphosphate unit
    b. cytosine + ribose + triphosphate unit
    c. thymine + deoxyribose + monophosphate unit
4. State the central dogma of molecular biology.
5. What two men elucidated the structure of DNA in 1953?
6. In naming the ring atoms of a pentose, what designation is used to distinguish them from atoms in the base?
7. What type of bonding helps to hold the two strands of DNA together as a double helix?
8. Each base pair in the DNA molecule lies at what angle to the long axis of the helix?
9. Which base in one strand of DNA is always hydrogen-bonded to adenine in the other strand?
10. One strand of the DNA molecule has a $5' \rightarrow 3'$ orientation. What is the orientation of the opposite strand?
11. As a result of specific base pairing in the DNA molecule, what is the relationship of the sequence of bases on one strand to the sequence of bases on the other strand?
12. What is the name of the DNA damage that is caused by ultraviolet light?
13. What is the term used to describe the mechanics of DNA replication?
14. Why is it necessary to produce faithful copies of the DNA through the process of DNA replication?
15. Which nitrogenous base is found only in RNA?
16. What is the sugar found in RNA molecules?
17. What disease is caused by the accumulation of mutations that cause uncontrolled cell growth and division?
18. List the classes of different RNA molecules.
19. Which form of RNA carries the needed amino acids to the ribosomes for use in protein synthesis?
20. Name the enzyme that catalyzes the transcription of DNA.
21. What sequence on DNA indicates the site at which the RNA polymerase should begin transcription?

22. In eukaryotic cells, the initial mRNA (primary transcript) carries the sequences for a protein but also contains noncoding sequences. What is the term for these noncoding sequences?

23. What is the term for the coding sequences that remain after mRNA splicing?

24. The three-letter genetic code contains 64 words. What is the term for these genetic words?

25. Who first proposed that different triplet codons may serve as code words for the same amino acid?

26. Which two amino acids have only a single codon?

27. When the abnormal gene for the formation of the β-chain in hemoglobin is translated into protein, a valine is substituted for which amino acid?

28. What genetic skin disorder is caused by a mutation in the DNA repair endonuclease gene or in other genes in the DNA repair pathway?

29. What are the terms used to describe the two subunits of the ribosome?

30. What is the name of the complexes of many ribosomes along a single mRNA?

31. What is the term used to describe a mutation that results in no change in the organism?

Vocabulary Quiz for Chapter Seventeen

1. The _____ is composed of a 5-carbon sugar, a nitrogenous base, and one, two, or three phosphoryl groups.

2. The term _____ describes the double stranded spiral nature of the DNA molecule.

3. The unidirectional flow of biological information (DNA → RNA → Protein) is called _____.

4. The class of RNA that carries the genetic code for a protein is _____.

5. The process of protein synthesis is called _____.

6. The platform on which protein synthesis occurs is called the _____.

7. Any change in the sequence of DNA is called a _____.

8. The enzyme that is responsible for transcription is _____.

9. A mutation that results in no change in the sequence of amino acids in the protein is called a _____.

18 *Carbohydrate Metabolism*

Learning Goals

1 Discuss the importance of ATP in cellular energy transfer processes.
2 Describe the three stages of catabolism of dietary proteins, carbohydrates, and lipids.
3 Discuss glycolysis in terms of its two major segments.
4 Looking at an equation representing any of the chemical reactions in glycolysis, describe the kind of reaction and the significance of that reaction to the pathway.
5 Discuss the practical and metabolic roles of fermentation reactions.
6 Name the regions of the mitochondria and the function of each region.
7 Describe the reaction that results in the conversion of pyruvate to acetyl CoA, describing the location of the reaction and the components of the pyruvate dehydrogenase complex.
8 Summarize the reactions of aerobic respiration.
9 Looking at an equation representing any of the chemical reactions in the citric acid cycle, describe the kind of reaction and the significance of that reaction to the pathway.
10 Describe the process of oxidative phosphorylation.
11 Discuss the biological function of gluconeogenesis.
12 Summarize the regulation of blood glucose levels by glycogenesis and glycogenolysis.

Introduction

Cells need a constant supply of energy to maintain essential life processes such as *active transport*, *biosynthesis*, and *mechanical work*. A supply of energy-rich food molecules, carbohydrates, protein, and fats is required to provide this needed cellular energy. Carbohydrates are the most readily used energy source, and glycolysis is the pathway for the first stages of carbohydrate degradation.

18.1 ATP: The Cellular Energy Currency

Learning Goal 1:	Discuss the importance of ATP in cellular energy transfer processes.

Catabolism is the degradation of fuel molecules. The energy released in catabolism is stored as *chemical bond energy*. The molecule used for the storage of chemical energy, and often called the universal energy currency, is **adenosine triphosphate (ATP).**

ATP is a **nucleotide** composed of the nitrogenous base adenine bonded in *N*-glycosidic linkage to the sugar ribose. Ribose is bonded to one (AMP), two (ADP), or three (ATP) phosphoryl groups. The molecule is a high-energy compound because the bonds holding the last two phosphoryl groups are high-energy bonds. This means that a large amount of energy is released when these bonds are hydrolyzed. Such high-energy bonds are indicated as squiggles ($\sim$).

ATP is an ideal "go-between," shuttling energy from exergonic (energy releasing) reactions to endergonic (energy requiring) reactions.

Example 1
What is the function of ATP?

Answer
Adenosine triphosphate (ATP) is a "go-between" molecule that can store or release chemical energy. The secret to the function of ATP as a go-between lies in its chemical structure. The molecule is a high-energy compound because of the bonds holding the terminal phosphoryl groups. When these bonds are hydrolyzed, they release a large amount of energy that can be used for cellular work. Then other pathways allow the resynthesis of ATP by providing the energy needed for the reaction between ADP + P_i which produces ATP.

Example 2
Write an equation showing the release of energy by the hydrolysis of ATP.

Answer
The hydrolysis of ATP to ADP and an inorganic phosphate group, abbreviated P_i releases 7.3 kcal/mole of energy:

$$\text{ATP} + \text{H}_2\text{O} \quad \rightarrow \quad \text{ADP} + P_i + 7.3 \text{ kcal/mole}$$

18.2 Overview of Catabolic Processes

Learning Goal 2:   **Describe the three stages of catabolism of dietary proteins, carbohydrates, and lipids.**

Stage I: Hydrolysis of Dietary Macromolecules into Small Subunits

The first stage of catabolism is the hydrolysis of dietary macromolecules into small subunits. Large polymeric molecules are degraded into their constituent subunits, which are taken into the cells. Polysaccharides are hydrolyzed to monosaccharides. Proteins are hydrolyzed to oligopeptides and amino acids, and fats are hydrolyzed into fatty acids, glycerol, and monoglycerides. All these molecules are then absorbed by the cells lining the intestine.

Stage II: Conversion of Monomers into a Form that Can Be Completely Oxidized

The second stage of catabolism is the conversion of monomers into a form that can enter one of the catabolic pathways and be degraded to yield energy. Sugars usually enter the glycolysis pathway. The amino groups are removed from the amino acids. Fatty acids are converted to acetyl CoA and glycerol. The glycerol is converted to glyceraldehyde-3-phosphate, which enters the glycolysis pathway.

Stage III: The Complete Oxidation of Nutrients and the Production of ATP

The final stage of catabolism is the complete oxidation of nutrients and the production of ATP. Pyruvate, the product of glycolysis, the carbon skeletons of many amino acids, and acetyl CoA from fatty acid metabolism all enter the citric acid cycle to be completely oxidized. The electrons and hydrogen atoms that are harvested in the citric acid cycle are used in the process of oxidative phosphorylation to produce ATP.

18.3 Glycolysis

Learning Goal 3:   Discuss glycolysis in terms of its two major segments.

An Overview

Glycolysis, the first stage of carbohydrate catabolism, was probably the first successful energy-harvesting pathway that evolved on earth. It is an anaerobic process, requiring no oxygen, and it is carried out by enzymes in the cytoplasm of the cell. The degradation of glucose by glycolysis yields chemical energy in the form of ATP. Four ATP molecules are produced by **substrate-level phosphorylation,** a reaction in which high-energy phosphoryl groups from one of the substrates in glycolysis are transferred to ADP to form ATP. However, the net yield of ATP is only 2, because 2 ATP are used in the early stages of glycolysis. Chemical energy in the form of reduced **nicotinamide adenine dinucleotide,** NADH, is also produced. Under aerobic conditions, the electrons are donated to an electron transport system for the production of ATP by **oxidative phosphorylation;** but under anaerobic conditions, NADH is used as a source of electrons in fermentation reactions. The final product of glycolysis is two pyruvate molecules. Under anaerobic conditions, pyruvate is used as a substrate in fermentation reactions.

Example 3
Describe the first stage of carbohydrate metabolism.

Answer
The pathway called glycolysis is the beginning stage for the use of glucose as a source of chemical energy. This pathway is anaerobic, which means that no oxygen is required. The glycolysis pathway releases only about 2% of the potential energy of the glucose molecule. In the pathway, glucose is finally transformed after many steps into two molecules of pyruvate.

Reactions of Glycolysis

Learning Goal 4:	Looking at an equation representing any of the chemical reactions in glycolysis, describe the kind of reaction and the significance of that reaction to the pathway.

The structures of the intermediates of glycolysis can be found on pages 545–548 of the text (Section 18.3). The following is a brief description of those reactions.

Reaction 1
Glucose is phosphorylated by the enzyme *hexokinase*. A phosphoryl group from ATP is transferred to C-6 of glucose, and the product is glucose-6-phosphate.

Reaction 2
The enzyme *phosphoglucose isomerase* isomerizes the glucose-6-phosphate to produce fructose-6-phosphate.

Reaction 3
The enzyme *phosphofructokinase* transfers a phosphoryl group from ATP to the C-1 hydroxyl of fructose-6-phosphate, producing fructose-1,6-bisphosphate.

Reaction 4
Aldolase splits the fructose-1,6-bisphosphate into glyceraldehyde-3-phosphate (G3P) and dihydroxyacetone phosphate (DHAP).

Reaction 5
Only G3P can be used in glycolysis, and the DHAP is isomerized by the enzyme *triose phosphate isomerase*. This produces a second molecule of G3P.

Reaction 6
Glyceraldehyde-3-phosphate dehydrogenase catalyzes the oxidation of the aldehyde group of G3P to a carboxyl group. The coenzyme in this reaction is NAD^+, which is reduced to NADH. Next, an inorganic phosphate group is transferred to the carboxylate group to produce 1,3-bisphosphoglycerate.

Reaction 7
Phosphoglycerate kinase catalyzes the transfer of a phosphoryl group from 1,3-bisphosphoglycerate to ADP. This is the first substrate-level phosphorylation of glycolysis to produce ATP.

Reaction 8
Phosphoglycerate mutase catalyzes the isomerization of 3-phosphoglycerate to 2-phosphoglycerate.

Reaction 9

Enolase catalyzes the dehydration of 2-phosphoglycerate, producing the energy-rich product phosphoenolpyruvate.

Reaction 10

Pyruvate kinase catalyzes the last substrate-level phosphorylation, in which a phosphoryl group from phosphoenolpyruvate is transferred to ADP to produce ATP. The final product of glycolysis is pyruvate.

Reactions 5 through 10 occur twice per glucose molecule; thus, the final products of glycolysis are 2 NADH, 2 pyruvate, and 4 ATP. However, since 2 ATP/glucose were invested early in glycolysis, the net yield is 2 ATP/glucose molecule.

Example 4

List the steps in glycolysis in which ATP molecules are hydrolyzed to provide energy for specific reactions.

Answer
1. Reaction 1 is a coupled reaction in which the enzyme *hexokinase* catalyzes the transfer of a phosphoryl group from ATP to glucose. The product is glucose-6-phosphate.
2. Reaction 3 is a coupled reaction in which the enzyme *phosphofructokinase* catalyzes the transfer of a phosphoryl group from ATP to fructose-6-phosphate. The product is fructose-1,6-bisphosphate.

Example 5

List the two steps of the anaerobic glycolysis pathway that result in the production of ATP molecules.

Answer
1. Reaction 7: This reaction is the first step of the pathway in which energy is harvested in the form of ATP. One of the phosphoryl groups of 1,3-bisphosphoglycerate is transferred to an ADP molecule in the first substrate-level phosphorylation of glycolysis. This reaction is catalyzed by the enzyme *phosphoglycerate kinase*.
2. Reaction 10: The final substrate-level phosphorylation in glycolysis involves the transfer of a phosphoryl group from phosphoenolpyruvate to ADP to form ATP. This reaction is catalyzed by the enzyme *pyruvate kinase*.

18.4 Fermentations

Learning Goal 5:   Discuss the practical and metabolic roles of fermentation reactions.

In order for glycolysis to continue to degrade glucose and produce ATP, the NADH must be reoxidized, and the pyruvate must be utilized and removed. Under anaerobic conditions, these two requirements are met through fermentation reactions.

Example 6
Explain how the NADH produced by glycolysis in reaction 6 is regenerated into NAD$^+$.

Answer
1. If the cell is functioning under aerobic conditions, the NADH will be reoxidized, and pyruvate will be completely oxidized by aerobic respiration.
2. Under anaerobic conditions, cells of different types employ a variety of fermentation reactions to accomplish the conversion of NADH to NAD$^+$.

Lactate Fermentation

Some cells, such as muscle cells and dairy bacteria, utilize lactate fermentation. Under anaerobic conditions, the enzyme *lactate dehydrogenase* reduces pyruvate to lactate. NADH is the reducing agent for this reaction, and thus NAD$^+$ is regenerated, while the pyruvate is used up.

This reaction occurs in muscle cells during strenuous exercise, when the body cannot provide sufficient oxygen for aerobic respiration to provide enough ATP. Lactate eventually builds up in the muscle to the extent that glycolysis can no longer proceed, and the muscle cells can no longer function. This point of exhaustion is called the **anaerobic threshold.**

Lactate fermentation is used by bacteria that produce yogurt and some cheeses. Lactate is one of the molecules that give these dairy products their tangy flavor.

Alcohol Fermentation

Under anaerobic conditions, yeast cells ferment sugars produced by fruit and grains to produce ethanol. First, *pyruvate decarboxylase* removes CO_2 from (decarboxylates) the pyruvate, producing CO_2 and acetaldehyde. Then, *alcohol dehydrogenase* catalyzes the reduction of acetaldehyde to ethanol, reoxidizing NADH in the process. Eventually, the stable fermentation end product, ethanol, builds up to a concentration that kills the yeast cells. It is characteristic of fermentations that the end product eventually builds up to levels that inhibit the cells carrying out the reaction.

Example 7
Summarize the lactate and ethanol fermentation pathways.

Answer
1. *Lactate fermentation:* This pathway occurs when you have exercised beyond the capacity of your lungs and circulatory system to deliver enough oxygen to the working muscles. The aerobic energy-generating pathways will no longer be able to supply enough ATP, but the muscles still demand energy. Under anaerobic conditions, pyruvate can be reduced by lactate dehydrogenase to form lactate.

Example 7 continued

> Simultaneously, NADH is oxidized to NAD$^+$, which is required for the continued anaerobic functioning of the glycolysis pathway.
>
> 2. *Alcohol fermentation:* Under anaerobic conditions in yeast (usually not humans!), the pyruvate formed by glycolysis is converted into ethanol and CO_2 in two enzyme-catalyzed steps. First, pyruvate is converted into acetaldehyde as CO_2 is released. Then the acetaldehyde is reduced by NADH. This produces ethanol and NAD$^+$ that allows anaerobic glycolysis to continue.

18.5 The Mitochondria

The energy-harvesting reactions that produce the greatest energy yield are aerobic reactions that occur in the **mitochondria.** Mitochondria are the organelles that serve as the cellular "powerplants." An organelle is a membrane-enclosed compartment within the cytoplasm that has a specialized function. In the mitochondria, the enzymes of the citric acid cycle strip electrons from substrates in the pathway. These are passed through the **electron transport system,** and the energy of the electrons is used to produce ATP. This process is **oxidative phosphorylation.**

Example 8

What is the function of the mitochondria?

Answer

The oxidative reactions of metabolism are responsible for most cellular ATP production. These reactions occur in metabolic pathways located in the mitochondria. This organelle in the cytoplasm is the power plant of the cell, producing most of the ATP for cellular processes. The mitochondria are responsible for the final oxidation of the acetyl group of acetyl CoA from glycolysis, fatty acid degradation, and amino acid catabolism. It also produces the majority of the ATP for the cell.

Learning Goal 6:	Name the regions of the mitochondria and the function of each region.

Structure and Function

Mitochondria are bounded by an **outer mitochondrial membrane** and an **inner mitochondrial membrane.** The region between the two membranes is known as the **intermembrane space,** and the region enclosed by the inner membrane is known as the **matrix space.**

The outer membrane is freely permeable to small molecules (less than 10^4 g/mol). Thus, metabolites to be oxidized via the citric acid cycle easily enter the intermembrane space through channel proteins.

The inner membrane is a highly folded, continuous structure. The individual folds are known as **cristae.** The inner mitochondrial membrane is virtually impermeable to most substances and contains three types of proteins. *Transport proteins* allow the transport of metabolites across the inner mitochondrial membrane into the matrix. *Electron transport system proteins* are involved in electron transfers, for which O_2 serves as the terminal electron acceptor. The third protein is a very large multiprotein complex known as **ATP synthase,** which is responsible for phosphorylation of ADP.

The enzymes of the citric acid cycle and those for the oxidation of fatty acids and amino acids are located in the matrix space.

Example 9
Describe the structure of the mitochondria.

Answer
Mitochondria are football-shaped organelles that are roughly the size of bacteria. This organelle has both an outer and an inner membrane. The region between the two membranes is called the intermembrane space, and the region enclosed by the inner membrane is known as the matrix space. The outer membrane is freely permeable to substances of less than 10^4 g/mol. This is because of the presence of a large number of transport proteins that form pores in the membrane.

The inner membrane is a continuous structure that is highly folded. These folds of the inner membrane are called cristae. This membrane is virtually impermeable to most substances. The inner membrane contains transport proteins to allow the transport of metabolites across the membrane. Protein complexes for electron transport, ATP synthesis, the citric acid cycle, and fatty acid oxidation are also located in the inner membrane or matrix space.

Origin of the Mitochondria

Mitochondria have several features that have led scientists to suspect that they may once have been free-living bacteria that were "captured" by eukaryotic cells. They have their own DNA and ribosomes; as a result, they can make many of their own proteins. They are also self-replicating; growing in size and dividing in a way similar to bacteria.

18.6 Conversion of Pyruvate to Acetyl CoA

Learning Goal 7:	Describe the reaction that results in the conversion of pyruvate to acetyl CoA, describing the location of the reaction and the components of the pyruvate dehydrogenase complex.

Under aerobic conditions, cells use oxygen and completely oxidize glucose to CO_2 in a metabolic pathway called the **citric acid cycle. Acetyl CoA** is the molecule that carries two-carbon fragments (acetyl groups) produced from pyruvate into the citric acid cycle.

The coenzyme A portion of acetyl CoA is derived from ATP and the vitamin pantothenic acid. The acetyl groups are linked to the thiol group of coenzyme A by a high-energy thioester bond. Acetyl CoA is an "activated" form of the acetyl group.

The reaction that converts pyruvate to acetyl CoA is carried out by the *pyruvate dehydrogenase complex*. Pyruvate is decarboxylated and oxidized and acetyl CoA is formed.

Acetyl CoA plays a central role in cellular metabolism. It is the product of the degradation of glucose, fatty acids, and some amino acids. It carries acetyl groups to the citric acid cycle for complete oxidation and the ultimate production of large amounts of ATP. It is also used for *anabolic* or biosynthetic reactions to produce cholesterol and fatty acids. Thus, acetyl CoA is the intermediate through which all energy sources (fats, proteins, and carbohydrates) are interconvertible.

18.7 An Overview of Aerobic Respiration

Learning Goal 8:   **Summarize the reactions of aerobic respiration.**

Aerobic respiration is the oxygen-requiring degradation of food molecules and production of ATP. Acetyl groups derived from the breakdown of sugars, amino acids, or lipids are completely oxidized to CO_2 in the reactions of the citric acid cycle. The electrons harvested in these oxidation reactions are used to reduce 3 NAD^+ and 1 FAD, producing 3 NADH and 1 $FADH_2$.

These electrons are then passed through an electron transport system that simultaneously pumps protons (H^+) into the H^+ reservoir in the mitochondrial intermembrane space. The energy of the H^+ reservoir is used by ATP synthase to produce ATP. The entire process is called **oxidative phosphorylation** because the energy of electrons from the oxidation of substrates is used to phosphorylate ADP and produce ATP.

18.8 The Citric Acid Cycle (The Krebs Cycle)

Reactions of the Citric Acid Cycle

Learning Goal 9:   **Looking at an equation representing any of the chemical reactions in the citric acid cycle, describe the kind of reaction and the significance of that reaction to the pathway.**

The following is a summary of the reactions of the **citric acid cycle.**

1. The acetyl group of acetyl CoA is transferred to oxaloacetate by the enzyme *citrate synthase*, forming citrate.

2. *Aconitase* catalyzes the dehydration of citrate, producing *cis*-aconitate. This same enzyme then adds a water molecule to the *cis*-aconitate, converting it to isocitrate. The result of these two steps is the isomerization of citrate to isocitrate.

3. *Isocitrate dehydrogenase* catalyzes the first oxidative reaction of the citric acid cycle. This is a complex reaction in which three things happen. First, the hydroxyl group of

isocitrate is oxidized to a ketone; then carbon dioxide is released; finally NAD^+ is reduced to NADH. The product of this oxidative decarboxylation reaction is α-ketoglutarate.

4. The *α-ketoglutarate dehydrogenase* complex, an enzyme complex very similar to the pyruvate dehydrogenase complex, mediates the next reaction. Once again, three chemical events occur. First, α-ketoglutarate loses a carboxylate group as CO_2; next, NAD^+ is reduced to NADH; and finally, coenzyme A combines with the product to form succinyl CoA. The bond between succinate and coenzyme A is a high-energy thioester linkage.

5. Succinyl CoA is converted to succinate by the enzyme *succinyl CoA synthase*, which removes the CoA group and uses the energy of the thioester bond to add an inorganic phosphate group to GDP, producing GTP. *Dinucleotide diphosphokinase* then shifts the phosphoryl group from GTP to ADP, producing ATP.

6. *Succinate dehydrogenase* then catalyzes the oxidation of succinate to fumarate. The oxidizing agent, flavin adenine dinucleotide (FAD), is reduced in this step.

7. *Fumarase* catalyzes the addition of H_2O to the double bond of fumarate, producing malate.

8. Finally, *malate dehydrogenase* reduces NAD^+ to NADH and oxidizes malate to oxaloacetate. Since the citric acid cycle began with the addition of an acetyl group to oxaloacetate, we have come full circle.

Example 10

Write the specific reactions in the citric acid cycle in which the coenzymes NAD^+ and FAD act as oxidizing agents.

Answer

1. The first oxidative step, Step 3, is catalyzed by isocitrate dehydrogenase:
   isocitrate + NAD^+ $\rightarrow$ α-ketoglutarate + CO_2 + NADH

2. The next step involves the oxidation of α-ketoglutarate into succinate:
   α-ketoglutarate + NAD^+ $\rightarrow$ succinate + CO_2 + NADH

3. In Step 6, succinate is oxidized to fumarate:
   succinate + FAD $\rightarrow$ fumarate + $FADH_2$

4. Then in the final step of the citric acid cycle, malate is oxidized to produce oxaloacetate:
   malate + NAD^+ $\rightarrow$ oxaloacetate + NADH

18.9 Oxidative Phosphorylation

Learning Goal 10: Describe the process of oxidative phosphorylation.

Oxidative phosphorylation is a series of reactions that couples the oxidation of NADH and $FADH_2$ to the phosphorylation of ADP to produce ATP.

Electron Transport Systems and the Hydrogen Ion Gradient

Embedded within the mitochondrial inner membrane is a series of electron carriers called the **electron transport system.** Prominent among these electron carriers are the cytochromes, which carry a heme group. These molecules are arranged within the membrane so that they pass electrons from one to the next. Some of these electron carriers are able to carry hydrogen atoms, while others can carry only electrons. Those that do not pass protons deposit them within the intermembrane space.

NADH carries electrons to the first carrier of the electron transport system, *NADH dehydrogenase*. There it is oxidized to NAD^+, donating a pair of hydrogen atoms. It then returns to the site of the citric acid cycle to be reduced again. The pair of electrons is passed to the next electron carrier, but the protons are pumped into the intermembrane compartment. The electrons are passed sequentially through the electron transport system and, at two additional points, protons from the matrix are pumped into the intermembrane compartment. In aerobic organisms, the **terminal electron acceptor** is molecular oxygen, O_2, and the product is water. $FADH_2$ donates its electrons to a carrier later in the electron transport system, so only four protons are pumped into the intermembrane space.

As the electron transport system continues to function, a high concentration of hydrogen ions accumulates in the intermembrane space, creating a concentration gradient across the inner membrane. Thus, the system generates a high concentration of protons in the intermembrane space; a lower concentration of protons is found in the matrix.

ATP Synthase and the Production of ATP

The last component needed for oxidative phosphorylation is a multiprotein complex called **ATP synthase,** or $\mathbf{F_0F_1}$ **complex.** The F_0 portion provides a channel in the membrane through which protons may pass. The F_1 portion phosphorylates ADP to produce ATP, using the energy of the proton gradient.

ATP synthase harvests the energy of this gradient and uses it to produce ATP. As the protons pass into the matrix through F_0, some of their energy is used by the enzymatic portion of ATP synthase (F_1) to catalyze the phosphorylation of ADP to ATP.

Example 11

What are the two different ways that ATP can be synthesized by the cell? Where are these pathways located, and how efficient are they?

Answer

1. ATP is synthesized by substrate-level phosphorylation and oxidative phosphorylation.
2. Substrate-level phosphorylation occurs in the cytoplasm of the cell and produces only a few ATP molecules per glucose molecule.
3. Oxidative phosphorylation occurs in the mitochondria and produces a large number of ATP molecules per glucose molecule.

273

Example 12

Explain how the electron transport system converts NADH back into NAD⁺, so that it may be used again by the citric acid cycle.

Answer

The NADH carries a hydride anion, which was originally from glucose or other food molecules, to the first carrier of the electron transport system. There it is oxidized to NAD⁺ and the first electron carrier, NADH dehydrogenase is reduced. NAD⁺ returns to be used again in the citric acid cycle.

Example 13

Write an equation to represent the last step of the electron transport system.

Answer

$$2\,H^+ \;+\; 2\,e^- \;+\; {}^{1}\!/_{2}\,O_2 \;\rightarrow\; H_2O$$

Summary of the Energy Yield

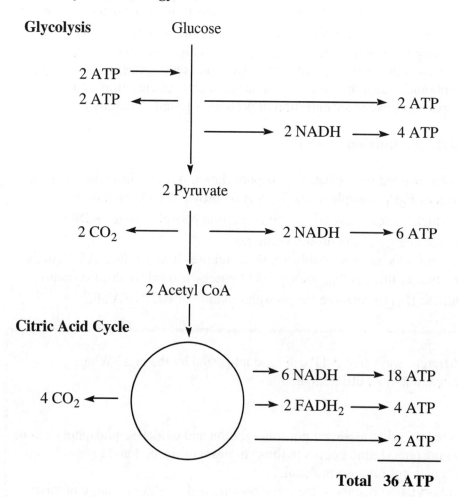

Glycolysis Glucose

2 ATP →

2 ATP ← → 2 ATP

→ 2 NADH → 4 ATP

2 Pyruvate

2 CO$_2$ ← → 2 NADH → 6 ATP

2 Acetyl CoA

Citric Acid Cycle

4 CO$_2$ ←

→ 6 NADH → 18 ATP

→ 2 FADH$_2$ → 4 ATP

→ 2 ATP

Total 36 ATP

18.10 Gluconeogenesis: The Synthesis of Glucose

Learning Goal 11: Discuss the biological function of gluconeogenesis.

During starvation and following extended exercise, the body must make glucose. **Gluconeogenesis** is the process by which glucose is produced from noncarbohydrate precursors. Lactate, all amino acids, except leucine and lysine, and glycerol from fats all serve as precursors for glucose biosynthesis. Gluconeogenesis, shown in Figure 18.14 in the text, appears to be the reverse of glycolysis. Certainly most of the intermediates of the two pathways are identical. However, steps 1, 3, and 10 of glycolysis are irreversible and must be bypassed by other enzymes. Glucose-6-phosphate is dephosphorylated by the enzyme glucose-6-phosphatase, found in the liver but not in muscle. Similarly, the phosphorylation of fructose-6-phosphate by phosphofructokinase is irreversible. Fructose bisphosphatase catalyzes the removal of the phosphoryl group from fructose-1,6-bisphosphate. Finally, step 10 of glycolysis is bypassed by two enzymes. Pyruvate carboxylase adds a carboxyl group to pyruvate by the addition of atmospheric CO_2, producing oxaloacetate. Then phosphoenolpyruvate carboxykinase removes the CO_2 and adds a phosphoryl group, producing phosphoenolpyruvate. The phosphoryl group donor is **guanosine triphosphate, GTP.**

In the Cori Cycle, lactate produced by working muscle is transported to the liver by the bloodstream. In the liver, the lactate is converted to pyruvate, which may be used to produce glucose by gluconeogenesis. The glucose produced in the liver may be degraded for energy or stored as glycogen.

18.11 Glycogen Synthesis and Degradation

Learning Goal 12: Summarize the regulation of blood glucose levels by glycogenesis and glycogenolysis.

The liver helps regulate the blood glucose level. One of the ways this is accomplished is through the uptake and storage of excess glucose as glycogen (**glycogenesis**). Alternatively, liver cells may degrade glycogen (**glycogenolysis**) and release glucose into the blood.

The Structure of Glycogen

Glycogen is a highly branched glucose polymer. The primary chain is linked by $\alpha(1\rightarrow4)$ glycosidic bonds. The branches are linked to the primary chain by $\alpha(1\rightarrow6)$ glycosidic bonds. Glycogen is stored in the cytoplasm of liver and muscle cells as **glycogen granules.** These granules are complexes of glycogen and the enzymes that carry out glycogen synthesis and degradation.

Glycogenolysis: Glycogen Degradation

Glycogenolysis is controlled by two hormones, **glucagon** and **epinephrine.** Glucagon is released from the pancreas in response to low blood glucose, and epinephrine is released from the adrenal glands in response to a threat or a stress. Both situations require an increase in blood glucose concentrations. This is accomplished by the fact that both glucagons and epinephrine regulate the activity of two enzymes in glycogen metabolism: *glycogen phosphorylase* and *glycogen synthase*. Glycogen phosphorylase, involved in glycogen degradation, is activated; glycogen synthetase, involved in glycogen synthesis, is inactivated. The result is an increase in blood glucose levels.

Example 14
Explain why glycogenolysis is important.

Answer
Glucose is the sole source of energy for mammalian red blood cells and the major source of energy for the brain. Neither of these can store glucose; thus, a constant supply must be available as blood glucose. This is provided by dietary glucose and by the production of glucose either by gluconeogenesis or by glycogenolysis, the degradation of glycogen molecules. Glycogen is a long, branched-chain polymer of glucose that is stored in the liver and skeletal muscles. Breakdown of glycogen in the liver mobilizes the glucose when hormonal signals register a need for increased levels of blood glucose.

Glycogenesis: Glycogen Synthesis

The hormone **insulin,** produced by the pancreas in response to high blood glucose levels, stimulates glycogenesis. It accelerates the uptake of glucose by most cells of the body. In the liver, insulin also promotes storage of the glucose by synthesis of glycogen. This is accomplished in two ways. First, insulin inhibits glycogen phosphorylase and thus inhibits glycogenolysis. Second, insulin stimulates glycogen synthesis directly by stimulating the enzymes glycogen synthase and glucokinase, which are involved in glycogen synthesis.

Glycogenesis and glycogenolysis are regulated by hormonal controls. When blood glucose levels are too high (**hyperglycemia**), insulin stimulates glucose uptake and storage. The net effect is the removal of glucose from the blood and its conversion into glycogen in the liver. Glucagon is produced in response to low blood glucose levels (**hypoglycemia**). By stimulating breakdown of glycogen, it raises blood glucose levels.

Glossary of Key Terms in Chapter Eighteen

acetyl CoA (18.6) a molecule composed of coenzyme A and an acetyl group; this intermediate provides acetyl groups for complete oxidation by aerobic respiration.
adenosine triphosphate (ATP) (18.1) a nucleotide composed of the purine adenine, the sugar ribose, and three phosphoryl groups; the primary energy storage and transport molecule used by the cells in cellular metabolism.
aerobic respiration (18.7) the oxygen requiring degradation of food molecules and production of ATP.
anabolism (18.1) all of the cellular energy-requiring pathways.
anaerobic threshold (18.4) the point at which the lactate concentration in the exercising muscle inhibits glycolysis and the muscle, deprived of energy, ceases to function.

ATP synthase (18.9) a multiprotein complex within the inner mitochondrial membrane that uses the energy of the proton (H^+) gradient to produce ATP.

catabolism (18.1) the degradation of fuel molecules and harvest of energy for cellular functions.

citric acid cycle (18.8) a cyclic biochemical pathway that is the final stage of degradation of carbohydrates, fats, and amino acids; it results from the complete oxidation of acetyl groups derived from these dietary fuels.

coenzyme A (18.6) a molecule derived from ATP and the vitamin pantothenic acid. Coenzyme A functions in the transfer of acetyl groups.

Cori Cycle (18.10) a metabolic pathway in which the lactate produced by working muscle is converted back to glucose by gluconeogenesis in the liver.

cristae (18.5) the highly folded inner membrane of the mitochondria.

electron transport system (18.9) the series of electron transport proteins embedded in the inner mitochondrial membrane that accept high-energy electrons from NADH and $FADH_2$ and transfer them in stepwise fashion to molecular oxygen (O_2).

fermentation (18.4) anaerobic catabolic reactions that occur with no net oxidation; pyruvate or an organic compound produced from pyruvate is reduced as NADH is oxidized.

F_0F_1 complex (18.9) an alternate term for the ATP synthase, the multiprotein complex in the inner mitochondrial membrane that uses the energy of the proton (H^+) gradient to produce ATP.

glucagon (18.11) a peptide hormone released by the pancreas in response to low blood glucose; it promotes glycogenolysis and thereby increases the concentration of blood glucose.

gluconeogenesis (18.10) the synthesis of glucose from noncarbohydrate precursors.

glycogen (18.11) a long, branched polymer of glucose stored in the liver and in muscles.

glycogen granule (18.11) a core of glycogen surrounded by enzymes responsible for glycogen synthesis and degradation.

glycogenesis (18.11) the metabolic pathway that results in the addition of glucose to growing glycogen polymers when blood glucose levels are high.

glycogenolysis (18.11) the biochemical pathway that results in the removal of glucose molecules from glycogen polymers when blood glucose levels are low.

glycolysis (18.3) the enzymatic pathway that converts a glucose molecule into two molecules of pyruvate; this anaerobic process harvests energy in the form of 2 ATP and 2 NADH.

hyperglycemia (18.11) blood glucose levels that are higher than normal.

hypoglycemia (18.11) blood glucose levels that are lower than normal.

inner mitochondrial membrane (18.5) the highly folded, impermeable membrane within the mitochondrion, which is the location of the electron transport system and ATP synthase.

insulin (18.11) a hormone released from the pancreas in response to high blood glucose levels; it promotes glycogenesis and fat storage.

intermembrane space (18.5) the region between the outer and inner mitochondrial membranes, which is the location of the proton reservoir that drives ATP synthesis.

matrix space (18.5) the region of the mitochondrion within the inner membrane; it is the location of the enzymes that carry out the reactions of the citric acid cycle.

mitochondria (18.5) the cellular "power plants" in which the reactions of the citric acid cycle, the electron transport system, and ATP synthase function to produce ATP.

nicotinamide adenine dinucleotide, NAD$^+$ (18.3) a molecule synthesized from the vitamin niacin and the nucleotide ATP, which serves as a carrier of hydride anions.

outer mitochondrial membrane (18.5) the membrane that surrounds the mitochondrion and separates it from the contents of the cytoplasm; it is highly permeable to small "food" molecules.

oxidative phosphorylation (18.9) a series of reactions that couples the oxidation of NADH and $FADH_2$ to the phosphorylation of ADP to produce ATP.

pyruvate dehydrogenase complex (18.6) a complex of all the enzymes and coenzymes required for the synthesis of CO_2 and acetyl CoA from pyruvate.

substrate-level phosphorylation (18.3) the production of ATP by the transfer of a phosphoryl group from the substrate of a reaction to ADP.

terminal electron acceptor (18.9) the final electron acceptor in an electron transport system that removes the low-energy electrons from the system; in aerobic organisms, the terminal electron acceptor is molecular oxygen.

Self Test for Chapter Eighteen

1. Which pathway is thought to be the first successful energy-generating pathway available to early organisms?

2. What molecule in a cell couples exergonic and endergonic reactions?

3. How many kcal/mole of energy are released by the hydrolysis of ATP?

4. Describe glycolysis in terms of its oxygen requirements.

5. What is the net yield of ATP produced by anaerobic glycolysis?

6. Which coenzyme must be reoxidized so that glycolysis can continue?

7. What is the final fermentation end product in yeast cells?

8. What is the final fermentation end product produced in muscle cells under anaerobic conditions?

9. Of the three classes of food molecules, which one is the most readily used by the human body?

10. What subunits are released by digestion of proteins?

11. Of the several high-energy compounds produced in cells, which one is the principal energy storage compound?

12. What reaction occurs in the first step of glycolysis?

13. Phosphofructokinase converts what substrate into fructose-1,6-bisphosphate?

14. Where in the cell does glycolysis occur?

15. During glycolysis, 1,3-bisphosphoglycerate is converted into 3-phosphoglycerate. Simultaneously, an ATP is produced. What is the general term for this kind of reaction?

16. Provide the missing product in the following reaction:

    ATP + water + glucose $\rightarrow$ ADP + 4 kcal/mole + ?

17. Hexokinase is used to convert what substrate into glucose-6-phosphate?

    a. glucose

    b. fructose

    c. glycogen

    d. pyruvate

    e. lactate

18. The enzyme triose phosphate isomerase converts dihydroxyacetone phosphate into what compound?

    a. pyruvate

    b. fructose-1,6-bisphosphate

    c. glucose-1,6-bisphosphate

    d. lactate

    e. glyceraldehyde-3-phosphate

19. Following strenuous exercise, the liver takes up excess lactate from the blood. Into what compound is this lactate converted?

    a. oxaloacetate

    b. glucose

    c. pyruvate

    d. acetyl CoA

    e. glycogen

20. True or false. Gluconeogenesis means the production of glycogen from a noncarbohydrate source.

21. True or false. The liver is the only tissue that contains glucose-6-phosphatase.

22. The oxidative reactions of metabolism provide most cellular energy. In what specific part of the cell do these reactions occur?

23. The number of mitochondria in a eukaryotic cell varies widely. What need is reflected by this variation?

24. To what compound is the acetyl group transferred in the first step of the citric acid cycle?

25. How many molecules of CO_2 are produced by the complete oxidation of an acetyl group by the citric acid cycle?

26. How many molecules of NADH are formed by the complete oxidation of an acetyl group by the citric acid cycle?

27. How many molecules of $FADH_2$ are produced by the complete oxidation of an acetyl group by the citric acid cycle?

28. In glycolysis, how many NADH are generated per glucose molecule?

29. In what organelle do the reactions of the Krebs cycle, fatty acid oxidation, and oxidative phosphorylation occur?

30. What is the term for the series of electron and proton carriers embedded within the mitochondrial inner membrane?

31. Which of the electron carriers in the electron transport system contain a heme unit?

32. When NADH dehydrogenase oxidizes NADH in the mitochondrial matrix, it can pass the electrons to the next carrier. What ion cannot be transferred?

33. Spanning the inner mitochondrial membrane is a protein complex (F_0) that serves as a channel protein. What specific particles pass through this channel?

34. Protruding into the mitochondrial matrix is a spherical protein complex (F_1). What is the enzymatic activity of F_1?

35. NADH carries electrons, originally from glucose, to the first carrier of the electron transport system. What compound is formed by the oxidation of NADH at this site?

36. After NADH is oxidized to NAD^+ in the first step of the electron transport system, what happens to the NAD^+?

37. In the last step of the electron transport system, the electrons have too little energy to accomplish any more work. In aerobic organisms, what is the terminal electron acceptor?

38. What is an alternative name for the F_0F_1 complex?

39. How many ATP are produced for each NADH generated in the mitochondria?

40. How many ATP are produced for each $FADH_2$ generated in the mitochondria?

Vocabulary Quiz for Chapter Eighteen

1. _____ is a nucleotide composed of the purine adenine, the sugar ribose, and a triphosphate group. It serves as the major energy storage form of the cell.

2. The enzymatic pathway that converts a glucose molecule into two molecules of pyruvate is called _____. This anaerobic process generates energy in the form of 2 ATP and 2 NADH.

3. _____ is the synthesis of glucose from noncarbohydrate precursors.

4. _____ is the production of ATP by the transfer of a phosphoryl group from the substrate of a reaction to ADP.

5. The degradation of fuel molecules and production of energy for cellular functions is called _____.

6. _____ is a long, branched polymer of glucose stored in the liver and in muscles.

7. _____ is the series of reactions that couples the oxidation of NADH and $FADH_2$ to the phosphorylation of ADP to produce ATP.

8. The _____ is the series of electron transport proteins embedded in the inner mitochondrial membrane that accept high energy electrons from NADH and $FADH_2$.

9. The cellular "power plants" in which the reactions of the citric acid cycle, the electron transport system, and ATP synthase function to produce ATP are called _____.

10. _____ is the multiprotein complex within the inner mitochondrial membrane that uses the energy of the proton gradient to produce ATP.

19 *Fatty Acid and Amino Acid Metabolism*

Learning Goals

1 Summarize the digestion and storage of lipids.
2 Describe the degradation of fatty acids by β-oxidation.
3 Explain the role of acetyl CoA in fatty acid metabolism.
4 Understand the role of ketone body production in β-oxidation.
5 Compare β-oxidation of fatty acids and fatty acid biosynthesis.
6 Describe the conversion of amino acids to molecules that can enter the citric acid cycle.
7 Explain the importance of the urea cycle and describe its essential steps.
8 Discuss the cause and effect of hyperammonemia.
9 Summarize the antagonistic effects of glucagon and insulin.

Introduction

Acetyl CoA is the key molecule of lipid metabolism. Fatty acids are degraded to acetyl CoA, which is oxidized by the citric acid cycle. On the other hand, acetyl CoA is also the precursor for the biosynthesis of fatty acids and cholesterol.

19.1 Lipid Metabolism in Animals

Learning Goal 1:   Summarize the digestion and storage of lipids.

Digestion and Absorption of Dietary Triglycerides

Fats are hydrophobic molecules and must be extensively processed before they can be digested and absorbed. The enzymes that hydrolyze lipids are called **lipases.** The most effective lipid digestion occurs in the small intestine, where **bile** causes formation of small lipid micelles. **Micelles** are aggregations of molecules having a polar region facing the aqueous exterior and an internal nonpolar region that dissolves lipid. Bile is produced in the liver and stored in the gallbladder. The presence of lipids in the small intestine stimulates secretion of bile into the duodenum.

Triglycerides are the major lipids in the micelles. A protein called **colipase** binds to the surface of the micelles and facilitates hydrolysis of the triglycerides by pancreatic lipases. The fatty acids and monoglycerides produced by hydrolysis are absorbed by cells of the intestinal epithelium.

Example 1

Explain the digestion and absorption of dietary fats.

Answer

Most dietary fats arrive in the duodenum in the form of fat globules. The globules stimulate the secretion of bile from the gallbladder. The bile salts emulsify the fat globules into tiny droplets. The protein colipase binds to the surface of the lipid droplets and helps pancreatic lipases adhere to the surface and hydrolyze the triglycerides into monoglycerides plus free fatty acids. These products are absorbed through the membranes of the intestinal epithelial cells.

In intestinal cells the monoglycerides and fatty acids are reassembled into triglycerides and combined with protein, producing **chylomicrons.** These chylomicrons eventually enter the blood, and the lipids are once again hydrolyzed to products that can be absorbed by cells of the body.

Example 2

Explain what happens to lipid digestion products after absorption.

Answer

The monoglycerides and fatty acids are reassembled into triglycerides, which are combined with protein to produce the class of plasma lipoproteins called chylomicrons. These are secreted into lymphatic vessels and eventually arrive in the bloodstream. Here triglycerides are again hydrolyzed, and the products are absorbed by the cells of the body.

Lipid Storage

Fatty acids are stored as triglycerides within **adipocytes** (fat cells) that compose **adipose tissue.** When the body demands energy, these triglycerides can be hydrolyzed, and the fatty acids are oxidized to generate ATP.

Example 3

Explain lipid storage in the human body.

Answer

Fatty acids are stored as triglycerides within fat droplets in the cytoplasm of adipocytes. These cells contain a large fat droplet, which accounts for nearly the entire volume of the cell.

19.2 Fatty Acid Degradation

An Overview of Fatty Acid Degradation

Fatty acids are degraded by a pathway called β-**oxidation,** which consists of a set of five reactions. Each trip through the β-oxidation pathway releases acetyl CoA and returns a fatty acyl CoA molecule that contains two fewer carbons. One $FADH_2$ (2 ATP) and one NADH (3 ATP) are produced for each cycle of β-oxidation. Acetyl CoA can enter the citric acid cycle, resulting in production of twelve more ATP. The remaining fatty acyl CoA cycles through steps 2–5 of β-oxidation until the fatty acid is completely converted to acetyl CoA.

Example 4
Give an overview of fatty acid degradation.

Answer
The β-oxidation pathway for fatty acid degradation occurs in the mitochondrial matrix. The fatty acids are converted into acetyl coenzyme A units by the enzymes of this pathway.

The Reactions of β-Oxidation

Learning Goal 2:   Describe the degradation of fatty acids by β-oxidation.

Learning Goal 3:   Explain the role of acetyl CoA in fatty acid metabolism.

Special transport mechanisms bring fatty acid molecules into the mitochondrial matrix, where β-oxidation occurs. The reactions of β-oxidation are summarized below.

1. *Activation:* A fatty acyl CoA molecule is formed between coenzyme A and the fatty acid. This reaction requires ATP, which is cleaved to AMP and pyrophosphate. The bond between the fatty acid and coenzyme A is a high-energy thioester bond.
2. *Oxidation:* A pair of hydrogen atoms is removed from the fatty acid, and FAD is reduced to $FADH_2$.
3. *Hydration:* The double bond produced in Step 2 undergoes a hydration reaction, and the β-carbon is hydroxylated.
4. *Oxidation:* The hydroxyl group of the β-carbon is now dehydrogenated, and NAD^+ is reduced to form NADH.
5. *Thiolysis:* A molecule of coenzyme A attacks the β-carbon, releasing acetyl CoA and a fatty acyl CoA that is two carbons shorter than the original fatty acid.

The fatty acyl CoA is further oxidized by cycling through Steps 2 through 5 until the fatty acid carbon chain is completely degraded to acetyl CoA. The acetyl CoA is completely oxidized in the citric acid cycle. The balance sheet for ATP production in the complete oxidation of a C16-fatty acid, palmitic acid, is found in Figure 19.6 of the text.

Example 5
Compare ATP production by β-oxidation of a fatty acid like palmitic acid to that produced by oxidation of an equivalent amount of glucose.

Example 5 continued

Answer
The complete oxidation of the C16-fatty acid palmitic acid produces 129 ATP molecules. This is 3.5 times more energy than results from the complete oxidation of the same amount of glucose.

19.3 Ketone Bodies

Learning Goal 4: Understand the role of ketone body production in β-oxidation.

If there is not a sufficient supply of citric acid cycle intermediates to allow the complete oxidation of the acetyl CoA produced in β-oxidation, acetyl CoA is converted to **ketone bodies,** such as β-hydroxybutyrate, acetoacetate, and acetone.

Example 6
Explain why ketone bodies are produced in the human body.

Answer
The β-oxidation of fatty acids produces a steady supply of acetyl CoA molecules. If glycolysis and β-oxidation are functioning at the same rate, there will be a steady supply of pyruvate, which can be converted into oxaloacetate. However, if the supply of oxaloacetate is too low to allow all of the acetyl CoA to enter the citric acid cycle, the acetyl CoA molecules are converted into ketone bodies. The ketone bodies are β-hydroxybutyrate, acetoacetate, and acetone.

Ketosis

Ketosis is the abnormal elevation of blood ketone body concentration. This may occur due to starvation, a diet that is extremely low in carbohydrates, or uncontrolled **diabetes mellitus.** In diabetes, this can lead to **ketoacidosis** because the ketone acids are relatively strong acids and dissociate to release H^+. This causes the pH of the blood to become acidic.

Ketogenesis

The first step in the pathway for ketone body synthesis is the fusion of two molecules of acetyl CoA to produce acetoacetyl CoA. This reaction is catalyzed by the enzyme thiolase. Acetoacetyl CoA reacts with a third acetyl CoA to yield β-hydroxy-β-methylglutaryl CoA (HMG-CoA). HMG-CoA is cleaved to yield acetoacetate and acetyl CoA. While some acetoacetate spontaneously loses carbon dioxide, producing acetone, most of it undergoes NADH-dependent reduction to produce β-hydroxybutyrate.

Acetoacetate and β-hydroxybutyrate, produced in the liver, are circulated to other tissues through the blood. There they may be reconverted to acetyl CoA and used to produce ATP.

Example 7
Explain ketogenesis.

Answer
The pathway for the production of ketone bodies begins with the reversal of the last step of the β-oxidation cycle. When oxaloacetate levels are low, the enzyme that mediates the last reaction of β-oxidation, thiolase, now mediates the fusion of two acetyl CoA molecules to produce acetoacetyl CoA. The acetoacetyl CoA reacts with another molecule of acetyl CoA to produce HMG-CoA. The HMG-CoA is cleaved to yield *acetoacetate* and acetyl CoA. Some of the acetoacetate spontaneously loses carbon dioxide to produce *acetone*. However, most of the acetoacetate is reduced to form *β-hydroxybutyrate*.

19.4 Fatty Acid Synthesis

Learning Goal 5:   Compare β-oxidation of fatty acids and fatty acid biosynthesis.

All organisms possess the ability to synthesize fatty acids. Fatty acid synthesis appears to be simply the reverse of β-oxidation. In fact, the fatty acid chain is constructed by the sequential addition of two carbon acetyl groups, as seen in Figure 19.9 of the text. However, there are several differences between the two processes, including intracellular location, acyl group carriers, enzymes involved, and the electron carriers used. Fatty acid biosynthesis occurs in the cytoplasm. The acyl group carrier is **acyl carrier protein (ACP).** The portion of ACP that binds the fatty acid is the **phosphopantetheine** group also found in coenzyme A. A multienzyme complex, *fatty acid synthase*, carries out fatty acid biosynthesis; NADPH is the reducing agent.

Example 8
Compare fatty acid synthesis with fatty acid degradation.

Answer
On first examination of fatty acid synthesis, it appears that it is simply the reverse of β-oxidation. Although fatty acid synthesis and breakdown are similar, there are several major differences:

1. The enzymes responsible for fatty acid biosynthesis are located in the cytoplasm of the cells of vertebrates, whereas those responsible for the degradation of fatty acids are found in the mitochondria.
2. The activated intermediates of fatty acid synthesis are bound to a thiol group of acyl carrier protein. Thus, the thioester intermediates of fatty acid synthesis are not derivatives of coenzyme A.
3. The seven steps of fatty acid biosynthesis are carried out by the multienzyme complex known as fatty acid synthase. The enzymes responsible for fatty acid degradation are not associated with one another as a complex.

Example 8 continued

4. NADH and $FADH_2$ are produced by fatty acid oxidation, whereas NADPH is the reducing agent for fatty acid biosynthesis.

19.5 Degradation of Amino Acids

Learning Goal 6: **Describe the conversion of amino acids to molecules that can enter the citric acid cycle.**

Amino acids obtained from the degradation of dietary protein may also be oxidized as energy sources. Amino acid degradation occurs mainly in the liver and takes place in two stages: (1) the removal of the α-amino group and (2) the degradation of the remaining carbon skeleton. In land mammals, the amino group is excreted in the urine as urea. The carbon skeletons of amino acids can be converted into a variety of compounds.

Example 9
When is body protein used for energy?

Answer
Carbohydrates are not the only source of energy. Dietary protein is digested to amino acids, which also may be used as an energy source. Most of the amino acids used for energy come from dietary protein. Only under starvation conditions, when stored glycogen and lipids have been greatly depleted, does the body begin to use its own muscle protein for energy.

Removal of α-Amino Groups: Transamination

A **transaminase** catalyzes the transfer of the α-amino group from an α-amino acid to an α-keto acid. This general reaction is summarized below.

The α-amino groups of many amino acids are transferred to α-ketoglutarate to produce the amino acid glutamate. This glutamate family of transaminases is especially important because the α-keto acid corresponding to glutamate is the citric acid cycle intermediate, α-ketoglutarate. The *glutamate transaminases* thus provide a direct link between amino acid degradation and the citric acid cycle. *Alanine transaminase* and *aspartate transaminase* are also important members of this family.

All transaminases require the prosthetic group pyridoxal phosphate, a coenzyme derived from vitamin B_6 (pyridoxine).

Example 10
Explain how the *transamination* of amino acids provides a direct linkage between amino acid degradation and the citric acid cycle.

Answer
Transaminases catalyze the transfer of the α-amino group from an α-amino acid to an α-keto acid. The α-amino groups of many amino acids are transferred to α-ketoglutarate to produce the amino acid glutamate. The glutamate family of transaminases is especially important because the α-keto acid corresponding to glutamate is the citric acid cycle intermediate, α-ketoglutarate. Thus, the glutamate transaminases, as well as others, provide a direct link to the citric acid cycle.

Removal of α-Amino Groups: Oxidative Deamination

Ammonium ion is now liberated from the glutamate by oxidative deamination, catalyzed by *glutamate dehydrogenase*:

$$
\begin{array}{l}
\text{COO}^- \\
| \\
\text{CH}\!-\!\text{NH}_3^+ \\
| \\
\text{CH}_2 \quad + \quad \text{NAD}^+ \quad + \quad \text{H}_2\text{O} \quad \rightleftharpoons \quad \text{NH}_4^+ \quad + \quad
\begin{array}{l}
\text{COO}^- \\
| \\
\text{C}\!=\!\text{O} \\
| \\
\text{CH}_2 \\
| \\
\text{CH}_2 \\
| \\
\text{COO}^-
\end{array}
\quad + \quad \text{NADH} \\
| \\
\text{CH}_2 \\
| \\
\text{COO}^-
\end{array}
$$

Glutamate   α-Ketoglutarate

The Fate of Amino Acid Carbon Skeletons

The carbon skeletons produced by these and other deamination reactions enter the energy-harvesting pathways at many steps, as seen in Figure 19.12 in the text.

Oxidative deamination of glutamate and deamination of other amino acids produce considerable quantities of ammonium ion. This must be incorporated into a biological molecule and removed from the body so that it does not reach toxic levels.

19.6 The Urea Cycle

Learning Goal 7:   Explain the importance of the urea cycle and describe its essential steps.

It is critically important to the survival of the organism to be able to excrete ammonium ions, regardless of the energy required. In humans and most terrestrial vertebrates, the means of ammonium ion removal is the urea cycle, which occurs in the liver. This cycle keeps excess ammonium ion out of the blood and allows the excretion of the excess in the form of urea.

Reactions of the Urea Cycle

The reactions of the urea cycle, summarized below, are shown in Figure 19.15 of the text.

Step 1. CO_2 and NH_4^+ react to form carbamoyl phosphate. This reaction requires ATP and H_2O.

Step 2. Carbamoyl phosphate condenses with the amino acid ornithine to produce the amino acid citrulline.

Step 3. Citrulline now condenses with aspartate to produce argininosuccinate. This reaction requires energy released by the hydrolysis of ATP.

Step 4. Argininosuccinate is cleaved to produce the amino acid arginine and the citric acid cycle intermediate fumarate.

Step 5. Finally, arginine is hydrolyzed, producing urea, which will be excreted, and ornithine, the original reactant in the cycle. Note that one of the amino groups of urea is the ammonium ion, and the second is derived from the amino acid aspartate.

Learning Goal 8:	Discuss the cause and effect of hyperammonemia.

A deficiency of urea cycle enzymes causes an elevation of the concentration of NH_4^+, a condition known as **hyperammonemia.** If there is a complete deficiency of one of the enzymes of the urea cycle, the result is death in early infancy. A partial deficiency results in less severe symptoms and can be treated with a low-protein diet.

Example 11

Give a summary of the reactions of the urea cycle.

Answer

1. *Step 1:* The first step involves the reaction of the waste product CO_2 with ammonium ions from the amino acids to form carbamoyl phosphate.

$$CO_2 + NH_4^+ + 2ATP + H_2O \longrightarrow H_2N-\overset{\overset{O}{\|}}{C}-O-\overset{\overset{O}{\|}}{\underset{\underset{O^-}{|}}{P}}-O^- + 2ADP + P_i + 3H^+$$

Carbamoyl phosphate

2. *Step 2:* The carbamoyl phosphate then condenses with the amino acid ornithine to produce citrulline (another amino acid).

288

Example 11 continued

3. *Step 3:* Citrulline condenses with aspartate to produce argininosuccinate.

4. *Step 4:* The argininosuccinate is cleaved to produce arginine and fumarate.

5. *Step 5:* Finally, arginine is hydrolyzed to generate urea and ornithine. The ornithine can then return to be used again in Step 2.

Example 12
Describe the genetic disorder that results from the deficiency of an enzyme in the urea cycle.

Answer
A deficiency of a urea cycle enzyme causes an elevation of the concentration of ammonium ion, a condition known as *hyperammonemia*. If there is a complete deficiency of one of these enzymes, the result is death in early infancy. If, however, there is a partial deficiency of one of the cycle's enzymes, the result can be retardation, convulsions, and vomiting. If the disorder is caught early, a diet low in protein may lead to less severe clinical symptoms.

19.7 The Effects of Insulin and Glucagon on Cellular Metabolism

Learning Goal 9:   Summarize the antagonistic effects of glucagon and insulin.

Insulin is a polypeptide hormone secreted by the β-cells of the islets of Langerhans in the pancreas in response to an increase in the blood glucose level. Insulin lowers the concentration of blood glucose by stimulating its storage as glycogen or triglycerides. In general, insulin activates biosynthetic processes and inhibits catabolism.

Insulin acts only on those *target cells* that possess a specific insulin receptor protein in their membranes. The major target cells for insulin are liver, adipose, and muscle cells. Insulin lowers blood glucose levels by altering cellular carbohydrate, protein, and lipid metabolism.

A second hormone, **glucagon,** is secreted by the α-cells of the islets of Langerhans in response to decreased blood glucose levels. The effects of glucagon are generally the opposite of the effects of insulin.

The effects of insulin and glucagon on a variety of metabolic reactions are summarized in the following table:

Actions	Insulin	Glucagon
Cellular glucose transport	Increased	No effect
Glycogen synthesis	Increased	Decreased
Glycogenolysis in liver	Decreased	Increased
Gluconeogenesis	Decreased	Increased
Amino acid uptake and protein synthesis	Increased	No effect
Inhibition of amino acid release and protein degradation	Decreased	No effect
Lipogenesis	Increased	No effect
Lipolysis	Decreased	Increased
Ketogenesis	Decreased	Increased

Glossary of Key Terms in Chapter Nineteen

acyl carrier protein (ACP) (19.4) the protein that forms a thioester linkage with fatty acids during fatty acid synthesis.

adipocyte (19.1) a fat cell.

adipose tissue (19.1) fatty tissue that stores most of the body lipids.

bile (19.1) micelles of lecithin, cholesterol, bile salts, protein, inorganic ions, and bile pigments that aid in lipid digestion by emulsifying fat droplets.

chylomicron (19.1) an aggregate of protein and triglycerides that carries triglycerides from the intestine to all body tissues via the blood stream.

colipase (19.1) a protein that aids in lipid digestion by binding to the surface of lipid droplets and facilitating binding of pancreatic lipase.

diabetes mellitus (19.3) the appearance of glucose in the urine caused by high blood glucose levels; the disease frequently results from the inability to produce insulin.

glucagon (19.7) a peptide hormone synthesized by the α-cells of the islets of Langerhans in the pancreas in response to low blood glucose levels.

insulin (19.7) a hormone released from the pancreas in response to high blood glucose levels; insulin stimulates glycogenesis, fat storage, and cellular uptake and storage of glucose from the blood.

hyperammonemia (19.6) a genetic disorder in one of the enzymes of the urea cycle that results in toxic or even fatal elevations of the concentration of ammonium ions in the body.

ketoacidosis (19.3) a drop in the pH of the blood caused by elevated ketone levels.

ketone bodies (19.3) acetone, acetoacetone, and β-hydroxybutyrate produced from fatty acids in the liver.

ketosis (19.3) an abnormal rise in the level of ketone bodies in the blood.

lipase (19.1) an enzyme that hydrolyzes the ester linkage between glycerol and the fatty acids of triglycerides.

micelle (19.1) an aggregation of molecules having a nonpolar and a polar region; the nonpolar regions of the molecules aggregate leaving the polar regions facing the surrounding water.

β-oxidation (19.2) the biochemical pathway that results in the oxidation of fatty acids and the production of acetyl CoA.

oxidative deamination (19.5) an oxidation-reduction reaction in which NAD^+ is reduced and the amino acid is deaminated.

phosphopantetheine (19.4) the portion of coenzyme A and the acyl carrier protein that is derived from the vitamin pantothenic acid; it forms a thioester linkage with fatty acids.

pyridoxal phosphate (19.5) a coenzyme derived from vitamin B_6 that is required for all transamination reactions.

transaminase (19.5) an enzyme that catalyzes the transfer of an amino group from one molecule to another.

transamination (19.5) a reaction in which an amino group is transferred from one molecule to another.

triglyceride (19.1) (also called triacylglycerol) a molecule composed of glycerol esterified to three fatty acids.

urea cycle (19.6) a cyclic series of reactions that detoxifies ammonium ions by incorporating them into urea, which is excreted from the body.

Self Test for Chapter Nineteen

1. The metabolism of fatty acids and lipids revolves around which specific compound?

2. Which compound is the starting material for the synthesis of half of the amino acids?

3. In what form do we find dietary fat when it arrives in the duodenum?

4. What mixture is composed of micelles of lecithin, cholesterol, protein, bile salts, inorganic ions, and bile pigments?

5. What is the term for an aggregate of molecules having polar and nonpolar regions?

6. Name the two major bile salts. (Refer to the textbook.)

7. Which protein molecule binds to the surface of lipid droplets and helps pancreatic lipases to adhere to the surface and hydrolyze the ester linkages?

8. Monoglycerides and fatty acids are reassembled into triglycerides and combined with protein in the cells of the small intestine. What class of lipoproteins is formed in this process?

9. What tissue stores most of the body's triglyceride molecules?

10. How many NADH and $FADH_2$ are produced in each round of β-oxidation?

11. What is another term for a fat cell?

12. The complete oxidation of a molecule of palmitic acid, using the β-oxidation cycle and the other oxidation pathways, produces how many molecules of ATP?

13. What is ketosis?

14. List three conditions that can produce ketosis.

15. In patients with uncontrolled diabetes, the very high concentration of ketone acids in the blood leads to what condition?

16. Where in the body are acetoacetate and β-hydroxybutyrate produced?

17. Which muscle derives more of its energy from the metabolic use of ketone bodies than it does from glucose?

18. What is the name of the multienzyme complex used for biosynthesis of fatty acids?

19. Where in the cell does fatty acid synthesis occur?

20. What lipoprotein complex transports triglycerides produced in the liver through the bloodstream?

21. What chemical group is shared in common between Coenzyme A and acyl carrier protein?

22. Where in the cell does the urea cycle begin?

23. What products are formed by the β-oxidation of butanoic acid?

24. What compound is produced in the first step of β-oxidation of fatty acids?

25. What compound is released in the final step of β-oxidation of fatty acids?

26. What are the major target cells for insulin?

27. The α-amino groups of a great many amino acids are transferred to α-ketoglutarate. What amino acid is produced in this reaction?

28. The glutamate transaminases provide a direct link between amino acid degradation and which other process?

29. One of the most important transaminases is aspartate transaminase, which catalyzes the transfer of the α-amino group of aspartate to α-ketoglutarate. What are the two new compounds produced in this reaction?

30. Where in the human body does the conversion of ammonium ions to urea occur?

Vocabulary Quiz for Chapter Nineteen

1. _____ is fatty tissue that stores most of the body lipids.

2. A(n) _____ is a molecule composed of glycerol esterified to three fatty acids.

3. _____ consists of micelles of lecithin, cholesterol, bile salts, protein, inorganic ions, and bile pigments that aid in lipid digestion by emulsifying fat droplets.

4. A(n) _____ is an aggregate of protein and triglycerides that carries triglycerides from the intestine to all body tissues via the blood stream.

5. The biochemical pathway that results in the oxidation of fatty acids and the production of acetyl CoA is called _____.

6. The molecule that serves as a carrier of acetyl groups in lipid and carbohydrate metabolism is _____.

7. _____ is a peptide hormone synthesized by the α-cells of the islets of Langerhans in the pancreas in response to low blood glucose levels.

8. The peptide hormone synthesized by the β-cells of the islets of Langerhans in the pancreas in response to high blood glucose levels is _____.

9. The _____ is a cyclic series of reactions that detoxifies ammonium ions by incorporating them into urea, which is excreted from the body.

10. Acetone, acetoacetone, and β-hydroxybutyrate produced from fatty acids in the liver via acetyl CoA are referred to as _____.

APPENDIX A

Chapter 1
Chemistry: Methods and Measurements
Solutions to the Odd-Numbered Problems

In-Chapter Questions and Problems

1.1   The scientific method is an organized way of doing science. It uses carefully planned experimentation to study our surroundings.

1.3   a. physical property
      b. chemical property
      c. physical property
      d. physical property
      e. physical property

1.5   Intensive property

1.7   a. pure substance
      b. heterogeneous mixture
      c. homogeneous mixture
      d. pure substance

1.9   Mass of aluminum, 5.40 g, and volume of aluminum, 2.00 cm^3, are data. The density, calculated from this data, constitutes the results (2.70 g/cm^3).

1.11 $155 \text{ lb} \times \dfrac{1 \text{ ton}}{2000 \text{ lb}} = 0.0775 \text{ ton}$

1.13   a. $1.0 \text{ L} \times \dfrac{10^3 \text{ mL}}{\text{L}} = 1.0 \times 10^3 \text{ mL}$

   b. $1.0 \text{ L} \times \dfrac{10^6 \text{ } \mu\text{L}}{\text{L}} = 1.0 \times 10^6 \text{ } \mu\text{L}$

   c. $1.0 \text{ L} \times \dfrac{1 \text{ kL}}{10^3 \text{ L}} = 1.0 \times 10^{-3} \text{ kL}$

   d. $1.0 \text{ L} \times \dfrac{10^2 \text{ cL}}{\text{L}} = 1.0 \times 10^2 \text{ cL}$

e. $1.0 \text{ L} \times \dfrac{10^{-1} \text{ daL}}{\text{L}} = 1.0 \times 10^{-1} \text{ daL}$

1.15   a. $0.50 \text{ in.} \times \dfrac{2.54 \text{ cm}}{1 \text{ in.}} \times \dfrac{1 \text{ m}}{10^2 \text{ cm}} = 1.3 \times 10^{-2} \text{ m}$

      b. $0.75 \text{ qt} \times \dfrac{0.946 \text{ L}}{1 \text{ qt}} = 0.71 \text{ L}$

      c. $56.8 \text{ g} \times \dfrac{1 \text{ lb}}{454 \text{ g}} \times \dfrac{16 \text{ oz}}{1 \text{ lb}} = 2.00 \text{ oz}$

      d. $1.5 \text{ cm}^2 \times \left(\dfrac{1 \text{ m}}{10^2 \text{ cm}}\right)^2 = 1.5 \text{ cm}^2 \times \dfrac{1 \text{ m}^2}{10^4 \text{ cm}^2} = 1.5 \times 10^{-4} \text{ m}^2$

1.17   a.   Three (all non-zero digits)

      b.   Three (all non-zero digits)

      c.   Four (zeros between non-zero digits are significant)

      d.   Two (trailing zero significant due to decimal)

      e.   Three (leading zeros are not significant)

1.19   a. 2.4×10^{-3}

      b. 1.80×10^{-2}

      c. 2.24×10^2

1.21   a.   8.09 (3 significant figures)

      b.   5.9 (2 significant figures)

      c.   20.19 (4 significant figures)

1.23   a.   51 (2 significant figures)

      b. 8.0×10^1 (2 significant figures)

      c. 1.6×10^2 (2 significant figures)

1.25   a.   61.404 rounds to 61.4

b. 6.1714 rounds to 6.17

c. 0.066494 rounds to 0.0665 (or, 6.65 x 10^{-2})

1.27 a. $°C = \dfrac{°F - 32}{1.8} = \dfrac{32 - 32}{1.8} = 0°C$

b. $K = °C + 273 = 0 + 273 = 273 \text{ K}$

1.29 g alcohol = 30.0 mL alcohol x $\dfrac{0.789 \text{ g alcohol}}{1 \text{ mL alcohol}}$ = 23.7 g alcohol

---

End-of-Chapter Questions and Problems

1.31 a. Chemistry is the study of matter and the changes that matter undergoes.
b. Matter is the material component of the universe.
c. Energy is the ability to do work.

1.33 a. gram (or kilogram)
b. liter
c. meter

1.35 Weight is the force exerted on a body by gravity; mass is a quantity of matter. Mass is an independent quantity while weight is dependent on gravity which may differ from location to location.

1.37 Density is mass per volume. Specific gravity is the ratio of the density of a substance to the density of water at 4°C.

1.39 The statement "stem-cell research has the potential to provide replacement 'parts' for the human body" is, at the time this book was written, a theory. Pay particular attention to the word "potential" in the statement. The meaning, and the scientific status of this statement would be quite different if the words "has the potential" were replaced by the word "can."

1.41 A physical property is a characteristic of a substance that can be observed without the substance undergoing a change in chemical composition.

1.43 Mixtures are composed of two or more substances. A homogeneous mixture has uniform composition while a heterogeneous mixture has non-uniform composition.

1.45 a. chemical reaction
b. physical change
c. physical change

1.47 a. physical property
      b. chemical property

1.49 a. pure substance
      b. pure substance
      c. mixture

1.51 a. homogeneous
      b. homogeneous
      c. homogeneous

1.53 a. extensive property
      b. extensive property
      c. intensive property

1.55 a. $2.0 \text{ lb} \times \dfrac{16 \text{ oz}}{1 \text{ lb}} = 32 \text{ oz}$

      b. $2.0 \text{ lb} \times \dfrac{1 \text{ t}}{2000 \text{ lb}} = 1.0 \times 10^{-3} \text{ t}$

      c. $2.0 \text{ lb} \times \dfrac{454 \text{ g}}{1 \text{ lb}} = 9.1 \times 10^{2} \text{ g}$

      d. $2.0 \text{ lb} \times \dfrac{454 \text{ g}}{1 \text{ lb}} \times \dfrac{10^{3} \text{ mg}}{1 \text{ g}} = 9.1 \times 10^{5} \text{ mg}$

      e. $2.0 \text{ lb} \times \dfrac{454 \text{ g}}{1 \text{ lb}} \times \dfrac{1 \text{ da}}{10^{1} \text{ g}} = 9.1 \times 10^{1} \text{ da}$

1.57 a. $3.0 \text{ g} \times \dfrac{1 \text{ lb}}{454 \text{ g}} = 6.6 \times 10^{-3} \text{ lb}$

      b. $3.0 \text{ g} \times \dfrac{1 \text{ lb}}{454 \text{ g}} \times \dfrac{16 \text{ oz}}{1 \text{ lb}} = 1.1 \times 10^{-1} \text{ oz}$

      c. $3.0 \text{ g} \times \dfrac{1 \text{ kg}}{10^{3} \text{ g}} = 3.0 \times 10^{-3} \text{ kg}$

      d. $3.0 \text{ g} \times \dfrac{10^{2} \text{ cg}}{1 \text{ g}} = 3.0 \times 10^{2} \text{ cg}$

      e. $3.0 \text{ g} \times \dfrac{10^{3} \text{ mg}}{1 \text{ g}} = 3.0 \times 10^{3} \text{ mg}$

1.59   a. $^{\circ}C = \dfrac{^{\circ}F - 32}{1.8} = \dfrac{50.0 - 32}{1.8} = 10.0^{\circ}C$

      b. $K = {^{\circ}C} + 273 = 10.0 + 273.15 = 283.2 \text{ K}$

1.61   a. $K = {^{\circ}C} + 273 = 20.0 + 273.15 = 293.2 \text{ K}$

      b.   If $^{\circ}C = \dfrac{^{\circ}F - 32}{1.8}$, then $1.8^{\circ}C = {^{\circ}F} - 32$

         and $^{\circ}F = 1.8^{\circ}C + 32$

         $^{\circ}F = [(1.8)(20.0)] + 32 = 68.0^{\circ}F$

1.63 $9 \text{ pt x } \dfrac{1 \text{ qt}}{1 \text{ pt}} \text{ x } \dfrac{0.946 \text{ L}}{1 \text{ qt}} = 4.257 \text{ L} \approx 4 \text{ L}$ (1 significant figure)

1.65   If $^{\circ}C = \dfrac{^{\circ}F - 32}{1.8}$, then $1.8^{\circ}C = {^{\circ}F} - 32$

         and $^{\circ}F = 1.8^{\circ}C + 32$

         $^{\circ}F = [(1.8)(38.5)] + 32 = 101^{\circ}F$

1.67   In order to compare the magnitude of the two lengths, we must convert them to a common unit.

      $5.0 \text{ cm x } \dfrac{1 \text{ in.}}{2.54 \text{ cm}} = 2.0 \text{ in.}$

      Therefore, 5 cm is shorter than 5 in.

1.69   a.   3
      b.   3
      c.   3
      d.   4
      e.   4
      f.   3

1.71   a. 3.87×10^{-3}

      b. 5.20×10^{-2}

      c. 2.62×10^{-3}

d. 2.43×10^1

e. 2.40×10^2

f. 2.41×10^0

1.73 a. $(23)(657) = 1.5 \times 10^4$

b. $0.00521 + 0.236 = 2.41 \times 10^{-1}$

c. $\dfrac{18.3}{3.0576} = 5.99$

d. $1157.23 - 17.812 = 1139.42$

e. $\dfrac{(1.987)(298)}{0.0821} = 7.21 \times 10^3$

1.75 a. 1.23×10^1

b. 5.69×10^{-2}

c. -1.527×10^3

d. 7.89×10^{-7}

e. 9.2×10^7

f. 5.280×10^{-3}

g. 1.279×10^0

h. -5.3177×10^2

1.77 $d = \dfrac{m}{V} = \dfrac{3.00 \times 10^2 \text{ g}}{50.0 \text{ mL}} = 6.00 \text{ g/mL}$

1.79 $1.50 \times 10^2 \text{ mL} \times \dfrac{7.20 \text{ g}}{1 \text{ mL}} = 1.08 \times 10^3 \text{ g}$

1.81 Specific gravity of urine $= \dfrac{\text{density of urine}}{\text{density of water at } 4\degree\text{C}}$

Multiplying both sides of the equation by the density of water at 4°C,

(Specific gravity of urine)(density of water at 4°C) = density of urine

(1.008)(1.000 g/mL) = density of urine = 1.008 g/mL

1.83 $d_{lead} = \dfrac{5.0 \times 10^1 \text{ g}}{6.36 \text{ cm}^3} = 7.9 \text{ g/cm}^3$

$d_{uranium} = \dfrac{75 \text{ g}}{3.97 \text{ cm}^3} = 19 \text{ g/cm}^3$

$d_{platinum} = \dfrac{m}{V} = \dfrac{2140 \text{ g}}{1.00 \times 10^2 \text{ cm}^3} = 21.4 \text{ g/cm}^3$

Lead has the lowest density and platinum has the greatest density.

1.85 $d = \dfrac{m}{V}$

Thus $V = \dfrac{m}{d} = \dfrac{10.0 \text{ g}}{0.791 \text{ g/mL}}$

$V = 12.6 \text{ mL}$

Chapter 2
The Structure of the Atom and the Periodic Table
Solutions to the Odd-Numbered Problems

In-Chapter Questions and Problems

2.1   a.   16 protons and 16 electrons (atomic number = 16)

   32 – 16 = 16 neutrons (mass number – atomic number)

   b.   11 protons and 11 electrons (atomic number = 11)

   23 – 11 = 12 neutrons (mass number – atomic number)

2.3   Step 1. Convert each percentage to a decimal fraction.

$$90.48 \ \% \ \ ^{20}_{10}Ne \ x \ \frac{1}{100\%} \ = \ 0.9048 \ ^{20}_{10}Ne$$

$$0.27 \ \% \ ^{21}_{10}Ne \ x \ \frac{1}{100\%} \ = \ 0.0027 \ ^{21}_{10}Ne$$

$$9.25 \ \% \ ^{22}_{10}Ne \ x \ \frac{1}{100\%} \ = \ 0.0925 \ ^{22}_{10}Ne$$

Step 2.

$$\begin{pmatrix} \text{contributions to} \\ \text{atomic mass by } ^{20}_{10}Ne \end{pmatrix} = \begin{pmatrix} \text{fraction of all Ne atoms} \\ \text{that are } ^{20}_{10}Ne \end{pmatrix} x \begin{pmatrix} \text{mass of a} \\ ^{20}_{10}Ne \text{ atom} \end{pmatrix}$$

$$= \ \ \ \ 0.9048 \ \ x \ \ 19.99 \ amu$$

$$= \ \ \ \ 18.087 \ amu$$

$$\begin{pmatrix} \text{contributions to} \\ \text{atomic mass by } ^{21}_{10}Ne \end{pmatrix} = \begin{pmatrix} \text{fraction of all Ne atoms} \\ \text{that are } ^{21}_{10}Ne \end{pmatrix} x \begin{pmatrix} \text{mass of a} \\ ^{21}_{10}Ne \text{ atom} \end{pmatrix}$$

$$= \ \ \ \ 0.0027 \ \ x \ \ 20.99 \ amu$$

$$= \ \ \ \ 0.057 \ amu$$

$$\begin{pmatrix} \text{contributions to} \\ \text{atomic mass by } ^{22}_{10}Ne \end{pmatrix} = \begin{pmatrix} \text{fraction of all Ne atoms} \\ \text{that are } ^{22}_{10}Ne \end{pmatrix} x \begin{pmatrix} \text{mass of a} \\ ^{22}_{10}Ne \text{ atom} \end{pmatrix}$$

$$= \ \ \ \ 0.9025 \ \ x \ \ 21.99 \ amu$$

$$= \quad 2.034 \text{ amu}$$

Step 3. The weighted average is:

atomic mass of naturally occurring neon = (contribution of Ne-20)
+ (contribution of Ne-21)
+ (contribution of Ne-22)

$$= 18.087 + 0.057 \text{ amu} + 2.034 \text{ amu}$$

$$= 20.18 \text{ amu}$$

2.5   Our understanding of the nucleus is based on the gold foil experiment performed by Geiger and interpreted by Rutherford. In this experiment, Geiger bombarded a piece of gold foil with alpha particles, and observed that some alpha particles passed straight through the foil, others were deflected and some simply bounced back. This led Rutherford to propose that the atom consisted of a small, dense nucleus (alpha particles bounced back), surrounded by a cloud of electrons (some alpha particles were deflected). The size of the nucleus is small when compared to the volume of the atom (most alpha particles were able to pass through the foil).

2.7   a. Zr (zirconium)
       b. 22.99
       c. Cr (chromium)
       d. At (astatine)

2.9   a. helium, atomic number = 2, mass = 4.00 amu
       b. fluorine, atomic number = 9, mass = 19.00 amu
       c. manganese, atomic number = 25, mass = 54.94 amu

2.11   a. Total electrons = 11, valence electrons = 1
       b. Total electrons = 12, valence electrons = 2
       c. Total electrons = 16, valence electrons = 6
       d. Total electrons = 17, valence electrons = 7
       e. Total electrons = 18, valence electrons = 8

2.13   a. Sulfur: $1s^2, 2s^2, 2p^6, 3s^2, 3p^4$
       b. Calcium: $1s^2, 2s^2, 2p^6, 3s^2, 3p^6, 4s^2$

2.15   a. [Ne] $3s^2, 3p^4$
       a. [Ar] $4s^2$

2.17   a. Ca^{2+} and Ar are isoelectronic
       b. Sr^{2+} and Kr are isoelectronic
       c. S^{2-} and Ar are isoelectronic
       d. Mg^{2+} and Ne are isoelectronic

e. P^{3-} and Ar are isoelectric

2.19 a. (smallest) F, N, Be (largest)
       b. (lowest) Be, N, F (highest)
       c. (lowest) Be, N, F, (highest)

End-of-Chapter Questions and Problems

2.21 a. Atomic number = 8, therefore 8 protons and 8 electrons.
          Mass number – atomic number = 16 – 8 = 8, therefore 8 neutrons.

       b. Atomic number = 15, therefore 15 protons and 15 electrons.
          Mass number – atomic number = 31 – 15 = 16, therefore 16 neutrons.

2.23

	Particle	Mass	Charge
a.	electron	5.4×10^{-4} amu	–1
b.	proton	1.00 amu	+1
c.	neutron	1.00 amu	0

2.25 a. An ion is a charged atom or group of atoms formed by the loss or gain of electrons.
       b. A loss of electrons by a neutral species results in a cation.
       c. A gain of electrons by a neutral species results in an anion.

2.27 From the periodic table, all isotopes of Rn have 86 protons. Isotopes differ in the number of neutrons.

2.29 a. 34
       b. 46

2.31 a. $^{1}_{1}H$ b. $^{14}_{6}C$

2.33

	Atomic Symbol	# Protons	# Neutrons	# Electrons	Charge
a)	$^{23}_{11}Na$	11	12	11	0
b)	$^{32}_{16}S^{2-}$	16	16	18	2–
c)	$^{16}_{8}O$	8	8	8	0
d)	$^{24}_{12}Mg^{2+}$	12	12	10	2+
e)	$^{39}_{19}K^{+}$	19	20	18	1+

2.35   a)   neutrons
        b)   protons
        c)   protons, neutrons
        d)   ion
        e)   nucleus, negative

2.37   The major postulates of Dalton's atomic theory include the following:
        • All matter consists of tiny particles called atoms.
        • Atoms cannot be created, divided, destroyed, or converted to any other type of atom.
        • All atoms of a particular element have identical properties.
        • Atoms of different elements have different properties.
        • Atoms combine in simple whole-number ratios.
        • Chemical change involves joining, separating, or rearranging atoms.

2.39   a.   Chadwick demonstrated the existence of the neutron in 1932. He accomplished this
              with a series of experiments using small nuclei as projectiles to study the nucleus.

        b.   Goldstein identified positive charge in the atom.

        c.   Crookes developed the cathode ray tube and discovered "cathode rays"; characterized
              electron properties.

2.41   Negative charge, affected by electric field, affected by magnetic field.

2.43   A cathode ray is the negatively charged particle formed in a cathode ray tube. It was
        characterized as an electron, with a mass of nearly zero and a charge of –1.

2.45   a.   sodium
        b.   potassium
        c.   magnesium
        d.   boron

2.47   Group IA (or 1) is known collectively as the alkali metals and consists of lithium,
        sodium, potassium, rubidium, cesium, and francium.

2.49   Group VIIA (or 17) is known collectively as the halogens and consists of fluorine,
        chlorine, bromine, iodine, and astatine.

2.51   a.   true
        b.   true

2.53   a.   The metals are: Na, Ni, Al
        b.   The representative metals are: Na, Al
        c.   The elements that tend to form positive ions are: Na, Ni, Al
        d.   The element that is inert is: Ar

2.55 a. one
      b. one
      c. three
      d. seven
      e. zero (or eight)
      f. zero (or two)

2.57 A principal energy level is designated n = 1, 2, 3, and so forth. It is similar to Bohr's orbits in concept. A sublevel is a part of a principal energy level and is designated $s, p, d,$ and f.

2.59 The s orbital represents the probability of finding an electron in a region of space surrounding the nucleus. A diagram of an s orbital is found in Figure 2.8 on p. 55.

2.61 Three p orbitals (p_x, p_y, p_z) can exist in a given principal energy level.

2.63 A $3p$ orbital is a higher energy orbital than a $2p$ orbital because it is a part of a higher energy principal energy level.

2.65 $2n^2 = 2(1)^2 = 2$ e$^-$ for $n = 1$
      $2n^2 = 2(2)^2 = 8$ e$^-$ for $n = 2$
      $2n^2 = 2(3)^2 = 18$ e$^-$ for $n = 3$

2.67 a. Al $1s^2, 2s^2, 2p^6, 3s^2, 3p^1$ (hence, a $3p$ orbital)
      b. Na $1s^2, 2s^2, 2p^6, 3s^1$ (hence, a $3s$ orbital)
      c. Sc $1s^2, 2s^2, 2p^6, 3s^2, 3p^6, 4s^2, 3d^1$ (hence, a $3d$ orbital)
      d. Ca $1s^2, 2s^2, 2p^6, 3s^2, 3p^6, 4s^2$ (hence, a $4s$ orbital)
      e. Fe $1s^2, 2s^2, 2p^6, 3s^2, 3p^6, 4s^2, 3d^6$ (hence, a $3d$ orbital)
      f. Cl $1s^2, 2s^2, 2p^6, 3s^2, 3p^5$ (hence, a $3p$ orbital)

2.69 a. Not possible; n=1 level can have only s-level orbitals.
      b. Possible; the electron configuration that is shown represents the carbon atom.
      c. Not possible; n=2 level can have only s- and p-orbitals.
      d. Not possible; an s-orbital cannot contain three electrons.

2.71 a. Li$^+$ (Li loses 1 electron to attain outermost octet)
      b. O^{2-} (O gains 2 electrons to attain outermost octet)
      c. Ca^{2+} (Ca loses 2 electrons to attain outermost octet)
      d. Br$^-$ (Br gains 1 electron to attain outermost octet)
      e. S^{2-} (S gains 2 electrons to attain outermost octet)
      f. Al^{3+} (Al loses 3 electrons to attain outermost octet)

2.73 a. O^{2-}, 10 e$^-$; Ne, 10 e$^-$; Isoelectronic
      b. S^{2-}, 18 e$^-$; Cl$^-$, e$^-$; Isoelectronic

2.75   a. Na$^+$; it has a noble gas electron configuration
       b. S^{2-}; it has a noble gas electron configuration
       c. Cl$^-$; it has a noble gas electron configuration

2.77   a. $1s^2, 2s^2, 2p^6, 3s^2, 3p^6$
       b. $1s^2, 2s^2, 2p^6$

2.79   a. (Smallest) F, O, N (Largest)
       b. (Smallest) Li, K, Cs (Largest)
       c. (Smallest) Cl, Br, I (Largest)

2.81   a. (Smallest) O, N, F (Largest)
       b. (Smallest) Cs, K, Li (Largest)
       c. (Smallest) I, Br, Cl (Largest)

2.83   A positive ion is always smaller than its parent atom because the positive charge of the nucleus is shared among fewer electrons in the ion. As a result, each electron is pulled closer to the nucleus and the volume of the ion decreases.

2.85   The fluoride ion has a completed octet of electrons and an electron configuration resembling its nearest noble gas.

Chapter 3
Structure and Properties of
Ionic and Covalent Compounds
Solutions to the Odd-Numbered Problems

In-Chapter Questions and Problems

3.1   a.

$$\text{Li} \cdot \; + \; :\!\ddot{Br}\cdot \; \longrightarrow \; \text{Li}^+ \; + \; :\!\ddot{Br}\!:^{-}$$

       b.

$$\cdot\text{Mg}\cdot \; + \; 2\,:\!\ddot{Cl}\cdot \; \longrightarrow \; \text{Mg}^{2+} \; + \; 2\,:\!\ddot{Cl}\!:^{-}$$

3.3   Yes; the electronegativity difference (C, 2.5 and N, 3.0) is 0.5. Two nonmetals with a nonzero electronegativity difference are joined by a polar covalent bond.

3.5   a. LiBr (one Li^+ and one Br^-)
       b. $CaBr_2$ (one Ca^{2+} and two Br^-)
       c. Ca_3N_2 (three Ca^{2+} and two N^{3-})

3.7   a. potassium cyanide
       b. magnesium sulfide
       c. magnesium acetate

3.9   a. $CaCO_3$
       b. $NaHCO_3$
       c. Cu_2SO_4

3.11   a. diboron trioxide
       b. nitrogen oxide
       c. iodine chloride
       d. phosphorus trichloride

3.13   a. P_2O_5
       b. SiO_2

3.15   a.   b.

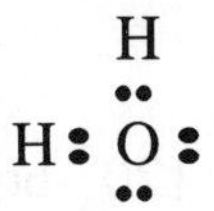

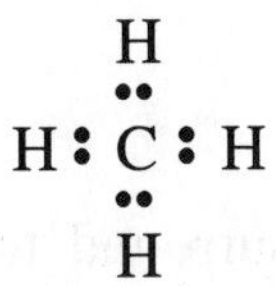

H
••
H : O :
••

H
••
H : C : H
••
H

3.17   a.   The bonded nuclei are closer together when a double bond exists, in comparison to a single bond.

b.   The bond strength increases as the bond order increases. Therefore, a double bond is stronger than a single bond.

3.19   a.

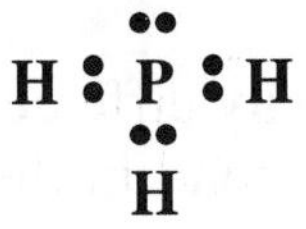

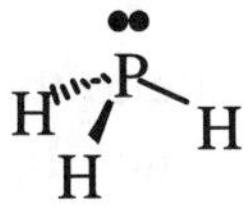

••
H : P : H
••
H

••
H''''P
H / \ H
H

Three groups and one lone pair of electrons surround the phosphorus atom; the structure is trigonal pyramidal (similar to the structure of ammonia).

b.

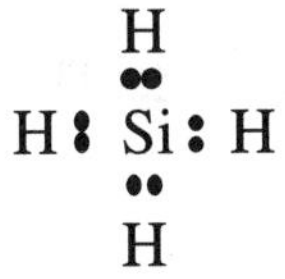

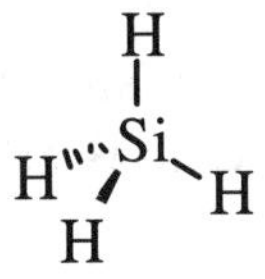

H
••
H : Si : H
••
H

H
|
H''''Si
H / \ H
H

Four groups surround the silicon atom; the structure is tetrahedral (similar to the structure of methane).

3.21   a.   O – S   Oxygen is more electronegative than sulfur; the bond is polar. The electrons are pulled toward the oxygen atom.

b.   C ≡ N   Nitrogen is more electronegative than carbon; the bond is polar. The electrons are pulled toward the nitrogen atom.

c.   Cl – Cl   There is no electronegativity difference between two identical atoms; the bond is nonpolar.

d.   I – Cl   Chlorine is more electronegative than iodine; the bond is polar. The electrons are pulled toward the chlorine atom.

3.23   a.

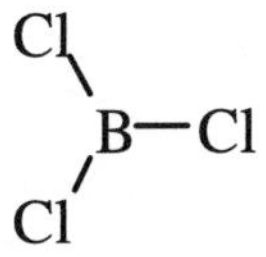

Cl
 \
  B—Cl
 /
Cl

Three groups surround the central atom forming 120° bond angles. Due to the symmetrical arrangement of the three B-Cl bonds, their polarities cancel and the molecule is nonpolar.

b.

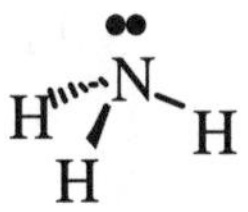

H⸍⸍⸍N⸜H
   /
  H

Three groups and a lone pair of electrons surround the central atom. Due to the effect of the lone pair, the molecule is polar.

c. H–Cl

The H-Cl bond is polar due to the electronegativity difference between hydrogen and chlorine. Since H-Cl is the only bond in the molecule, the molecule is polar.

d.

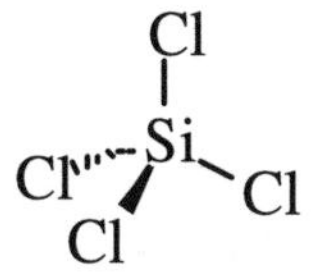

    Cl
    |
Cl⸍⸍⸍Si⸜Cl
   /
  Cl

Four groups, all equivalent, surround the central atom. The structure is tetrahedral and the molecule is nonpolar.

3.25   a. H_2O is polar → higher melting and boiling point.
            C_2H_4 is nonpolar

       b. CO is polar → higher melting and boiling point.
            CH_4 is nonpolar

       c. NH_3 is polar → higher melting and boiling points.
            N_2 is nonpolar

       d. Cl_2 is nonpolar.
            ICl is polar → higher melting and boiling points.

End-of-Chapter Questions and Problems

3.27   a. Ionic
       b. Covalent
       c. Covalent
       d. Covalent

3.29   a.   covalent (2 nonmetals)
        b.   covalent (2 nonmetals)
        c.   covalent (2 nonmetals)
        d.   ionic (metals and nonmetals)

3.31   a.

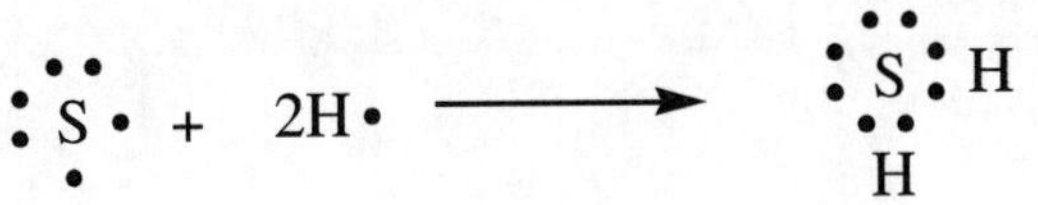

$$:\overset{..}{\underset{.}{S}}\cdot \ + \ 2H\cdot \ \longrightarrow \ :\overset{..}{\underset{..}{S}}:H$$
$$\qquad\qquad\qquad\qquad\qquad\qquad\ \ H$$

       b.

$$\cdot \underset{.}{P}\cdot \ + \ 3H\cdot \ \longrightarrow \ H:\overset{..}{P}:H$$
$$\qquad\qquad\qquad\qquad\qquad\qquad\ \ \overset{..}{H}$$

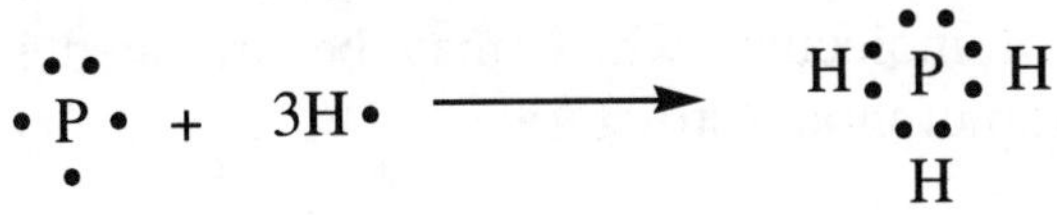

3.33   He has two valence electrons (electron configuration $1s^2$) and a complete N=1 level. It has a stable electron configuration, with no tendency to gain or lose electrons, and satisfies the octet rule (2 e$^-$ for period 1). Hence, it is nonreactive.

       He $:$

3.35   a.   Copper (II) ion (or cupric ion)
        b.   Iron (II) ion (or ferrous ion)
        c.   Iron (III) ion (or ferric ion)

3.37   a. K^+

        b. Br^-

3.39   a. SO_4^{2-}

        b. NO_3^-

3.41   a.   NaCl   [one 1+ cancels one 1–]
        b. $MgBr_2$ [one 2+ cancels two 1–]

3.43   a.   AgCN [one 1+ cancels out 1–]
        b. NH_4Cl [one 1+ cancels out 1–]

3.45   a.   Magnesium chloride
        b.   Aluminum chloride

3.47   a.   Nitrogen dioxide
        b.   Sulfur trioxide

3.49 a. Al_2O_3
      b. Li_2S

3.51 a. CO or CO_2
      b. H_2S

3.53 a. sodium hypochlorite
      b. sodium chlorite

3.55 Ionic solid state compounds exist in regular, repeating, three-dimensional structures; the crystal lattice. The crystal lattice is made up of positive and negative ions. Solid state covalent compounds are made up of molecules which may be arranged in a regular crystalline pattern or in an irregular (amorphous) structure.

3.57 The boiling points of ionic solids are generally much higher than those of covalent solids.

3.59 KCl would be expected to exist as a solid at room temperature; it is an ionic compound, and ionic compounds are characterized by high melting points.

3.61 Water will have a higher boiling point. Water is a polar molecule with strong intermolecular attractive forces, whereas carbon tetrachloride is a nonpolar molecule with weak intermolecular attractive forces. More energy, hence, a higher temperature is required to overcome the attractive forces among the water molecules.

3.63 a. H•   b. He⦂   c.  •C•   d. 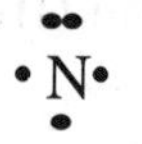 •N•

3.65 a. Li^+ b. Mg^{2+} c. 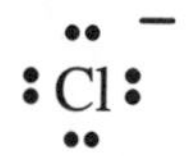 :Cl:⁻   d. 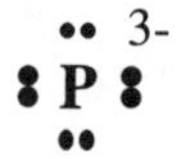 ⦂P⦂³⁻

3.67 a. NCl_3

      :Cl: N :Cl:
         :Cl:

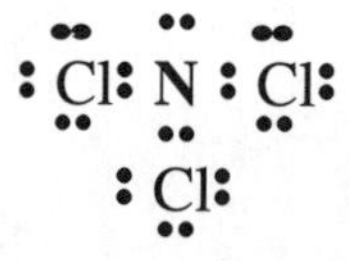

      One nitrogen atom has 5 valence electrons and three chlorine atoms contribute 21 valence electrons (3 x 7 valence electrons/atom of chlorine).

      This produces a total of **26 valence electrons.**
      This structure satisfies the octet rule for N and Cl.

b.   CH₃OH

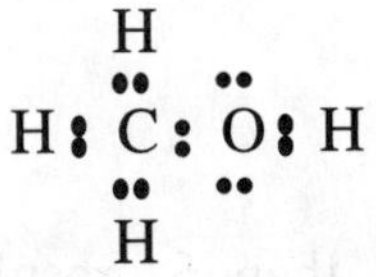

```
        H
        ••   ••
H ⦂ C ⦂ O ⦂ H
        ••   ••
        H
```

One carbon atom has 4 valence electrons, one oxygen atom has 6 valence electrons, and four hydrogen atoms contribute 4 valence electrons (one for each hydrogen atom).

This produces a total of **14 valence electrons.**
This structure satisfies the octet rule for C and O.

c. CS_2

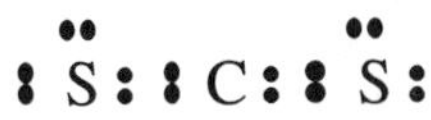

```
  ••         ••
⦂ S ⦂⦂ C ⦂⦂ S ⦂
```

One carbon atom contributes 4 valence electrons, two sulfur atoms contribute 12 valence electrons (six from each sulfur atom) for a total of **16 valence electrons.**
This structure satisfies the octet rule for C and S.

3.69   a.   b.   c.

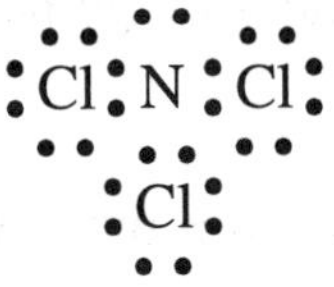

```
  ••    ••
⦂Cl⦂N⦂Cl⦂
  ••   •• ••
     ⦂Cl⦂
       ••
```

```
        H
        ••
H ⦂ C ⦂ O ⦂ H
        ••   ••
        H
```

```
  ••         ••
  S ⦂⦂ C ⦂⦂ S
  ••         ••
```

pyramidal,   tetrahedral around C,   linear
polar   angular around O   nonpolar
water soluble   polar   not water soluble
                            water soluble

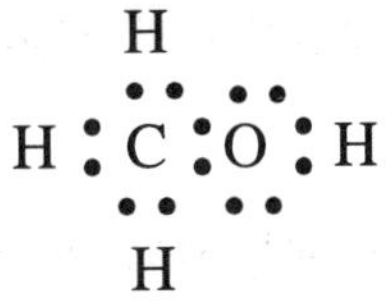

3.71

```
     H   H
     ••  ••   ••
H ⦂ C ⦂ C ⦂ O ⦂ H
     ••  ••   ••
     H   H
```

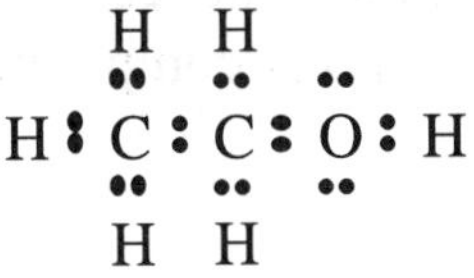

3.73

3.75   a. S and O   Polar covalent; O is more electronegative than S, resulting in an unequal sharing of electrons in the bond.

      b. Si and P   Polar covalent; P is more electronegative than Si, resulting in the unequal sharing of electrons in the bond.

      c. Na and Cl   Ionic; the bond is between a metal and a non-metal.

      d. Na and O   Ionic; the bond is between a metal and a non-metal.

      e. Ca and Br   Ionic; the bond is between a metal and a non-metal.

3.77   a. S and O $:\ddot{\text{O}}:\ddot{\text{S}}::\ddot{\text{O}}$

      b. Si and P $[:\text{Si}\equiv\text{P}:]^-$

    c), d), and e) are ionic compounds.

3.79   A molecule containing no polar bonds <u>must</u> be nonpolar. A molecule containing polar bonds may or may not itself be polar. It depends upon the number and arrangement of the bonds. For example:

- If a molecule contains only one bond, and that bond is polar, the molecule must be polar.
- If the molecule contains more than one polar bond, the molecule will be nonpolar if the arrangement of the bonds causes their effects to cancel. If not, the molecule will be polar.
- If lone pairs of electrons are present, their effect must be considered as well.
- If a molecule contains no polar bonds, it cannot be a polar molecule.

3.81   Polar compounds have strong intermolecular attractive forces. Higher temperatures are needed to overcome these forces and convert the solid to a liquid; hence, we predict higher melting points for polar compounds when compared to nonpolar compounds.

3.83   Yes. Many ionic compounds are water soluble.

Chapter 4
Calculations and the Chemical Equation
Solutions to the Odd-Numbered Problems

In-Chapter Questions and Problems

4.1 $\dfrac{26.98 \text{ amu Al}}{1 \text{ atom Al}} \times \dfrac{1.661 \times 10^{-24} \text{ g Al}}{1 \text{ amu Al}} \times \dfrac{6.022 \times 10^{23} \text{ atoms Al}}{1 \text{ mol Al}} = 26.98 \ \dfrac{\text{g Al}}{\text{mol Al}}$

4.3   a.   O atoms $= 2.50 \text{ mol O} \times \dfrac{6.022 \times 10^{23} \text{ atoms O}}{1 \text{ mol O}}$

O atoms $= 1.51 \times 10^{24}$ oxygen atoms

     b.   O atoms $= 2.50 \text{ mol O}_2 \times \dfrac{6.022 \times 10^{23} \text{ molecules O}_2}{1 \text{ mol O}_2} \times \dfrac{2 \text{ atoms O}}{1 \text{ molecule O}_2}$

O atoms $= 3.01 \times 10^{24}$ oxygen atoms

4.5   g He $= 3.50 \text{ mol He} \times \dfrac{4.00 \text{ g He}}{1 \text{ mol He}} = 14.0$ g He

4.7   a.

3 atoms of hydrogen x	1.01 amu/atom =	3.03 amu
1 atom of nitrogen x	14.01 amu/atom =	14.01 amu
		17.04 amu

The mass of a single unit of NH_3 is 17.04 amu/formula unit. Therefore, the mass of 1 mole of formula units is 17.04 grams or 17.04 g/mol.

     b.

6 atoms of carbon x	12.01 amu/atom =	72.06 amu
12 atoms of hydrogen x	1.01 amu/atom =	12.12 amu
6 atoms of oxygen x	16.00 amu/atoms =	96.00 amu
		180.18 amu

The mass of a single unit of $C_6H_{12}O_6$ is 180.18 amu/formula unit. Therefore, the mass of 1 mole of formula units is 180.18 grams or 180.18 g/mol.

     c.

1 atom of cobalt x	58.93 amu/atom =	58.93 amu
2 atoms of chlorine x	35.45 amu/atom =	70.90 amu
12 atoms of hydrogen x	1.01 amu/atom =	12.12 amu
6 atoms of oxygen x	16.00 amu/atom =	96.00 amu
		237.95 amu

The mass of a single unit of $CoCl_2 \cdot 6H_2O$ is 237.95 amu/formula unit. Therefore, the mass of 1 mole of formula units is 237.95 grams or 237.95 g/mol.

4.9   a. DR
       b. SR
       c. DR
       d. D

4.11   a. $KCl(aq) + AgNO_3(aq) \rightarrow KNO_3(aq) + AgCl(s)$

       The solubility rules predict that silver chloride is insoluble; a precipitation reaction occurs.

       b. $CH_3COOK(aq) + AgNO_3(aq) \rightarrow$ no reaction

       The solubility rules predict that both potential products, potassium nitrate (KNO_3) and silver acetate (CH_3COOAg) are soluble; no precipitation reaction occurs.

4.13 $Ca \rightarrow Ca^{2+} + 2\ e^-$ (oxidation ½ reaction)
       $S\ +\ 2\ e^- \rightarrow S^{2-}$ (reduction ½ reaction)
       $Ca\ +\ S \rightarrow CaS$ (complete reaction)

4.15   a. $4Fe(s) + 3O_2(g) \rightarrow 2Fe_2O_3(s)$

       b. $2C_6H_6(l) + 15O_2(g) \rightarrow 12CO_2(g) + 6H_2O(g)$

4.17   a. $5.00\ mol\ H_2 \times \dfrac{18.02\ g\ H_2O}{1\ mol\ H_2O} = 90.1\ g\ H_2O$

       b. $25.0\ g\ LiCl \times \dfrac{1\ mol\ LiCl}{42.39\ g\ LiCl} = 0.590\ mol\ LiCl$

4.19   a. $1\ mol\ C_2H_5OH \times \dfrac{3\ mol\ O_2}{1\ mol\ C_2H_5OH} = 3\ mol\ O_2$

       b. $1\ mol\ C_2H_5OH \times \dfrac{3\ mol\ O_2}{1\ mol\ C_2H_5OH} \times \dfrac{32\ g\ O_2}{1\ mol\ O_2} = 96.00\ g\ O_2$

4.21   Iron(III) oxide is Fe_2O_3

       a. $4Fe(s) + 3O_2(g) \rightarrow 2Fe_2O_3(s)$

b. $5.00 \text{ g Fe}_2\text{O}_3 \times \dfrac{1 \text{ mol Fe}_2\text{O}_3}{159.7 \text{ g Fe}_2\text{O}_3} \times \dfrac{4 \text{ mol Fe}}{2 \text{ mol Fe}_2\text{O}_3} \times \dfrac{55.85 \text{ g Fe}}{1 \text{ mol Fe}} = 3.50 \text{ g Fe}$

4.23 a. Step 1. Write down information about the reaction:

$$\text{Sn(s)} + 2 \text{ HF(aq)} \rightarrow \text{SnF}_2\text{(s)} + \text{H}_2\text{(g)}$$
$$100.0 \text{ g} \quad \text{(excess)}$$

Step 2. Convert the mass of Sn to moles of Sn:

$$100.0 \text{ g Sn} \times \dfrac{1 \text{ mol Sn}}{118.69 \text{ g Sn}} = 0.8425 \text{ mol Sn}$$

Step 3. The reaction states that one mole of Sn will react to form one mole of SnF_2, so the mole ratio is 1:1. Use this conversion factor to calculate the mass of product:

$$0.8425 \text{ mol Sn} \times \dfrac{1 \text{ mol SnF}_2}{1 \text{ mol Sn}} \times \dfrac{156.69 \text{ g SnF}_2}{1 \text{ mol SnF}_2} = 132.0 \text{ g SnF}_2$$

b. $\% \text{ yield} = \dfrac{\text{actual yield}}{\text{theoretical yield}} \times 100 \%$

$$\% \text{ yield} = \dfrac{5.00 \text{ g}}{132.00 \text{ g}} \times 100 \% = 3.79 \% \text{ yield}$$

End-of-Chapter Questions and Problems

4.25 Examples of other packaging units include a <u>ream</u> of paper (500 sheets of paper), a <u>six-pack</u> of soft drinks, a <u>case</u> of canned goods (24 cans), to name a few.

4.27 a. The average molar mass of Si (silicon) is 28.09 g.
      b. The average molar mass of Ag (silver) is 107.9 g.

4.29 The mass of Avogadro's number of argon atoms is the same as the average molar mass, 39.95 g.

4.31

$$2.00 \text{ mol Ne} \quad \times \quad \dfrac{20.18 \text{ g Ne}}{1 \text{ mol Ne}} \quad = \quad 40.36 \text{ g Ne}$$

4.33 $\dfrac{1.66 \times 10^{-24} \text{ g He}}{1 \text{ amu}} \times \dfrac{4.00 \text{ amu}}{1 \text{ atom He}} \times \dfrac{6.022 \times 10^{23} \text{ atoms He}}{1 \text{ mol He}} = 4.00 \dfrac{\text{g He}}{\text{mol He}}$

4.35   a. $20.0 \text{ g He} \times \dfrac{1 \text{ mol He}}{4.00 \text{ g He}} = 5.00 \text{ mol He}$

      b. $0.040 \text{ kg Na} \times \dfrac{10^3 \text{g Na}}{1 \text{ kg Na}} \times \dfrac{1 \text{ mol Na}}{22.99 \text{ g Na}} = 1.7 \text{ mol Na}$

      c. $3.0 \text{ g Cl}_2 \times \dfrac{1 \text{ mol Cl}_2}{70.90 \text{ g Cl}_2} = 4.2 \times 10^{-2} \text{ mol Cl}_2$

4.37   A molecule is a single unit comprised of atoms joined by covalent bonds. An ion-pair is composed of positive and negatively charged ions joined by electrostatic attraction, the ionic bond. The ion pairs, unlike the molecule, do not form single units; the electrostatic charge is directed to other ions in a crystal lattice, as well.

4.39   a. $1 \text{ atom Na} \times \dfrac{22.99 \text{ amu Na}}{1 \text{ atom Na}} = 22.99 \text{ amu Na}$

         $1 \text{ atom Cl} \times \dfrac{35.45 \text{ amu Cl}}{1 \text{ atom Cl}} = \underline{35.45 \text{ amu Cl}}$

                                                      58.44 amu NaCl

The mass of a single unit of NaCl is 58.44 amu/formula unit. Therefore, the mass of a mole of NaCl formula units is 58.44 g/mol.

      b. $2 \text{ atoms Na} \times \dfrac{22.99 \text{ amu Na}}{1 \text{ atom Na}} = 45.98 \text{ amu Na}$

         $1 \text{ atom S} \times \dfrac{32.06 \text{ amu S}}{1 \text{ atom S}} = 32.06 \text{ amu S}$

         $4 \text{ atoms O} \times \dfrac{16.00 \text{ amu O}}{1 \text{ atom O}} = \underline{64.00 \text{ amu O}}$

                                                142.04 amu Na_2SO_4

The mass of a single unit of Na_2SO_4 is 142.04 amu/formula unit. Therefore, the mass of a mole of Na_2SO_4 formula units is 142.04 g/mol.

      c. $3 \text{ atoms Fe} \times \dfrac{55.85 \text{ amu Fe}}{1 \text{ atom Fe}} = 167.55 \text{ amu Fe}$

         $2 \text{ atoms P} \times \dfrac{33.97 \text{ amu P}}{1 \text{ atom P}} = 61.94 \text{ amu P}$

$$8 \text{ atoms O} \times \frac{16.00 \text{ amu O}}{1 \text{ atom O}} = \underline{128.00 \text{ amu O}}$$

$$357.49 \text{ amu Fe}_3(PO_4)_2$$

The mass of a single unit of $Fe_3(PO_4)_2$ is 357.49 amu/formula unit. Therefore, the mass of a mole of $Fe_3(PO_4)_2$ formula units is 357.49 g/mol.

4.41 $\dfrac{2 \text{ mol O}}{1 \text{ mol O}_2} \times \dfrac{16.00 \text{ g O}}{1 \text{ mol O}} = 32.00 \text{ g/mol O}_2$

4.43

$$1 \text{ atom Cu} \quad \times \quad \frac{63.55 \text{ amu Cu}}{1 \text{ atom Cu}} = \quad 63.55 \text{ amu Cu}$$

$$1 \text{ atom S} \quad \times \quad \frac{32.07 \text{ amu S}}{1 \text{ atom S}} = \quad 32.07 \text{ amu S}$$

$$9 \text{ atom O} \quad \times \quad \frac{16.00 \text{ amu O}}{1 \text{ atom O}} = \quad 144.00 \text{ amu O}$$

$$10 \text{ atom H} \quad \times \quad \frac{1.008 \text{ amu H}}{1 \text{ atom H}} = \quad 10.08 \text{ amu H}$$

$$\overline{\qquad\qquad\qquad}$$

$$249.70 \text{ amu CuSO}_4 \cdot 5 \text{ H}_2\text{O}$$

The molar mass of $CuSO_4 \cdot 5 H_2O$ is 249.70 grams.

4.45   a.   The formula weight of NaCl is 58.44 g/mol.

$$15.0 \text{ g NaCl} \times \frac{1 \text{ mol NaCl}}{58.44 \text{ g NaCl}} = 0.257 \text{ mol NaCl}$$

b.   The formula weight of Na_2SO_4 is 142.04 g/mol.

$$15.0 \text{ g Na}_2\text{SO}_4 \times \frac{1 \text{ mol Na}_2\text{SO}_4}{142.04 \text{ g Na}_2\text{SO}_4} = 0.106 \text{ mol Na}_2\text{SO}_4$$

4.47   a.   The formula weight of H_2O is 18.02 g/mol.

$$1.000 \text{ mol H}_2\text{O} \times \frac{18.02 \text{ g H}_2\text{O}}{1 \text{ mol H}_2\text{O}} = 18.02 \text{ g H}_2\text{O}$$

b.   The formula weight of NaCl is 58.44 g/mol.

$$2.000 \text{ mol NaCl} \times \frac{58.44 \text{ g NaCl}}{1 \text{ mol NaCl}} = 116.9 \text{ g NaCl}$$

4.49   a.   The formula weight of He is 4.00 g/mol.

$$10.0 \text{ mol He} \times \frac{4.00 \text{ g He}}{1 \text{ mol He}} = 40.0 \text{ g He}$$

       b.   The formula weight of $H_2 = 2.016$ g/mol.

$$1.00 \times 10^2 \text{ mol } H_2 \times \frac{2.016 \text{ g } H_2}{1 \text{ mol } H_2} = 2.02 \times 10^2 \text{g H}$$

4.51   a.   The formula weight of KBr is 119.01 g/mol.

$$50.0 \text{ g KBr} \times \frac{119.01 \text{ g KBr}}{1 \text{ mol KBr}} = 0.420 \text{ mol KBr}$$

       b.   The formula weight of $MgSO_4$ is 120.37 g/mol.

$$50.0 \text{g } MgSO_4 \times \frac{120.37 \text{ g } MgSO_4}{1 \text{ mol } MgSO_4} = 0.415 \text{ mol } MgSO_4$$

4.53   a.   The formula weight of CS_2 is 76.13 g/mol.

$$50.0 \text{ g } CS_2 \times \frac{1 \text{ mol } CS_2}{76.13 \text{ g } CS_2} = 6.57 \times 10^{-1} \text{ mol } CS_2$$

       b.   The formula weight of $Al_2(CO_3)_3$ is 233.99 g/mol.

$$50.0 \text{ g } Al_2(CO_3)_3 \times \frac{1 \text{ mol } Al_2(CO_3)_3}{233.99 \text{ g } Al_2(CO_3)_3} = 2.14 \times 10^{-1} \text{ mol } Al_2(CO_3)_3$$

4.55   The subscript tells us the number of atoms or ions contained in one unit of the compound.

4.57   a.   Heating an alkaline earth metal carbonate, for example:

$$MgCO_3(s) \xrightarrow{\Delta} MgO(s) + CO_2(g)$$

       b.   The replacement of copper by zinc in copper sulfate,

$$Zn(s) + CuSO_4(aq) \rightarrow ZnSO_4(aq) + Cu(s)$$

4.59   Reaction of two soluble substances to form an insoluble product. For example:

$$2NaOH(aq) + FeCl_2(aq) \rightarrow Fe(OH)_2(s) + 2NaCl(aq)$$

4.61   The symbol Δ over the reaction arrow means that heat is necessary for the reaction to occur.

4.63   If we change the subscript we change the identity of the compound.

4.65   Reactants are found on the left side of the reaction arrow.

4.67   a. $2C_2H_6(g) + 7O_2(g) \rightarrow 4CO_2(g) + 6H_2O(g)$

       b. $6K_2O(s) + P_4O_{10}(s) \rightarrow 4K_3PO_4(s)$

       c. $MgBr_2(aq) + H_2SO_4(aq) \rightarrow 2HBr(g) + MgSO_4(aq)$

4.69   a. $Ca(s) + F_2(g) \rightarrow CaF_2(s)$

       b. $2Mg(s) + O_2(g) \rightarrow 2MgO(s)$

       c. $3H_2(g) + N_2(g) \rightarrow 2NH_3(g)$

4.71   a. $2C_4H_{10}(g) + 13O_2(g) \rightarrow 10H_2O(g) + 8CO_2(g)$

       b. $Au_2S_3(s) + 3H_2(g) \rightarrow 2Au(s) + 3H_2S(g)$

       c. $Al(OH)_3(s) + 3HCl(aq) \rightarrow AlCl_3(aq) + 3H_2O(l)$

       d. $(NH_4)_2Cr_2O_7(s) \rightarrow Cr_2O_3(s) + N_2(g) + 4H_2O(g)$

       e. $C_2H_5OH(l) + 3O_2(g) \rightarrow 2CO_2(g) + 3H_2O(g)$

4.73   a. $N_2(g) + 3H_2(g) \rightarrow 2NH_3(g)$

       b. $HCl(aq) + NaOH(aq) \rightarrow NaCl(aq) + H_2O(l)$

       c. $2HNO_3(aq) + Ca(OH)_2(aq) \rightarrow Ca(NO_3)_2(aq) + 2H_2O(l)$

       d. $2C_4H_{10}(g) + 13O_2(g) \rightarrow 10H_2O(g) + 8CO_2(g)$

4.75   The formula weight of B_2O_3 is 69.62 g/mol and the formula weight of B_2H_6 is 27.67 g/mol.

$$20 \text{ g } B_2H_6 \times \frac{1 \text{ mol } B_2H_6}{27.67 \text{ g } B_2H_6} \times \frac{1 \text{ mol } B_2O_3}{1 \text{ mol } B_2H_6} \times \frac{69.62 \text{ g } B_2O_3}{1 \text{ mol } B_2O_3} = 50.3 \text{ g } B_2O_3$$

4.77   a. $N_2(g) + 3H_2(g) \rightarrow 2NH_3(g)$

   b.   Three moles of H_2 will react with one mole of N_2, according to the coefficients in the balanced equation.

   c.   One mole of N_2 will produce two moles of the product NH_3, according to the coefficients in the balanced equation.

   d. $140.0 \text{ g N}_2 \times \dfrac{1 \text{ mol N}_2}{28.02 \text{ g N}_2} \times \dfrac{3 \text{ mol H}_2}{1 \text{ mol N}_2} = 1.50 \text{ mol H}_2$

   e. $1.50 \text{ mol H}_2 \times \dfrac{2 \text{ mol NH}_3}{3 \text{ mol H}_2} \times \dfrac{17.03 \text{ g NH}_3}{1 \text{ mol NH}_3} = 17.0 \text{ g NH}_3$

4.79   a.   5 C atoms x 12.01 amu/atom C =   60.05 amu
         11 H atoms x 1.008 amu/atom H =   11.09 amu
         1 N atom x 14.01 amu/atom N =   14.01 amu
         2 O atoms x 16.00 amu/atom O =   32.00 amu
         1 S atom x 32.06 amu/atom S =   32.06 amu
                                          149.21 amu

   The mass of a single unit of $C_5H_{11}NO_2S$ is 149.21 amu/formula unit. Therefore the mass of a mole of $C_5H_{11}NO_2S$ formula units is 149.21 g/mol.

   b. $1 \text{ mol C}_5H_{11}NO_2S \times \dfrac{2 \text{ mol O atoms}}{1 \text{ mol C}_5H_{11}NO_2S} \times \dfrac{6.02 \times 10^{23} \text{ O atoms}}{1 \text{ mol O atoms}}$

         $= 1.20 \times 10^{24} \text{ O atoms}$

   c. $1 \text{ mol C}_5H_{11}NO_2S \times \dfrac{2 \text{ mol O atoms}}{1 \text{ mol C}_5H_{11}NO_2S} \times \dfrac{16.00 \text{ g O}}{1 \text{ mol O atoms}} = 32.00 \text{ g O}$

   d. $50.0 \text{ g C}_5H_{11}NO_2S \times \dfrac{1 \text{ mol C}_5H_{11}NO_2S}{149.21 \text{ g C}_5H_{11}NO_2S} \times \dfrac{2 \text{ mol O atoms}}{1 \text{ mol C}_5H_{11}NO_2S} \times$

         $\dfrac{16.00 \text{ g O}}{1 \text{ mol O atoms}} = 10.7 \text{ g O}$

4.81   The balanced equation is:

   $2C_2H_2(g) + 5O_2(g) \rightarrow 4CO_2(g) + 2H_2O(g)$

   The formula weight of C_2H_2 is 26.04 g/mol.

$$20.0 \text{ kg } C_2H_2 \times \frac{10^3 \text{ g } C_2H_2}{1 \text{ kg } C_2H_2} \times \frac{1 \text{ mol } C_2H_2}{26.04 \text{ g } C_2H_2} \times \frac{5 \text{ mol } O_2}{2 \text{ mol } C_2H_2} \times \frac{32.00 \text{ g } O_2}{1 \text{ mol } O_2}$$

$$= 6.14 \times 10^4 \text{ g } O_2$$

4.83 Step 1. Write down information about the reaction:

$$N_2O_4(l) + 2 N_2H_4(l) \rightarrow 3 N_2(g) + 4 H_2O(g)$$

1.00 kg   (excess)

Step 2. Convert the mass of N_2O_4 to moles of N_2O_4:

$$1.00 \text{ kg } \times \frac{10^3 \text{ g } N_2O_4}{1 \text{ kg } N_2O_4} \times \frac{1 \text{ mol } N_2O_4}{92.02 \text{ g } N_2O_4} = 10.9 \text{ mol } N_2O_4$$

Step 3. The reaction states that the ratio of moles of N_2O_4 to N_2 is 1:3. Use this conversion factor to calculate the mass of N_2:

$$10.9 \text{ mol } N_2O_4 \times \frac{2 \text{ mol } N_2}{1 \text{ mol } N_2O_4} \times \frac{28.02 \text{ g } N_2}{1 \text{ mol } N_2} = 9.13 \times 10^2 \text{ g } N_2$$

4.85 In problem 4.83 we found that the theoretical yield of nitrogen was 9.13×10^2 g. Since only 75.0% was actually obtained, corresponding to a decimal fraction of 0.750,

$$9.13 \times 10^2 \text{ g } N_2 \times 0.750 = 6.85 \times 10^2 \text{ g } N_2$$

Chapter 5
Energy, Rate, and Equilibrium
Solutions to the Odd-Numbered Problems

In-Chapter Questions and Problems

5.1   Intensive property; it is independent of quantity.

5.3   Electrical energy is converted to light energy.

5.5   a. Exothermic. The reaction <u>produces</u> energy used to heat our homes.

      b. Exothermic. $\Delta H°$ is <u>negative</u>, meaning that energy is <u>released</u> by the reaction.

      c. Exothermic. 18.3 kcal of energy is shown as a <u>product</u> of the reaction.

5.7   He(g) has a greater entropy than Na(s). Gases have a greater degree of disorder than solids.

5.9   A positive value for ΔH predicts nonspontaneity; furthermore, a decrease in entropy (a negative value for ΔS) also predicts nonspontaneity. Therefore, the reaction must be nonspontaneous.

5.11 $Q = m_w \times \Delta T_w \times SH_w$

Solving for ΔT_w

$$\Delta T_w = \frac{Q}{m_w \times SH_w}$$

Substituting,

$$\Delta T_w = \frac{6.5 \times 10^2 \text{ cal}}{(50.0 \text{ g})(1.00 \text{ cal/g°C})} = 13°C$$

5.13 6.5×10^2 cal $\times \dfrac{4.18 \text{ J}}{1 \text{ cal}} = 2.7 \times 10^3$ J

5.15 $Q = m_w \times \Delta T_w \times SH_w$

$$Q = 1.00 \times 10^3 \text{ g} \times 3.0°C \times \frac{1.00 \text{ cal}}{g \cdot °C}$$

$Q = 3.0 \times 10^3$ cal in a 1.0 gram sample of candy

and,

3.0×10^3 cal $\times \dfrac{1 \text{ nutritional Cal}}{10^3 \text{ cal}} = 3.0$ nutritional Calories in a 1.0 gram sample of candy

and,

$$\dfrac{3.0 \text{ nutritional Cal}}{1.0 \text{ g candy}} \times \dfrac{454 \text{ g candy}}{1 \text{ lb candy}} \times \dfrac{1 \text{ lb candy}}{16 \text{ oz candy}} \times \dfrac{2.5 \text{ oz}}{1 \text{ candy bar}} =$$

$$\dfrac{2.1 \times 10^2 \text{ nutritional Cal}}{\text{candy bar}}$$

5.17   Heat energy produced by the friction of striking the match provides the activation energy necessary for this combustion process.

5.19   If the enzyme catalyzed a process needed to sustain life, the substance interfering with that enzyme would be classified as a poison.

5.21   At rush hour, approximately the same number of passengers enter and exit the train at any given stop. Throughout the trip, the number of passengers on the train may be essentially unchanged, but the identity of the individual passengers is continually changing.

5.23   Measure the concentrations of products and reactants at a series of times until no further concentration change is observed.

5.25   a.   A would decrease; the system would shift to remove excess B. B reacts with A to form products C and D.

       b.   A would increase; addition of excess C shifts the equilibrium to the left producing both A and B.

       c.   A would decrease; A would react with B to compensate for the loss of some D. The equilibrium shifts to the right.

       d.   A would remain the same. The presence of a catalyst increases the rate of attainment of equilibrium, but does not affect the equilibrium position.

End-of-Chapter Questions and Problems

5.27   joule

5.29   An exothermic reaction is one in which energy is released during chemical change.

5.31 The temperature of the water (or solution) is measured in a calorimeter. If the reaction being studied is exothermic, released energy heats the water and the temperature increases. In an endothermic reaction, heat flows from the water to the reaction and the water temperature decreases.

5.33 Double-walled containers, used in calorimeters, provide a small airspace between the part of the calorimeter (inside wall) containing the sample solution and the outside wall, contacting the surroundings. This makes heat transfer more difficult.

5.35 The first law of thermodynamics, the law of conservation of energy, states that the energy of the universe is constant.

5.37 Enthalpy is a measure of heat energy.

5.39 $Q = m_w \times \Delta T_w \times SH_w$

$$Q = 2.00 \times 10^2 \text{ g H}_2\text{O} \times 6.00°\text{C} \times \frac{1.00 \text{ cal}}{\text{g H}_2\text{O } °\text{C}}$$

$$Q = 1.20 \times 10^3 \text{ cal}$$

5.41 $1.20 \times 10^3 \text{ cal} \times \dfrac{4.18 \text{ J}}{1 \text{ cal}} = 5.02 \times 10^3 \text{ J}$

5.43 a. Entropy increases. Conversion of a solid to a liquid results in an increase in disorder of the substance. Solids retain their shape while liquids will flow and their shape is determined by their container.

b. Entropy increases. Conversion of a liquid to a gas results in an increase in disorder of the substance. Gas particles move randomly with very weak interactions between particles, much weaker than those interactions in the liquid state.

5.45 Isopropyl alcohol quickly evaporates (liquid → gas) after being applied to the skin. Conversion of a liquid to a gas requires heat energy. The heat energy is supplied by the skin. When this heat is lost, the skin temperature drops.

5.47 Decomposition of leaves and twigs to produce soil.

5.49 The activated complex is the arrangement of reactants in an unstable transition state as a chemical reaction proceeds. The activated complex must form in order to convert reactants to products.

5.51 A catalyst increases the rate of a reaction without itself undergoing change.

5.53   Assume a generalized exothermic reaction. In the catalyzed reaction, the energy released
        by the reaction is the same, but the activation energy is much lower. Equilibrium
        concentrations of products and reactants are the same in both the catalyzed and
        uncatalyzed reactions.

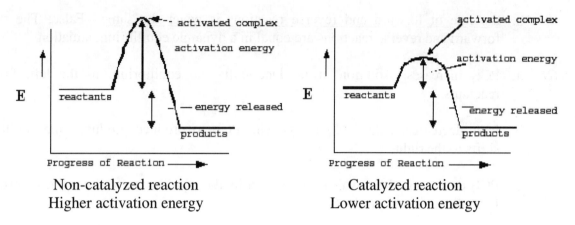

Non-catalyzed reaction   Catalyzed reaction
Higher activation energy   Lower activation energy

5.55   Enzymes are biological catalysts. The enzyme lysozyme catalyzes a process that results
        in the destruction of the cell walls of many harmful bacteria. This helps to prevent disease
        in organisms.

        The breakdown of foods to produce material for construction and repair of body tissue, as
        well as energy, is catalyzed by a variety of enzymes. For example, amylase begins the
        hydrolysis of starch in the mouth.

5.57   An increase in concentration of reactants means that there are more molecules in a certain
        volume. The probability of collision is enhanced because each molecule travels a shorter
        distance before meeting another molecule. The rate is proportional to the number of
        collisions per unit time.

5.59   A catalyst speeds up a chemical reaction by facilitating the formation of the activated
        complex, thus lowering the activation energy, the energy barrier for the reaction.

5.61   LeChatelier's principle states that when a system at equilibrium is disturbed, the
        equilibrium shifts in the direction that minimizes the disturbance.

5.63   A dynamic equilibrium has fixed concentrations of all reactants and products—these
        concentrations do not change with time. However, the process is dynamic because
        products and reactants are continuously being formed and consumed. The concentrations
        do not change because the rates of production and consumption are equal.

5.65   a. Equilibrium shifts to the left. Increasing the temperature increases the energy, a
            product of the reaction.

        b. No change; the number of moles of gaseous products and reactants are equal.

c. No change; a catalyst has no effect on the equilibrium position of the reaction.

5.67 a. <u>a slow reaction is an incomplete reaction</u>—False; A slow reaction may go to completion, but take a longer period of time.

b. <u>the rate of forward and reverse reactions is never the same</u>—False; The rate of forward and reverse reactions are equal in a dynamic equilibrium situation.

5.69 a. PCl_3 increases. Addition of product shifts the equilibrium to the left, favoring reactants.

b. PCl_3 decreases. Added Cl_2 reacts with PCl_3 to produce products; the equilibrium shifts to the right.

c. PCl_3 decreases. Removal of product shifts the equilibrium to the right, favoring the formation of more product.

d. PCl_3 decreases. Decreasing the temperature removes heat from the system. Heat is a product; therefore the equilibrium shifts to the right.

e. PCl_3 remains the same. Addition of a catalyst has no effect on the equilibrium position.

5.71 To determine the effect of pressure on equilibrium concentrations, focus on the number of moles of substances in the gaseous state. An increase in pressure would shift the equilibrium to the side of the reaction that has the least number of moles of gas. For *each* mole of *reactant* (in the gaseous state), *two* moles of *product* (in the gaseous state) are formed. Therefore, this reaction equilibrium would shift to the *left* upon an increase in pressure, and the concentration of H_2, a product, would <u>decrease</u>.

5.73 False. The position of equilibrium is not affected by a catalyst, only the rate at which equilibrium is attained.

5.75 Removing the cap allows CO_2 to escape into the atmosphere. This corresponds to the removal of product (CO_2):

$$CO_2 \text{ (l)} \rightleftharpoons CO_2 \text{ (g)}$$

The equilibrium shifts to the right, dissolved CO_2 is lost, and the beverage goes "flat".

Chapter 6
States of Matter
Gases, Liquids, and Solids
Solutions to the Odd-Numbered Problems

In-Chapter Questions and Problems

6.1   a. $725 \text{ mm Hg} \times \dfrac{1 \text{ atm}}{760 \text{ mm Hg}} = 0.954 \text{ atm}$

       b. $29.0 \text{ cm Hg} \times \dfrac{10 \text{ mm Hg}}{1 \text{ cm Hg}} \times \dfrac{1 \text{ atm}}{760 \text{ mm Hg}} = 0.382 \text{ atm}$

       c. $555 \text{ torr} \times \dfrac{1 \text{ atm}}{760 \text{ torr}} = 0.730 \text{ atm}$

6.3   a. $P_i V_i = P_f V_f$

          $P_i = \dfrac{P_f V_f}{V_i}$

          $P_i = \dfrac{(5.0 \text{ atm})(7.5 \text{ L})}{1.0 \text{ L}} = 37.5 \text{ atm} \approx 38 \text{ atm (2 significant figures)}$

       b. $P_f = \dfrac{P_i V_i}{V_f}$

          $P_f = \dfrac{(5.0 \text{ atm})(1.0 \text{ L})}{0.20 \text{ L}} = 25 \text{ atm}$

6.5   Initial temperature: $25°C + 273 = 298 \text{ K}$

       a. $100°C + 273 = 373 \text{ K}$

          $\dfrac{V_i}{T_i} = \dfrac{V_f}{T_f}$

          $V_f = \dfrac{V_i T_f}{T_i}$

$$V_f = \frac{(3.00 \text{ L})(373 \text{ K})}{298 \text{ K}} = 3.76 \text{ L}$$

b. $°C = \dfrac{5}{9}(°F - 32) = \dfrac{5}{9}(150 - 32) = 66°C$

$K = °C + 273 = 66 + 273 = 339 \text{ K}$

$$\frac{V_i}{T_i} = \frac{V_f}{T_f}$$

$$V_f = \frac{V_i T_f}{T_i}$$

$$V_f = \frac{(3.00 \text{ L})(339 \text{ K})}{298 \text{ K}} = 3.41 \text{ L}$$

c. $\dfrac{V_i}{T_i} = \dfrac{V_f}{T_f}$

$$V_f = \frac{V_i T_f}{T_i}$$

$$V_f = \frac{(3.00 \text{ L})(273 \text{ K})}{298 \text{ K}} = 2.75 \text{ L}$$

6.7 $\quad \dfrac{P_i V_i}{T_i} = \dfrac{P_f V_f}{T_f}$

$$P_f = \frac{P_i V_i T_f}{V_f T_i}$$

$P_i = 760 \text{ torr} \times \dfrac{1 \text{ atm}}{760 \text{ torr}} = 1.00 \text{ atm}$

$V_i = 2.00 \text{ L}$

$T_f = T_i = 25.0°C + 273 = 298 \text{ K}$

$V_f = 10.0 \text{ L}$

$$P_f = \frac{(1.00 \text{ atm})(2.00 \text{ L})(298 \text{ K})}{(10.0 \text{ L})(298 \text{ K})} = 0.200 \text{ atm}$$

6.9

$$\frac{V_i}{n_i} = \frac{V_f}{n_f}$$

$$n_f = \frac{n_i V_f}{V_i}$$

$$n_f = \frac{(1.00 \text{ mol H}_2)(100.0 \text{ L})}{22.4 \text{ L}} = 4.46 \text{ mol H}_2$$

6.11 $PV = nRT$

Solving for volume, V,

$$V = \frac{nRT}{P}$$

$$n = 10.0 \text{ g N}_2 \times \frac{1 \text{ mol N}_2}{28.00 \text{ g N}_2} = 0.357 \text{ mol N}_2$$

$$R = 0.0821 \frac{\text{L} \cdot \text{atm}}{\text{K} \cdot \text{mol}}$$

$$T = 30°\text{C} + 273 = 303 \text{ K}$$

$$P = 750 \text{ torr} \times \frac{1 \text{ atm}}{760 \text{ torr}} = 0.987 \text{ atm}$$

Substituting,

$$V = \frac{(0.357 \text{ mol N}_2)(0.0821 \text{ L} \cdot \text{atm/K} \cdot \text{mol})(303 \text{ K})}{0.987 \text{ atm}}$$

$$V = 9.00 \text{ L}$$

6.13 $PV = nRT$

Solving for number of moles of N_2, n,

$$n = \frac{PV}{RT}$$

$P = 1.00 \text{ atm}$ $\qquad$ $V = 5.00 \text{ L}$ $\qquad$ $R = 0.0821 \dfrac{\text{L} \cdot \text{atm}}{\text{K} \cdot \text{mol}}$ $\qquad$ $T = 273 \text{ K}$

Substituting,

$$n = \frac{(1.00 \text{ atm})(5.00 \text{ L})}{(0.0821 \text{ L} \cdot \text{atm/K} \cdot \text{mol})(273 \text{ K})} = 0.223 \text{ mol } N_2$$

6.15 Intermolecular forces in liquids are considerably stronger than intermolecular forces in gases. Particles are, on average, much closer together in liquids and the strength of attraction is inversely proportional to the distance of separation.

6.17 Evaporation is the conversion of a liquid to a gas at a temperature lower than the boiling point of the liquid. Condensation is the conversion of a gas to a liquid at a temperature lower than the boiling point of the liquid.

6.19 Solids are essentially incompressible because the average distance of separation among particles in the solid state is small. There is literally no space for the particles to crowd closer together.

End-of-Chapter Questions and Problems

6.21 In all cases, gas particles are much further apart than similar particles in the liquid or solid state. In most cases, particles in the liquid state are, on average, farther apart that those in the solid state. Water is the exception; liquid water's molecules are closer together than they are in the solid state.

6.23 Pressure is a force/unit area. Gas particles are in continuous, random motion. Collisions with the walls of the container results in a force (mass x acceleration) on the walls of the container. The sum of these collisional forces constitutes the pressure exerted by the gas.

6.25 Gases are easily compressed simply because there is a great deal of space between particles; they can be pushed closer together (compressed) because the space is available.

6.27 Gases exhibit more ideal behavior at low pressures. At low pressures, gas particles are more widely separated and therefore the attractive forces between particles are less. The ideal gas model assumes negligible attractive forces between gas particles.

6.29 The kinetic molecular theory states that the average kinetic energy of the gas particles increases as the temperature increases. Kinetic energy is proportional to (velocity)^2. Therefore, as the temperature increases the gas particle velocity increases and the rate of mixing increases as well.

6.31 The volume of the balloon is directly proportional to the pressure exerted on its inner walls. As the contents of the balloon cool, the average kinetic energy of the particles decreases, the pressure drops, and the balloon contracts.

6.33 Volume will decrease according to Boyle's law. Volume is inversely proportional to the pressure exerted on the gas.

6.35 The Kelvin scale is the only scale that is directly proportional to molecular motion, and it is the motion that determines the physical properties of gases.

6.37 Examine each effect separately:

- Volume and temperature are <u>directly</u> proportional; increasing T <u>increases</u> V.
- Volume and pressure are <u>inversely</u> proportional; decreasing P <u>increases</u> V.

Therefore, both variables work together to <u>increase</u> the volume.

6.39 A volume of 5 L (ordinate) corresponds to a pressure of 1 atm (abscissa).

6.41 A volume of 2 L (ordinate) corresponds to a pressure of 2.5 atm (abscissa).

$PV = k$

$(2.5 \text{ atm})(2 \text{ L}) = 5 \text{ L-atm} = k$

6.43 $P_i = 1.00$ atm $\qquad P_f = ?$ atm

$V_i = 20.9$ L $\qquad V_f = 4.00$ L

$P_i V_i = P_f V_f$

$P_f = \dfrac{P_i V_i}{V_f}$

$P_f = \dfrac{(1.00 \text{ atm})(20.9 \text{ L})}{4.00 \text{ L}}$

$P_f = 5.23$ atm

6.45 No. The volume is proportional to the temperature in K, not Celsius.

6.47 $V_i = 2.00$ L $\qquad V_f = ?$ L

$T_i = 250°C$ $\qquad T_f = 500°C$

$$\frac{V_i}{T_i} = \frac{V_f}{T_f}$$

$$V_f = \frac{V_i T_f}{T_i}$$

$$V_f = \frac{(2.00 \text{ L})(500\,^\circ\text{C} + 273 \text{ K})}{(250\,^\circ\text{C} + 273 \text{ K})}$$

$$V_f = 2.96 \text{ L}$$

The change in volume, $\qquad \Delta V = V_f - V_i$

$$\Delta V = 2.96 \text{ L} - 2.00 \text{ L}$$

$$\Delta V = 0.96 \text{ L}$$

6.49 $\quad \dfrac{P_i V_i}{T_i} = \dfrac{P_f V_f}{T_f}$

$$P_f V_f T_i = P_i V_i T_f$$

$$V_f = \frac{P_i V_i T_f}{P_f T_i}$$

6.51 $\quad P_i = 1.00$ atm $\qquad P_f = 125$ atm

$\qquad V_i = 2.25$ L $\qquad V_f = ?$ L

$\qquad T_i = 16°\text{C} \qquad T_f = 20°\text{C}$

Using the equation derived in question 6.49,

$$V_f = \frac{P_i V_i T_f}{P_f T_i}$$

and substituting:

$$V_f = \frac{(1.00 \text{ atm})(2.25 \text{ L})(20\,^\circ\text{C} + 273 \text{ K})}{(125 \text{ atm})(16\,^\circ\text{C} + 273 \text{ K})}$$

$$V_f = \frac{(1.00 \text{ atm})(2.25 \text{ L})(293 \text{ K})}{(125 \text{ atm})(289 \text{ K})}$$

$$V_f = 1.82 \times 10^{-2} \text{ L}$$

6.53 $n_i = 1.00 \text{ g He} \times \dfrac{1 \text{ mol He}}{4.00 \text{ g He}} = 0.25 \text{ mol He}$

$V_i = 1.00 \text{ L}$

$n_f = 6.00 \text{ g He} \times \dfrac{1 \text{ mol He}}{4.00 \text{ g He}} = 1.50 \text{ mol He}$

$V_f = ? \text{ L}$

$$\frac{V_i}{n_i} = \frac{V_f}{n_f}$$

$$V_f = \frac{V_i n_f}{n_i}$$

$$V_f = \frac{(1.00 \text{ L})(1.50 \text{ mol He})}{(0.25 \text{ mol He})} = 6.00 \text{ L}$$

6.55 No. One mole of an ideal gas will occupy exactly 22.4 L; however, there is no completely ideal gas and careful measurement will show a different volume.

6.57 Standard temperature is 273 K.

6.59 $PV = nRT$
$P = 5.0 \text{ atm}$
$V = 4.0 \text{ L}$
$T = 32\,^{\circ}\text{C} + 273 = 305 \text{ K}$
$R = 0.0821 \text{ L} \cdot \text{atm/K} \cdot \text{mol}$

$$n = \frac{PV}{RT} = \frac{(5.0 \text{ atm})(4.0 \text{ L})}{(0.0821 \text{ L} \cdot \text{atm/K} \cdot \text{mol})(305 \text{ K})} = 0.799 \text{ mol} \approx 0.80 \text{ mol}$$

6.61 $PV = nRT$

$n = 44.0 \text{ g CO}_2 \times \dfrac{1 \text{ mol}}{44.0 \text{ g CO}_2} = 1.00 \text{ mol CO}_2$

$T = 273 \text{ K}$

$P = 1.00$ atm

$R = 0.0821$ L $\cdot$ atm/K $\cdot$ mol

$$V = \frac{nRT}{P} = \frac{(1.00 \text{ mol})(0.0821 \text{ L} \cdot \text{atm/K} \cdot \text{mol})(273 \text{ K})}{(1.00 \text{ atm})} = 22.4 \text{ L}$$

6.63 $PV = nRT$

$P = 725$ mm Hg x $\dfrac{1 \text{ atm}}{760 \text{ mm Hg}} = 0.954$ atm

$V = 7.55$ L

$T = 45^{\circ}\text{C} + 273 = 318$ K

$R = 0.0821$ L $\cdot$ atm/K $\cdot$ mol

$$n = \frac{PV}{RT} = \frac{(0.945 \text{ atm})(7.55 \text{ L})}{(0.0821 \text{ L} \cdot \text{atm/K} \cdot \text{mol})(318 \text{ K})} = 0.276 \text{ mol}$$

6.65 $P_i = 750$ torr $P_f = 1.00$ atm

$V_i = 65.0$ mL $V_f = ?$ L

$T_i = 22^{\circ}\text{C}$ $T_f = 273$ K

$$\frac{P_i V_i}{T_i} = \frac{P_f V_f}{T_f}$$

$$V_f = \frac{P_i V_i T_f}{P_f T_i}$$

$$V_f = \frac{\left(750 \text{ torr x } \dfrac{1 \text{ atm}}{760 \text{ torr}}\right)\left(65.0 \text{ mL x } \dfrac{1 \text{ L}}{10^3 \text{ mL}}\right)(273 \text{ K})}{(1.00 \text{ atm})(22^{\circ}\text{C} + 273 \text{ K})}$$

$$V_f = \frac{(0.987 \text{ atm})(6.5 \times 10^{-2} \text{ L})(273 \text{ K})}{(1.00 \text{ atm})(295 \text{ K})}$$

$V_f = 5.94 \times 10^{-2}$ L

6.67   The molar volume of any gas at STP is 22.4 L, as shown by the following calculation:

$PV = nRT$

$$V = \frac{nRT}{P}$$

$$n = 1.00 \text{ mol} \qquad R = 0.0821 \frac{\text{L} \cdot \text{atm}}{\text{K} \cdot \text{mol}} \qquad T = 273 \text{ K} \qquad P = 1.00 \text{ atm}$$

$$V = \frac{(1.00 \text{ mol})(0.0821 \text{ L} \cdot \text{atm/K} \cdot \text{mol})(273 \text{ K})}{1.00 \text{ atm}}$$

$$V = 22.4 \text{ L}$$

6.69 Dalton's law states that the total pressure of a mixture of gases is the sum of the partial pressures of the component gases.

6.71 $P_\text{T} = P_{\text{N}_2} + P_{\text{F}_2} + P_{\text{He}}$

$P_\text{T} = 0.40 \text{ atm} + 0.16 \text{ atm} + 0.18 \text{ atm}$

$P_\text{T} = 0.74 \text{ atm}$

6.73 The vapor pressure of a liquid increases as the temperature of the liquid increases.

6.75 Viscosity is the resistance to flow caused by intermolecular attractive forces. Complex molecules may become entangled and not slide smoothly across one another.

6.77 All molecules exhibit London forces.

6.79 Only methanol exhibits hydrogen bonding. Methanol has an oxygen atom bonded to a hydrogen atom, a necessary condition for hydrogen bonding.

6.81 a. ionic solids—high melting temperature, brittle

b. covalent solids—high melting temperature, hard

6.83 Beryllium. Metallic solids are good electrical conductors. Carbon forms covalent solids that are poor electrical conductors.

6.85 Mercury. Mercury is a liquid at room temperature, whereas chromium is a solid at room temperature. Liquids have higher vapor pressures than solids.

Chapter 7
Solutions
Solutions to the Odd-Numbered Problems

In-Chapter Questions and Problems

7.1   A chemical analysis must be performed in order to determine the identity of all components, a qualitative analysis. If only one component is found, it is a pure substance; two or more components indicates a true solution.

7.3   After the container of soft drink is opened, CO_2 diffuses into the surrounding atmosphere; consequently the partial pressure of CO_2 over the soft drink decreases and the equilibrium

$$CO_2 \, (g) \rightleftharpoons CO_2 \, (aq)$$

shifts to the left, lowering the concentration of CO_2 in the soft drink.

7.5 $\% \, (W/V) = \dfrac{\text{grams of solute}}{\text{milliliters of solution}} \times 100\%$

$\% \, (W/V) = \dfrac{10.0 \text{ g NaCl}}{\left(0.0600 \text{ L} \times \dfrac{10^3 \text{ mL}}{1 \text{ L}} \right)} \times 100\%$

$\% \, (W/V) = 16.7 \, \% \text{ NaCl}$

7.7 $\% \, (W/V) = \dfrac{\text{grams of solute}}{\text{milliliters of solution}} \times 100\%$

$\% \, (W/V) = \dfrac{15.0 \text{ g KCl}}{\left(0.200 \text{ L} \times \dfrac{10^3 \text{ mL}}{1 \text{ L}} \right)} \times 100\%$

$\% \, (W/V) = 7.50 \, \% \text{ KCl}$

7.9 $\% \, (W/V) = \dfrac{\text{grams of solute}}{\text{milliliters of solution}} \times 100\%$

$\% \, (W/V) = \dfrac{20.0 \text{ g O}_2}{\left(78.0 \text{ L} \times \dfrac{10^3 \text{ mL}}{1 \text{ L}} \right)} \times 100\%$

$\% \, (W/V) = 2.56 \times 10^{-2} \, \% \text{ oxygen}$

7.11 $\% (W/W) = \dfrac{\text{grams solute}}{\text{grams solution}} \times 100\%$

$\% (W/W) = \dfrac{20.0 \text{ g O}_2}{(20.0 \text{ g O}_2 + 80.0 \text{ g N}_2)} \times 100\%$

$\% (W/W) = 20.0 \%$ oxygen

7.13 $ppt = \dfrac{\text{grams solute}}{\text{grams solution}} \times 10^3 \quad ppt = \dfrac{20.0 \text{ g O}_2}{20.0 \text{ g O}_2 + 80.0 \text{ g N}_2} \times 10^3 \quad ppt = 2.00 \times 10^2 \; ppt$

and

$ppm = \dfrac{\text{grams solute}}{\text{grams solution}} \times 10^6 \quad ppm = \dfrac{20.0 \text{ g O}_2}{20.0 \text{ g O}_2 + 80.0 \text{ g N}_2} \times 10^6 \quad ppm = 2.00 \times 10^5 \; ppm$

7.15 $M_{\text{HCl}} = \dfrac{\text{mol HCl}}{\text{L}_{\text{solution}}}$

Solving for mol HCl,

mol HCl $= (M_{\text{HCl}})(\text{L}_{\text{solution}})$

mol HCl $= (0.250 \; M)\left(5.00 \times 10^2 \text{ mL} \times \dfrac{1 \text{ L}}{10^3 \text{ mL}} \right)$

mol HCl $= 0.125$ mol HCl

7.17 $(M_1)(V_1) = (M_2)(V_2)$

$M_1 = 12.0 \; M$

$M_2 = 2.0 \; M$

$V_2 = 1.0 \times 10^2 \text{ mL} \times \dfrac{1 \text{ L}}{10^3 \text{ mL}} = 1.0 \times 10^{-1} \text{ L}$

Solving for the initial volume of 12 M HCl, V_1,

$V_1 = \dfrac{(M_2)(V_2)}{(M_1)} = \dfrac{(2.0 \; M)(1.0 \times 10^{-1} \text{ L})}{(12 \; M)}$

$V_1 = 1.7 \times 10^{-2}$ L (or 17 mL) of 12 M HCl

To prepare the solution, dilute 1.7×10^{-2} L of 12 M HCl with sufficient water to produce 1.0×10^2 mL of total solution.

7.19 Pure water has the higher freezing point. The presence of a solute decreases the freezing point.

7.21 The water flows from the cucumber to the surrounding solution. This is why the "cucumber" becomes a "pickle."

7.23 The water flows from the cells of the sailor to the fluid (now a more concentrated salt solution) surrounding the cells.

7.25 The Lewis structures of these two compounds are:

$$:C\equiv O: \qquad \text{and} \qquad \ddot{O}=C=\ddot{O}$$

Since oxygen is more electronegative than carbon, all carbon-oxygen bonds are polar. However, only carbon monoxide has a dipole moment. Hence, only carbon monoxide is polar. Carbon dioxide is linear and symmetrical. Consequently, carbon dioxide is nonpolar. Polar carbon monoxide is more soluble in water.

7.27

$$\frac{1.54 \text{ meq Na}^+}{1 \text{ L}} \times \frac{1 \text{ eq Na}^+}{10^3 \text{ meq Na}^+} \times \frac{1 \text{ mol charge}}{1 \text{ eq Na}^+} \times \frac{1 \text{ mol Na}^+}{1 \text{ mol charge}} = 1.54 \times 10^{-2} \frac{\text{mol Na}^+}{\text{L}}$$

End-of-Chapter Questions and Problems

7.29 $\% \text{ (W/V)} = \dfrac{\text{grams of solute}}{\text{milliliters of solution}} \times 100\%$

a. $\% \text{ (W/V)} = \dfrac{20.0 \text{ g NaCl}}{\left(1.00 \text{ L soln} \times \dfrac{10^3 \text{ mL soln}}{1 \text{ L soln}}\right)} \times 100\%$

$\% \text{ (W/V)} = 2.00 \% \text{ NaCl}$

b. $\% \text{ (W/V)} = \dfrac{33.0 \text{ g C}_6\text{H}_{12}\text{O}_6}{5.00 \times 10^2 \text{ mL soln}} \times 100\%$

$\% \text{ (W/V)} = 6.60 \% \text{ C}_6\text{H}_{12}\text{O}_6$

7.31 $\% \, (W/W) = \dfrac{\text{grams solute}}{\text{grams solution}} \times 100\%$

a. $\% \, (W/W) = \dfrac{21.0 \text{ g NaCl}}{1.00 \times 10^2 \text{ g soln}} \times 100\%$

$\% \, (W/W) = 21.0 \, \% \text{ NaCl}$

b.   Use the density of the sodium chloride solution as a conversion factor to calculate the volume of the sodium chloride solution.

$5.00 \times 10^2 \text{ mL soln} \times \dfrac{1.12 \text{ g soln}}{1 \text{ mL soln}} = 5.60 \times 10^2 \text{ g soln}$

Then,

$\% \, (W/W) = \dfrac{21.0 \text{ g NaCl}}{5.60 \times 10^2 \text{ g soln}} \times 100\%$

$\% \, (W/W) = 3.75 \, \% \text{ NaCl}$

7.33 $\% \, (W/W) = \dfrac{\text{grams solute}}{\text{grams solution}} \times 100\%$

Solve for g solute,

$\text{g solute} = \dfrac{\% \, (W/W)(\text{g soln})}{100\%}$

a. $\text{g NaCl} = \dfrac{[0.900\% \, (W/W)](2.50 \times 10^2 \text{g soln})}{100\%}$

$\text{g NaCl} = 2.25 \text{ g NaCl}$

b.   Assume that the density of the solution is 1.00 g/mL; then

$\text{g CH}_3\text{COONa} = \dfrac{[1.25\% \, (W/W)](2.50 \times 10^2 \text{g soln})}{100\%}$

$\text{g CH}_3\text{COONa} = 3.13 \text{ g NaC}_2\text{H}_3\text{O}_2$

7.35 $\%(W/V) = \dfrac{\text{grams of solute}}{\text{milliliters of solution}} \times 100\%$

$$\%(W/V) = \frac{14.6 \text{ g KNO}_3}{75.0 \text{ mL of solution}} \times 100\%$$

$$\%(W/V) = 19.47\% \text{ KNO}_3 \approx 19.5\% \text{ KNO}_3$$

7.37 $\quad \%(W/V) = \dfrac{\text{grams of solute}}{\text{milliliters of solution}} \times 100\%$

$$\text{Grams of solute} = \frac{\%(W/V)(\text{milliliters of solution})}{100\%}$$

$$\text{Grams of sugar} = \frac{1.00 \%(W/V)(100 \text{ mL of solution})}{100\%}$$

$$\text{Grams of sugar} = 1.00 \text{ g sugar}$$

7.39 $\quad 1.0 \text{ mg Cu}^{2+} \times \dfrac{1 \text{ g}}{10^3 \text{ mg}} = 1.0 \times 10^{-3} \text{ g Cu}^{2+}$

$$0.50 \text{ kg} \times \frac{10^3 \text{ g}}{1 \text{ kg}} = 5.0 \times 10^2 \text{ g solution}$$

$$\text{ppt Cu}^{2+} = \frac{\text{g Cu}^{2+}}{\text{g solution}} \times 10^3 \text{ ppt} = \frac{1.0 \times 10^{-3} \text{ g Cu}^{2+}}{5.0 \times 10^2 \text{ g solution}} \times 10^3 \text{ ppt}$$

$$\text{ppt Cu}^{2+} = 2.0 \times 10^{-3} \text{ ppt}$$

7.41 $\quad$ a. $\quad M_{\text{NaCl}} = \dfrac{\text{mol NaCl}}{\text{L solution}}$

$$M_{\text{NaCl}} = \frac{\left(20.0 \text{ g NaCl} \times \dfrac{1 \text{ mol NaCl}}{58.44 \text{ g NaCl}} \right)}{1.00 \text{ L solution}}$$

$$M_{\text{NaCl}} = 0.342 \; M \text{ NaCl}$$

b. $\quad M_{\text{C}_6\text{H}_{12}\text{O}_6} = \dfrac{\text{mol C}_6\text{H}_{12}\text{O}_6}{\text{L solution}}$

$$M_{\text{C}_6\text{H}_{12}\text{O}_6} = \frac{\left(33.0 \text{ g C}_6\text{H}_{12}\text{O}_6 \times \dfrac{1 \text{ mol C}_6\text{H}_{12}\text{O}_6}{180.0 \text{ g C}_6\text{H}_{12}\text{O}_6} \right)}{\left(5.00 \times 10^2 \text{ mL solution} \times \dfrac{1 \text{ L soln}}{10^3 \text{ mL soln}} \right)}$$

$$M_{C_6H_{12}O_6} = 0.367 \ M \ C_6H_{12}O_6$$

7.43 $$M = \dfrac{\text{mol solute}}{\text{L solution}}$$

Solve for mol solute,   mol solute = (M)(L solution)

a.   mol NaCl = $(0.100 \ M)\left(2.50 \times 10^2 \ \text{mL soln} \times \dfrac{1 \ \text{L soln}}{10^3 \ \text{mL soln}} \right)$

mol NaCl = 2.50×10^{-2} mol NaCl

and, 2.50×10^{-2} mol NaCl x $\dfrac{58.44 \ \text{g NaCl}}{1 \ \text{mol NaCl}}$ = 1.46 g NaCl

b.   mol $C_6H_{12}O_6$ = $(0.200 \ M)\left(2.50 \times 10^2 \ \text{mL soln} \times \dfrac{1 \ \text{L soln}}{10^3 \ \text{mL soln}} \right)$

mol $C_6H_{12}O_6$ = 5.00×10^{-2} mol $C_6H_{12}O_6$

and, 5.00×10^{-2} mol $C_6H_{12}O_6$ x $\dfrac{180.0 \ \text{g } C_6H_{12}O_6}{1 \ \text{mol } C_6H_{12}O_6}$ = 9.00 g $C_6H_{12}O_6$

7.45 $$M_{C_{12}H_{22}O_{11}} = \dfrac{\text{mol } C_{12}H_{22}O_{11}}{\text{L solution}}$$

$$M_{C_{12}H_{22}O_{11}} = \dfrac{\left(50.0 \ \text{g } C_{12}H_{22}O_{11} \times \dfrac{1 \ \text{mol } C_{12}H_{22}O_{11}}{342.3 \ \text{g } C_{12}H_{22}O_{11}} \right)}{L}$$

$$M_{C_{12}H_{22}O_{11}} = 0.146 \ M \ C_{12}H_{22}O_{11}$$

7.47 $(M_1)(V_1) = (M_2)(V_2)$

$M_1 = 1.00 \ M$ $M_2 = 0.100 \ M$

$V_1 = ?$ $V_2 = 0.500 \ L$

Solve for V_1,

$$V_1 = \dfrac{(M_2)(V_2)}{M_1}$$

Substitute,

$$V_1 = \frac{(0.100)(0.500)}{1.00}$$

$$V_1 = 5.00 \times 10^{-2} \text{ L}$$

7.49 $(M_1)(V_1) = (M_2)(V_2)$

Solve for M_1,

$$M_1 = \frac{(M_2)(V_2)}{V_1}$$

Substitute,

$$M_1 = \frac{(2.00)(5.000 \times 10^{-1} \text{ L})}{5.00 \times 10^{-2} \text{ L}} = 20.0 \ M$$

7.51   A colligative property is a solution property that depends on the concentration of solute particles rather than the identity of the particles.

7.53   Salt is an ionic substance that dissociates in water to produce positive and negative ions. These ions (or particles) lower the freezing point of water. If the concentration of salt particles is large, the freezing point may be depressed below the surrounding temperature, and the ice would melt.

7.55   Chemical properties depend on the identity of the substance, whereas colligative properties depend on concentration, not identity.

7.57   Raoult's law states that when a solute is added to a solvent, the vapor pressure of the solvent decreases in proportion to the concentration of the solute.

7.59   One mole of $CaCl_2$ produces three moles of particle in solution whereas one mole of NaCl produces two moles of particles in solution. Therefore, a one molar $CaCl_2$ solution contains a greater number of particles than a one molar NaCl solution and will produce a greater freezing point depression.

Overview of questions 7.61 – 7.65:

The *molar* concentration of both solutions is identical. Sodium chloride is an ionic compound that dissociates in water to produce two ions for each ion pair. Therefore, the molarity of particles is 2 x 0.50 M = 1.0 M. Sucrose, in contrast, is a covalent, nondissociating solute; the number of particles and molecules is identical. Consequently, the molarity of *particles* is 0.50 *M*. Armed with this information, we can now answer each question.

7.61 The NaCl solution would melt at a *lower* temperature since it has a larger number of particles per liter. So, the sucrose solution, with fewer particles per liter, would have the higher melting point. (Recall that solute particles depress the freezing point.)

7.63 The NaCl solution would have the lower vapor pressure because it has the larger number of particles per liter. The sucrose solution, with fewer particles per liter, would have the higher vapor pressure.

7.65 a.   A → B (more dilute to more concentrated solution)

     b.   A → B (A is more dilute, 0.10 *M* particles and B is more concentrated, 0.10 *M* x 2 particles)

7.67 a. $CaCl_2$ is 3 x 0.15 *M* = 0.45 *M* particles
           NaCl is 2 x 0.15 *M* = 0.30 *M* particles; 0.45 *M* > 0.30 *M*; hypertonic

     b.   Glucose is 1 x 0.35 *M* = 0.35 *M* particles
           NaCl is 2 x 0.15 *M* = 0.30 *M* particles; 0.35 *M* > 0.30 *M*; hypertonic

7.69 Water is often termed the "universal solvent" because it is a polar molecule and will dissolve, at least to some extent, most ionic and polar covalent compounds. The majority of our body mass is water and this water is an important part of the nutrient transport system due to its solvent properties. This is true in other animals and plants as well. Because of its ability to hydrogen bond, water has a high boiling point and a low vapor pressure. Also, water is abundant and easily purified.

7.71 The shelf life is a function of the stability of the ammonia-water solution. The ammonia can react with the water to convert to the extremely soluble and stable ammonium ion. Also, ammonia and water are polar molecules. Polar interactions, particularly hydrogen bonding, are strong and contribute to the long-term solution stability.

7.73 In the Lewis structure of water, there are two lone pairs of electrons on the oxygen atom. Remember that this causes the geometry, or shape, of the water molecule to be bent.

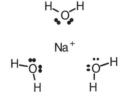

The partial (−) side of the water molecule attracts the positive sodium ion.

Several water molecules "hydrate" each sodium ion.

7.75 Polar; like dissolves like (H_2O is polar)

7.77 In dialysis, sodium ions move from a region of high concentration to a region of low concentration. If we wish to remove (transport) sodium ions from the blood, they can move to a region of lower concentration, the dialysis solution.

7.79 Elevated concentrations of sodium ion in the blood may occur whenever large amounts of water are lost. Diarrhea, diabetes, and certain high-protein diets are particularly problematic.

Chapter 8
Acids and Bases
Solutions to the Odd-Numbered Problems

In-Chapter Questions and Problems

8.1   a. $HF(aq) + H_2O \text{ (l)} \rightleftharpoons H_3O^+ (aq) + F^- (aq)$

      b. $NH_3(aq) + H_2O \text{ (l)} \rightleftharpoons NH_4^+(aq) + OH^-(aq)$

8.3   a.   HF and F^-; H_2O and H_3O^+

      b. NH_3 and NH_4^+; H_2O and OH^-

8.5   a. NH_4^+; based on Figure 8.1

      b. H_2SO_4; based on Figure 8.1

8.7 $[H_3O^+] [OH^-] = 1.0 \times 10^{-14}$
      $$[OH^-] = \frac{1.0 \times 10^{-14}}{1.0 \times 10^{-3}}$$
      $[OH^-] = 1.0 \times 10^{-11} \, M$

8.9   Referring to the discussion of the decimal-based system, a solution of sodium hydroxide producing $[OH^-] = 1.0 \times 10^{-2} \, M$ corresponds to a pH of 12.00.

8.11 A pH of 8.50 is a non-integer. The calculation of the $[H_3O^+]$ is most easily accomplished with the aid of a calculator.

      $pH = -\log [H_3O^+]$

      and, $[H_3O^+] = 10^{-pH}$

      On the calculator:
          Enter 8.50
          Press "change sign" key
          Press 10^X key
          The result is $[H_3O^+] = 3.2 \times 10^{-9} \, M$

8.13 The acid-base reaction is:

$$HCl(aq) + NaOH(aq) \rightarrow NaCl(aq) + H_2O(l)$$

$$(M_{acid})(V_{liters\ acid}) = (M_{base})(V_{liters\ base})$$

Solving for M_{base}, $\qquad M_{base} = \dfrac{(M_{acid})(V_{liters\ acid})}{(V_{liters\ base})}$

Substituting

$$M_{base} = \dfrac{(0.2000\ M)\left(20.00\ ml \times \dfrac{1\ L}{10^3\ mL}\right)}{\left(40.00\ ml \times \dfrac{1\ L}{10^3\ mL}\right)} = 0.1000\ M\ NaOH$$

8.15   The equilibrium reaction is:

$$CO_2 + H_2O \rightleftharpoons H_2CO_3 \rightleftharpoons H_3O^+ + HCO_3^-$$

An increase in the partial pressure of CO_2 is a stress on the left side of the equilibrium. The equilibrium will shift to the right in an effort to decrease the concentration of CO_2. This will cause the molar concentration of H_2CO_3 to increase.

8.17   The equilibrium reaction is:

$$CO_2 + H_2O \rightleftharpoons H_2CO_3 \rightleftharpoons H_3O^+ + HCO_3^-$$

In Question 8.15, the equilibrium shifts to the right. Therefore the molar concentration of H_3O^+ should increase.

In Question 8.16, the equilibrium shifts to the left. Therefore the molar concentration of H_3O^+ should decrease.

End-of-Chapter Questions and Problems

8.19   a.   An Arrhenius acid is a substance that dissociates, producing hydrogen ions.

       b.   A Brønsted-Lowry acid is a substance that behaves as a proton donor.

8.21   The Brønsted-Lowry theory provides a broader view of acid-base theory than does the Arrhenius theory. Brønsted-Lowry emphasizes the role of the solvent in the dissociation process.

8.23   a. $HNO_2\ (aq) + H_2O\ (l) \rightleftharpoons H_3O^+\ (aq) + NO_2^-\ (aq)$

b. $HCN\ (aq) + H_2O\ (l) \rightleftharpoons H_3O^+\ (aq) + CN^-\ (aq)$

8.25   a. HNO_2 and NO_2^-; H_2O and H_3O^+

        b. HCN and CN^-; H_2O and H_3O^+

8.27   a. Weak

        b. Weak

        c. Weak

8.29   a. CN^- and HCN; NH_3 and NH_4^+

        b. CO_3^{2-} and HCO_3^-; Cl^- and HCl

8.31   Concentration refers to the quantity of acid or base contained in a specified volume of solvent. Strength refers to the degree of dissociation of the acid or base.

8.33   a. H_3O^+ is a Brønsted acid
        b. OH^- is a Brønsted base
        c. H_2O can behave as both acid and base

8.35   a. H_2CO_3 is a Brønsted acid
        b. HCO_3^- can behave as both acid and base
        c. CO_3^{2-} is a Brønsted base

8.37 $CN^- + H^+ \rightarrow HCN$
       (base)   (conjugate acid)

8.39 $HI \rightarrow H^+ + I^-$
       (acid)   (conjugate base)

8.41 $[H_3O^+][OH^-] = 1.0 \times 10^{-14}$

Solving for $[H_3O^+]$, $[H_3O^+] = \dfrac{1 \times 10^{-14}}{[OH^-]}$

a. Substituting $[OH^-] = 1.0 \times 10^{-7}\ M$

$$[H_3O^+] = \frac{1 \times 10^{-14}}{1.0 \times 10^{-7}} = 1.0 \times 10^{-7} M$$

b. Substituting $[OH^-] = 1.0 \times 10^{-3} \, M$

$$[H_3O^+] = \frac{1 \times 10^{-14}}{1.0 \times 10^{-3}} = 1.0 \times 10^{-11} M$$

8.43   a. Neutral, $[H_3O^+] = 1.0 \times 10^{-7} \, M$

   b. Basic, $[H_3O^+]$ is less than $1.0 \times 10^{-7} \, M$

8.45   a. $pH = -\log [H_3O^+]$

   $pH = -\log [1.0 \times 10^{-7}]$

   $pH = 7.00$

   b. $pH = -\log [H_3O^+]$

   Since $[H_3O^+][OH^-] = 1.0 \times 10^{-14}$

   and, $[H_3O^+] = \dfrac{1 \times 10^{-14}}{[OH^-]}$

   Substituting, $[H_3O^+] = \dfrac{1 \times 10^{-14}}{1 \times 10^{-9}}$

   $[H_3O^+] = 1.0 \times 10^{-5} \, M$

   and, $pH = -\log [1.0 \times 10^{-5}]$

   $pH = 5.00$

8.47   a. $pH = -\log [H_3O^+]$

   and, $[H_3O^+] = 10^{-pH}$

   On the calculator
   Enter 1.00
       Press "change sign" key
       Press 10^x key
       The result is $[H_3O^+] = 1.0 \times 10^{-1} \, M$

   and, $[H_3O^+][OH^-] = 1.0 \times 10^{-14}$

solving for [OH⁻]

$$[OH^-] = \frac{1.0 \times 10^{-14}}{[H_3O^+]} = \frac{1.0 \times 10^{-14}}{1.0 \times 10^{-1}} = 1.0 \times 10^{-13} M$$

b. $pH = -\log [H_3O^+]$

and, $[H_3O^+] = 10^{-pH}$

On the calculator:
    Enter 9.00
    Press "change sign" key
    Press 10^x key
    The result is $[H_3O^+] = 1.0 \times 10^{-9} M$

and, $[H_3O^+][OH^-] = 1.0 \times 10^{-14}$

solving for [OH⁻]

$$[OH^-] = \frac{1.0 \times 10^{-14}}{[H_3O^+]} = \frac{1.0 \times 10^{-14}}{1.0 \times 10^{-9}} = 1.0 \times 10^{-5} M$$

8.49   a. $pH = -\log [H_3O^+]$

and, $[H_3O^+] = 10^{-pH}$

On the calculator:
    Enter 1.30
    Press "change sign" key
    Press 10^x key
    The result is $[H_3O^+] = 5.0 \times 10^{-2} M$

and, $[H_3O^+][OH^-] = 1.0 \times 10^{-14}$

solving for [OH⁻]

$$[OH^-] = \frac{1.0 \times 10^{-14}}{[H_3O^+]} = \frac{1.0 \times 10^{-14}}{5.0 \times 10^{-2}} = 2.0 \times 10^{-13} M$$

b. $pH = -\log [H_3O^+]$

and, $[H_3O^+] = 10^{-pH}$

On the calculator:
   Enter 9.70
   Press "change sign" key
   Press 10^x key
   The result is $[H_3O^+] = 2.0 \times 10^{-10} M$

and, $[H_3O^+][OH^-] = 1.0 \times 10^{-14}$

solving for $[OH^-]$

$$[OH^-] = \frac{1.0 \times 10^{-14}}{[H_3O^+]} = \frac{1.0 \times 10^{-14}}{2.0 \times 10^{-10}} = 5.0 \times 10^{-5} M$$

8.51   A neutralization reaction is one in which an acid and a base react to produce water and a salt (a "neutral" solution).

8.53   a. $pH = -\log [H_3O^+]$

and, $[H_3O^+] = 10^{-pH}$

On the calculator:
   Enter 6.00
   Press "change sign" key
   Press 10^x key
   The result is $[H_3O^+] = 1.0 \times 10^{-6} M$

and, $[H_3O^+] [OH^-] = 1.0 \times 10^{-14}$

solving for $[OH^-]$
$$[OH^-] = \frac{1.0 \times 10^{-14}}{[H_3O^+]} = \frac{1.0 \times 10^{-14}}{1.0 \times 10^{-6}} = 1.0 \times 10^{-8} M$$

   b. $pH = -\log [H_3O^+]$

and, $[H_3O^+] = 10^{-pH}$

On the calculator:
   Enter 5.20
   Press "change sign" key
   Press 10^x key
   The result is $[H_3O^+] = 6.3 \times 10^{-6} M$

and, $[H_3O^+][OH^-] = 1.0 \times 10^{-14}$

solving for $[OH^-]$

$$[OH^-] = \frac{1.0 \times 10^{-14}}{[H_3O^+]} = \frac{1.0 \times 10^{-14}}{6.3 \times 10^{-6}} = 1.6 \times 10^{-9} M$$

c. $pH = -\log [H_3O^+]$

and, $[H_3O^+] = 10^{-pH}$

On the calculator:
  Enter 7.80
  Press "change sign" key
  Press 10^x key
  The result is $[H_3O^+] = 1.6 \times 10^{-8} M$

and, $[H_3O^+][OH^-] = 1.0 \times 10^{-14}$

solving for $[OH^-]$

$$[OH^-] = \frac{1.0 \times 10^{-14}}{[H_3O^+]} = \frac{1.0 \times 10^{-14}}{1.6 \times 10^{-8}} = 6.3 \times 10^{-7} M$$

8.55 One pH unit difference corresponds to a tenfold (10^1) difference in concentration of H_3O^+ or OH^-.

a. $4 - 2 = 2$ pH units, therefore $(10^1)^2$ or 1×10^2

b. $11 - 7 = 4$ pH units, therefore $(10^1)^4$ or 1×10^4

c. $12 - 2 = 10$ pH units, therefore $(10^1)^{10}$ or 1×10^{10}

8.57 $pH = -\log [H_3O^+]$

and, $[H_3O^+] = 10^{-pH}$

a. pH = 5.0
   $[H_3O^+] = 10^{-5.0}$
   $[H_3O^+] = 1 \times 10^{-5}$

b. pH = 12.0

$[H_3O^+] = 10^{-12.0}$

$[H_3O^+] = 1 \times 10^{-12}$

c. pH = 5.5

$[H_3O^+] = 10^{-5.5}$

On the calculator:

Enter 5.50

Press "change sign" key

Press 10^x key

The result is $[H_3O^+] = 3.2 \times 10^{-6}$

8.59 pH $= -\log[H_3O^+]$

a. $[H_3O^+] = 1.0 \times 10^{-6}\,M$

$pH = -\log[1.0 \times 10^{-6}]$

pH = 6.00

b. $[H_3O^+] = 1.0 \times 10^{-8}\,M$

$pH = -\log[1.0 \times 10^{-8}]$

pH = 8.00

c. $[H_3O^+] = 5.6 \times 10^{-4}\,M$

$pH = -\log[5.6 \times 10^{-4}]$

pH = 3.25

8.61 pH $= -\log[H_3O^+]$

pH $= -\log 7.5 \times 10^{-4}$

pH $= 3.12$

8.63 $[H_3O^+][OH^-] = 1.0 \times 10^{-14}$

$[H_3O^+] = \dfrac{1.0 \times 10^{-14}}{[OH^-]} = \dfrac{1.0 \times 10^{-14}}{[5.5 \times 10^{-4}]} = 1.8 \times 10^{-11}$

$pH = -\log[H_3O^+]$

$pH = -\log 1.8 \times 10^{-11}$

pH = 10.74

8.65 a. NH_3 is a weak base.

NH_4Cl is a salt formed from NH_3.

Therefore, NH_3 and NH_4Cl can form a buffer solution.

b. HNO_3 is a strong acid; strong acids are not suitable for buffer preparation; they are completely dissociated.
Therefore, HNO_3 and KNO_3 cannot form a buffer solution.

8.67 a. A buffer solution contains components (a weak acid and its salt or a weak base and its salt) that enable the solution to resist large changes in pH when acids or bases are added.

b. Acidosis is a medical condition characterized by higher-than-normal levels of CO_2 in the blood and lower-than-normal blood pH.

8.69 a. Addition of strong acid is equivalent to adding H_3O^+. This is a stress on the right side of the equilibrium and the equilibrium will shift to the left. Consequently the $[CH_3COOH]$ increases.

b. Water, in this case, is a solvent and does not appear in the equilibrium expression. Hence, it does not alter the position of the equilibrium.

Chapter 9
The Nucleus and Radioactivity
Solutions to the Odd-Numbered Problems

In-Chapter Questions and Problems

9.1 The nucleus

9.3 a. $^{85}_{36}\text{Kr} \rightarrow {}^{85}_{37}\text{Rb} + {}^{0}_{-1}\text{e}$

Always double-check your answer. Make sure that the sum of the atomic numbers is the same on both sides of the equation, and that the sum of the mass numbers are the same on both sides of the equation.
Atomic numbers: $36 = 37 + (-1)$
Mass numbers: $85 = 85 + 0$

b. $^{226}_{88}\text{Ra} \rightarrow {}^{4}_{2}\text{He} + {}^{222}_{86}\text{Rn}$

9.5 The half-life of sodium-24 is 15 hours. The number (n) of half-lives elapsed is:

$$n = 2.5 \text{ days x } \frac{24 \text{ hours}}{1 \text{ day}} \text{ x } \frac{1 \text{ half-life}}{15 \text{ hours}} = 4 \text{ half-lives}$$

Then,

$$100 \text{ ng } \xrightarrow{\text{first half-life}} 50 \text{ ng}$$
$$\xrightarrow{\text{second half-life}} 25.0 \text{ ng}$$
$$\xrightarrow{\text{third half-life}} 12.5 \text{ ng}$$
$$\xrightarrow{\text{fourth half-life}} 6.3 \text{ ng}$$

Therefore, 6.3 ng of sodium-24 remain after 2.5 days.

9.7 The half-life of technetium-99m is 6 hours. The number (n) of half-lives elapsed is:

$$n = 12 \text{ hours x } \frac{1 \text{ half-life}}{6 \text{ hours}} = 2 \text{ half-lives}$$

Then, assume that the original amount is x g.

$$x \text{ g} \xrightarrow{\substack{\text{first} \\ \text{half-life}}} \frac{x}{2} \text{ g} \xrightarrow{\substack{\text{second} \\ \text{half-life}}} \frac{x}{4} \text{ g}$$

Therefore, 1/4 of the radioisotope remains after 2 half-lives, 12 hours.

9.9   Isotopes with short half-lives release their radiation rapidly. There is much more radiation per unit time observed with short half-life substances; hence, the signal is stronger and the sensitivity of the procedure is enhanced.

9.11   The rem takes into account the relative biological effect of the radiation in addition to the quantity of radiation. This provides a more meaningful estimate of potential radiation damage to human tissue.

End-of-Chapter Questions and Problems

9.13   Natural radioactivity is the spontaneous decay of a nucleus to produce high-energy particles or rays.

9.15   Alpha particles contain two protons and two neutrons; hence they have a charge of +2.

9.17   A beta particle is an electron with a –1 charge.

9.19   A positron has a positive charge and a beta particle has a negative charge.

9.21
   - charge, $\alpha = +2$, $\beta = -1$
   - mass, $\alpha = 4$ amu, $\beta = 0.000549$ amu
   - velocity, $\alpha = 10\%$ of the speed of light, $\beta = 90\%$ of the speed of light

9.23   Chemical reactions involve joining, separating and rearranging atoms; valence electrons are critically involved. Nuclear reactions involve changes in nuclear composition only.

9.25 $^{4}_{2}\text{He}$

9.27 $^{235}_{92}\text{U}$

9.29   Mass number – atomic number = number of neutrons;
       So $^{1}_{1}\text{H}$, $1 - 1 = 0$ neutrons, $^{2}_{1}\text{H}$, $2 - 1 = 1$ neutron, $^{3}_{1}\text{H}$ $3 - 1 = 2$ neutrons.

9.31 $^{15}_{7}\text{N}$

9.33 $^{60}_{27}\text{Co} \rightarrow {}^{60}_{28}\text{Ni} + {}^{0}_{-1}\beta + \gamma$

9.35   a. $^{23}_{11}\text{Na} + {}^{2}_{1}\text{H} \rightarrow {}^{24}_{11}\text{Na} + {}^{1}_{1}\text{H}$

       b. $^{238}_{92}\text{U} + {}^{14}_{7}\text{N} \rightarrow {}^{246}_{99}\text{Np} + 6\,{}^{1}_{0}\text{n}$

       c. $^{24}_{10}\text{Ne} \rightarrow {}^{0}_{-1}\beta + {}^{24}_{11}\text{Na}$

9.37 $\quad^{209}_{83}\text{Bi} + \,^{54}_{24}\text{Cr} \rightarrow \,^{262}_{107}\text{Bh} + \,^{1}_{0}\text{n}$

9.39 $\quad^{27}_{12}\text{Mg} \rightarrow \,^{27}_{13}\text{Al} + \,^{0}_{-1}\text{e}$

9.41 $\quad^{12}_{7}\text{N} \rightarrow \,^{12}_{6}\text{C} + \,^{0}_{1}\text{e}$

9.43 Natural radioactivity is a spontaneous process; artificial radioactivity is nonspontaneous and results from a nuclear reaction that produces an unstable nucleus.

9.45
- Nuclei for light atoms tend to be most stable if their neutron/proton ratio is close to 1.
- Nuclei with more than 84 protons tend to be unstable.
- Isotopes with a "magic number" of protons or neutrons (2, 8, 20, 50, 82, or 126 protons or neutrons) tend to be stable.
- Isotopes with even numbers of protons or neutrons tend to be more stable.

9.47 The half-life of iodine-131 is 8.1 days. The number (n) of half-lives elapsed is:

$$n = 24 \text{ days } \times \frac{1 \text{ half-life}}{8.1 \text{ days}} = 3.0 \text{ half-lives} \quad \textit{(2 significant figures)}$$

Then,

$$3.2 \text{ mg} \xrightarrow[\text{half-life}]{\text{first}} 1.6 \text{ mg} \xrightarrow[\text{half-life}]{\text{second}} 0.80 \text{ mg} \xrightarrow[\text{half-life}]{\text{third}} 0.40 \text{ mg}$$

Therefore, 0.40 mg of iodine-131 remains after 24 days.

9.49 The half-life of iron-59 is 45 days. The number (n) of half-lives elapsed is:

$$n = 135 \text{ days } \times \frac{1 \text{ half - life}}{45 \text{ days}} = 3.0 \text{ half-lives}$$

Then,

$$100 \text{ mg} \xrightarrow[\text{half-life}]{\text{first}} 50 \text{ mg} \xrightarrow[\text{half-life}]{\text{second}} 25 \text{ mg} \xrightarrow[\text{half-life}]{\text{third}} 13 \text{ mg}$$

Therefore, 13 mg of iron-59 remains after 135 days.

9.51
$$200 \text{ μg} \xrightarrow[\text{half-life}]{\text{first}} 100 \text{ μg} \xrightarrow[\text{half-life}]{\text{second}} 50 \text{ μg} \xrightarrow[\text{half-life}]{\text{third}} 25 \text{ μg}$$

So, $3 \text{ half-lives} \times \dfrac{67 \text{ hours}}{1 \text{ half - life}} = 201 \text{ hours}$

9.53 Fission splits nuclei to produce energy.

9.55   a. The fission process involves the breaking down of large, unstable nuclei into smaller, more stable nuclei. This process releases energy in the form of heat and/or light.

       b. The heat generated during the fission process could be used to generate steam, which is then used to drive a turbine to create electricity.

9.57 $^{3}_{1}H + ^{1}_{1}H \rightarrow ^{4}_{2}He + energy$

9.59   A "breeder" reactor creates the fuel which can be used by a conventional fission reactor during its fission process.

9.61   The reaction in a fission reactor that involves neutron production and causes subsequent reactions accompanied by the production of more neutrons in a continuing process.

9.63   High operating temperatures

9.65 $^{108}_{47}Ag + ^{4}_{2}He \rightarrow ^{112}_{49}In$

       $^{112}_{49}In$ is the intermediate isotope of indium.

9.67   a. Technetium-99m is used to study the heart (cardiac output, size, and shape), kidney (follow-up procedure for kidney transplant), and liver and spleen (size, shape, presence of tumors).

       b. Xenon-133 is used to locate regions of reduced ventilation and presence of tumors in the lung.

9.69   Radiation therapy provides sufficient energy to destroy molecules critical to the reproduction of cancer cells.

9.71   Background radiation, radiation from natural sources, is emitted by the sun as cosmic radiation, and from naturally radioactive isotopes found throughout our environment.

9.73   Level decreases. Radiation level is inversely proportional to the square of the distance from the source.

9.75   Positive effect. Potential damage is often directly proportional to the time of exposure.

9.77   Positive effect. The operator of the robotic device could be located far from the source, no physical contact with the source is necessary, and barriers of lead or other shielding can isolate the control and the robot.

9.79   Yes. Concrete has a higher density than wood, and thus serves as a better radiation shield.

9.81 A film badge detects gamma radiation by darkening photographic film in proportion to the amount of radiation exposure over time. Badges are periodically collected and evaluated for their level of exposure. This mirrors the level of exposure of the personnel wearing the badges.

9.83 Relative biological effect is a measure of the damage to biological tissue caused by different forms of radiation.

Chapter 10
An Introduction to Organic Chemistry:
The Saturated Hydrocarbons
Solutions to the Odd-Numbered Problems

In-Chapter Questions and Problems

10.1   The student could test the solubility of the substance in water and in an organic solvent, such as hexane. Solubility in hexane would suggest an organic substance; whereas solubility in water would indicate an inorganic compound. The student could also determine the melting and boiling points of the substance. If the melting and boiling points are very high, an inorganic substance would be suspected.

10.3

Hexane   2-Methylpentane   3-Methylpentane

2,3-Dimethylbutane   2,2-Dimethylbutane

10.5

a.

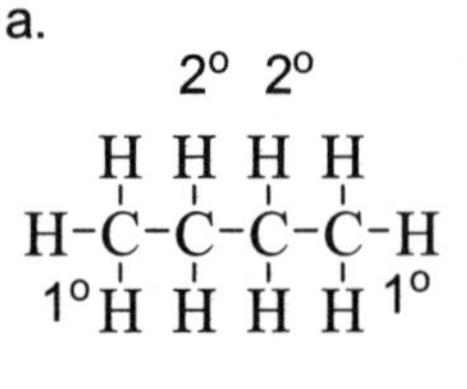

2° 2°

H H H H
| | | |
H-C-C-C-C-H
| | | |
1^oH H H H 1°

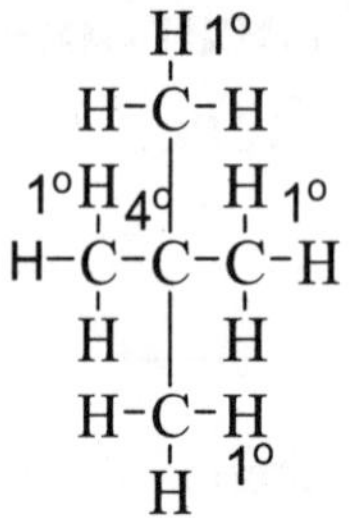

b.

H 1°
|
H-C-H
|
1^oH 4° H 1°
| | |
H-C-C-C-H
| | |
H | H
|
H-C-H
| 1°
H

c.

1^oH
|
H-C-H
|
H H 3° H H 1°
| | | | |
H-C-C-C-C-H
| | | | |
1^oH 3^oH H H
|
H-C-H
| 1°
H

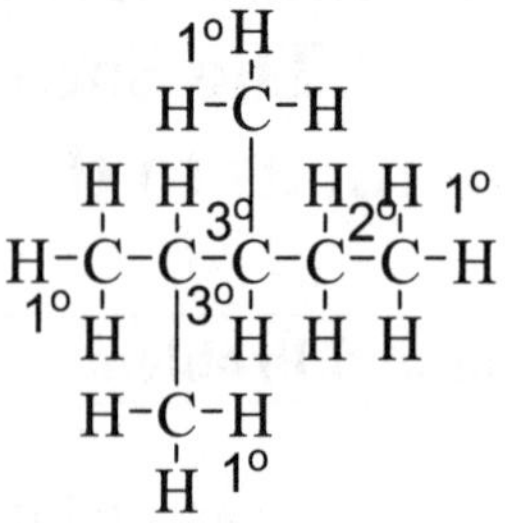

10.7   a. 2,3-Dimethylbutane
       b. 2,2-Dimethylpentane
       c. Dimethylpropane
       d. 1,2,3-Tribromopropane

10.9   a. The straight-chain isomers of molecular formula C_4H_9Br:

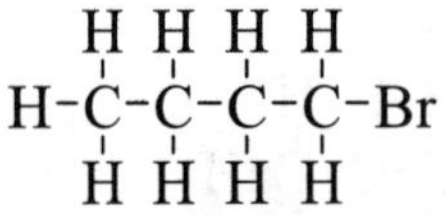

H H H H
| | |
H-C-C-C-C-Br
| | | |
H H H H

1-Bromobutane

H H Br H
| | | |
H-C-C-C-C-H
| | | |
H H H H

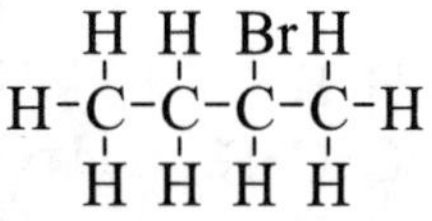

2-Bromobutane

b. The linear isomers of molecular formula $C_4H_8Br_2$:

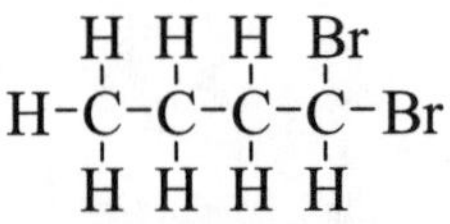

H H H Br
| | | |
H-C-C-C-C-Br
| | | |
H H H H

1,1-Dibromobutane

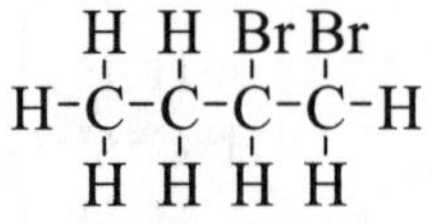

H H Br Br
| | | |
H-C-C-C-C-H
| | | |
H H H H

1,2-Dibromobutane

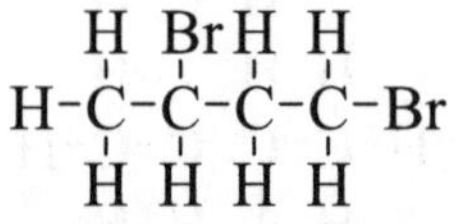

H Br H H
| | | |
H-C-C-C-C-Br
| | | |
H H H H

1,3-Dibromobutane

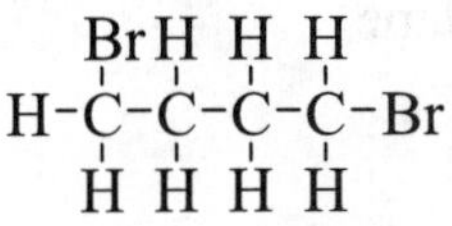

Br H H H
| | | |
H-C-C-C-C-Br
| | | |
H H H H

1,4-Dibromobutane

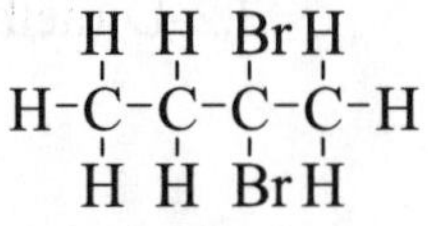

H H Br H
| | | |
H-C-C-C-C-H
| | | |
H H Br H

2,2-Dibromobutane

H Br Br H
| | | |
H-C-C-C-C-Br
| | | |
H H H H

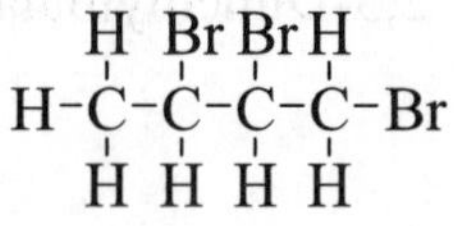

2,3-Dibromobutane

360

10.11 a. *trans*-1-Bromo-2-ethylcyclobutane
       b. *trans*-1,2-Dimethylcyclopropane
       c. Propylcyclohexane

10.13 a. The combustion of cyclobutane:

$$\square + 6O_2 \longrightarrow 4CO_2 + 4H_2O + \text{heat energy}$$

b. The monobromination of propane will produce two products, as shown in the
following two equations:

$$CH_3CH_2CH_3 + Br_2 \xrightarrow{\text{Light or heat}} CH_3CH_2CH_2Br + HBr$$

$$CH_3CH_2CH_3 + Br_2 \xrightarrow{\text{Light or heat}} CH_3CHBrCH_3 + HBr$$

c. The complete combustion of ethane:

$$2CH_3CH_3 + 7O_2 \rightarrow 4CO_2 + 6H_2O + \text{heat energy}$$

d. The monochlorination of butane will produce two products, as shown in the following
two equations:

$$CH_3CH_2CH_2CH_3 + Cl_2 \xrightarrow{\text{Light or heat}} CH_3CH_2CH_2CH_2Cl + HCl$$

$$CH_3CH_2CH_2CH_3 + Cl_2 \xrightarrow{\text{Light or heat}} CH_3CH_2CHClCH_3 + HCl$$

10.15 The products of the reaction in Problem 10.13b are 1-bromopropane and 2-
bromopropane. The products of the reactions in Problem 10.13d are 1-chlorobutane and
2-chlorobutane.

End-of-Chapter Questions and Problems

10.17 The number of organic compounds is nearly limitless because carbon forms stable
covalent bonds with other carbon atoms in a variety of different patterns. In addition,
carbon can form stable bonds with other elements and functional groups, producing many
families of organic compounds, including alcohols, aldehydes, ketones, esters, ethers,
amines, and amides. Finally, carbon can form double or triple bonds with other carbon
atoms to produce organic molecules with different properties.

10.19 The allotropes of carbon include graphite, diamond, and buckminsterfullerene.

10.21 Because ionic substances often form three-dimensional crystals made up of many positive and negative ions, they generally have much higher melting and boiling points than covalent compounds.

10.23 a. LiCl > H_2O > CH_4

b. NaCl > C_3H_8 > C_2H_6

10.25 a. LiCl would be a solid; H_2O would be a liquid; and CH_4 would be a gas.

b. NaCl would be a solid; both C_3H_8 and C_2H_6 would be gases.

10.27 a. Water-soluble inorganic compounds are good electrolytes because they dissociate into ions in water. The ions conduct an electrical charge.
b. Inorganic compounds exhibit ionic bonding.
c. Organic compounds have lower melting points.
d. Inorganic compounds are more likely to be water soluble.
e. Organic compounds are flammable.

10.29

a.

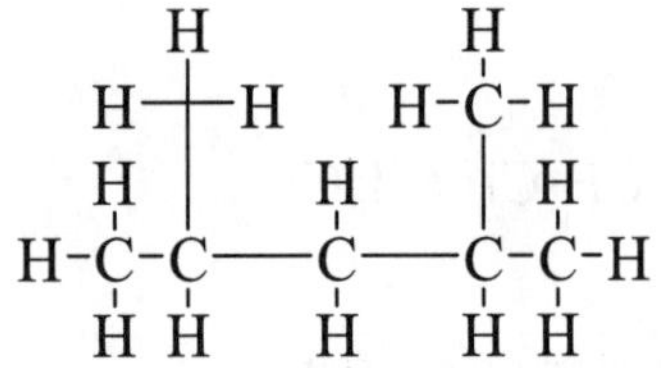

```
        H              H
  H――H        H-C-H
  H  |      H   |  H
  |  |      |   |  |
H-C-C――――C――――C-C-H
  H  H    H    H  H
```

b.

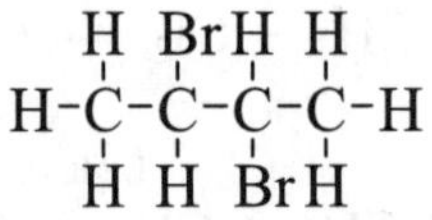

```
  H  Br H  H
  |  |  |  |
H-C-C-C-C-H
  |  |  |  |
  H  H  Br H
```

10.31

a.

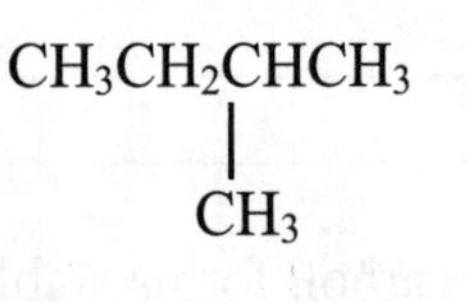

CH₃CH₂CHCH₃
      |
    CH₃

b.

        CH₃   CH₃
        | |
CH₃CH₂CHCH₂CH₂CHCH₃
        |
        CH₃

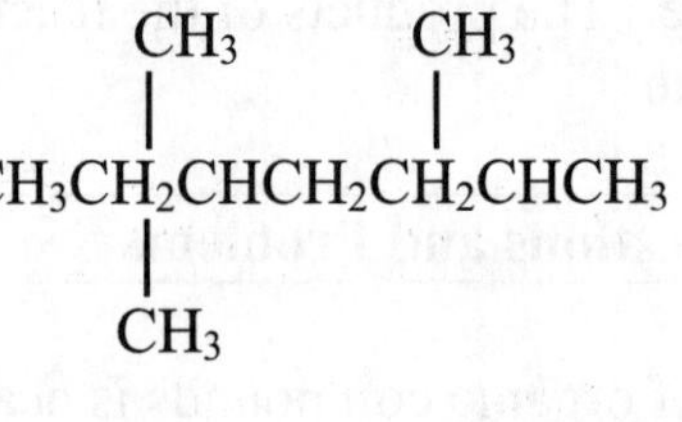

10.33

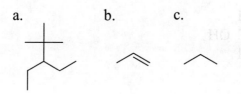

10.35

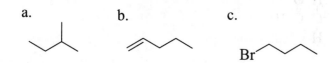

10.37 Structure b is not possible because there are five bonds to carbon-2. Structure d is not possible because there are five bonds to carbon-3. Structure e is not possible because there are five bonds to carbon-2 and carbon-3 and only three bonds to carbon-4. Structure f is not possible because there are five bonds to carbon-3.

10.39

a.
```
            H
            |
          H-C-H
     H      |   H H H
     |      |   | | |
   H-C------C---C-C-C-H
     |      |   | | |
     H H    |   H H
          H-C-H
            |
            H
```

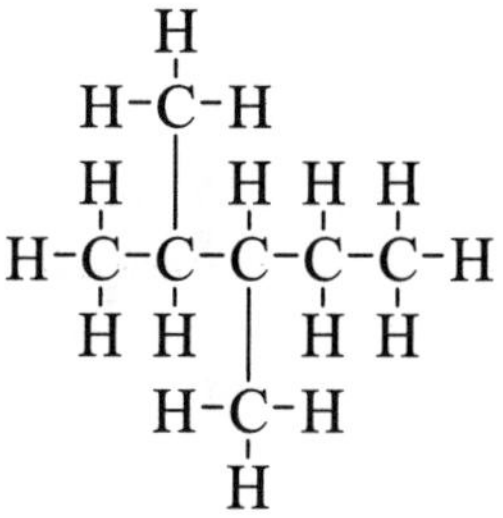

b.
```
     H H H H H
     | | | | |
   H-C-C-C-C-C-H
     | | | | |
     H H H H H
```

c.
```
            H
            |
          H-C-H
            |
          H-C-H
     H H    |   H H H
     | |    |   | | |
   H-C-C----C---C-C-C-H
     | |    |   | | |
     H H    H   H H H
```

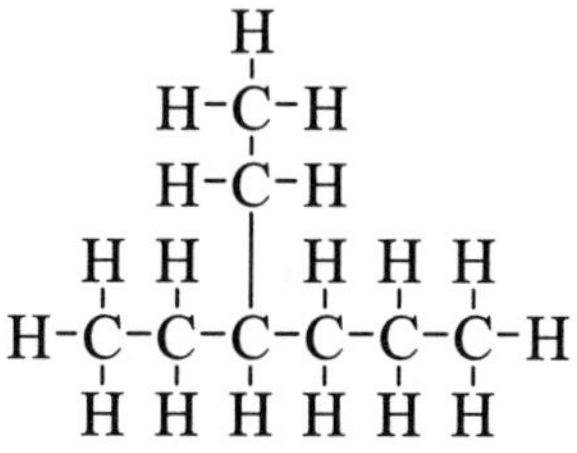

d.
```
            H
            |
          H-C-H
     H      |   H H H H H
     |      |   | | | | |
   H-C------C---C-C-C-C-C-H
     |      |   | | | | |
     H H    |   H H H H
          H-C-H
            |
            H
```

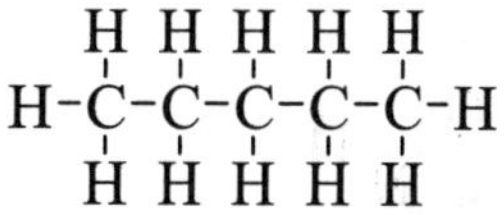

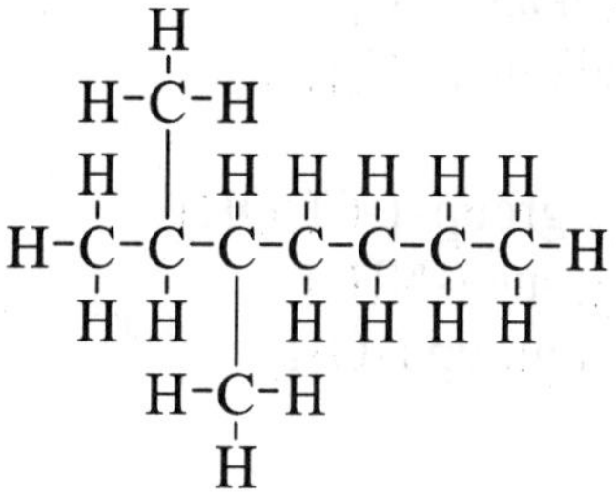

10.41 An alcohol-Ethanol:

$$
\begin{array}{cc}
\text{H} & \text{H} \\
| & | \\
\text{H}-\text{C}-\text{C}-\text{OH} \\
| & | \\
\text{H} & \text{H}
\end{array}
$$

An aldehyde-Ethanal:

$$
\begin{array}{cc}
\text{H} & \text{O} \\
| & \| \\
\text{H}-\text{C}-\text{C}-\text{H} \\
| \\
\text{H}
\end{array}
$$

A ketone-Propanone:

$$
\begin{array}{ccc}
\text{H} & \text{O} & \text{H} \\
| & \| & | \\
\text{H}-\text{C}-\text{C}-\text{C}-\text{H} \\
| & & | \\
\text{H} & & \text{H}
\end{array}
$$

A carboxylic acid-Ethanoic acid:

$$
\begin{array}{cc}
\text{H} & \text{O} \\
| & \| \\
\text{H}-\text{C}-\text{C}-\text{OH} \\
| \\
\text{H}
\end{array}
$$

An amine-Ethanamine:

$$
\begin{array}{cc}
\text{H} & \text{O} \\
| & \| \\
\text{H}-\text{C}-\text{C}-\text{N} \begin{array}{c} \text{H} \\ \text{H} \end{array} \\
| \\
\text{H}
\end{array}
$$

10.43 a. An alkane: C_nH_{2n+2}
       b. An alkyne: C_nH_{2n-2}
       c. An alkene: C_nH_{2n}
       d. A cycloalkane: C_nH_{2n}
       e. A cycloalkene: C_nH_{2n-2}

10.45 a. Carboxyl group (-COOH)
       b. Amino group (-NH$_2$)
       c. Hydroxyl group (-OH)

10.47

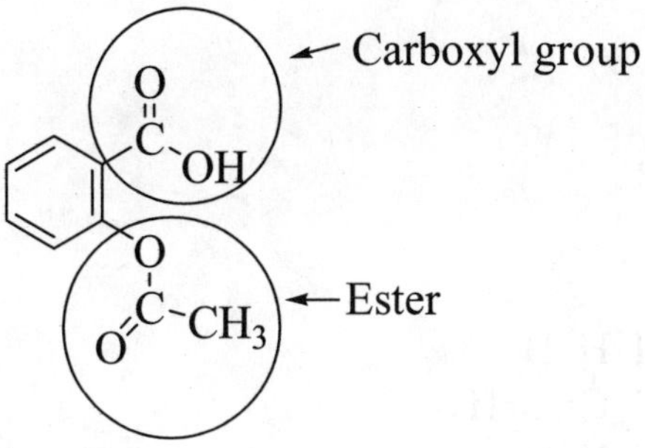

Aspirin
Acetylsalicylic acid

10.49 Hydrocarbons are nonpolar molecules, and hence are not soluble in water.

10.51 a. heptane > hexane > butane > ethane
b. $CH_3CH_2CH_2CH_2CH_2CH_2CH_2CH_2CH_3$ > $CH_3CH_2CH_2CH_2CH_3$ > $CH_3CH_2CH_3$

10.53 a. Heptane and hexane would be liquid at room temperature; butane and ethane would be gases.
b. $CH_3CH_2CH_2CH_2CH_2CH_2CH_2CH_2CH_3$ and $CH_3CH_2CH_2CH_2CH_3$ would be liquids at room temperature; $CH_3CH_2CH_3$ would be a gas.

10.55 a. 2-Bromobutane:

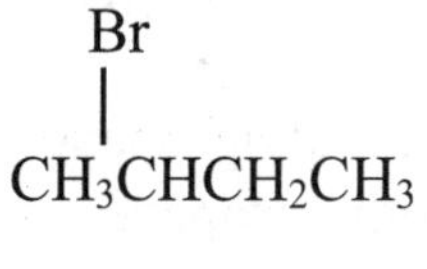

$$\begin{array}{c} Br \\ | \\ CH_3CHCH_2CH_3 \end{array}$$

b. 2-Chloro-2-methylpropane:

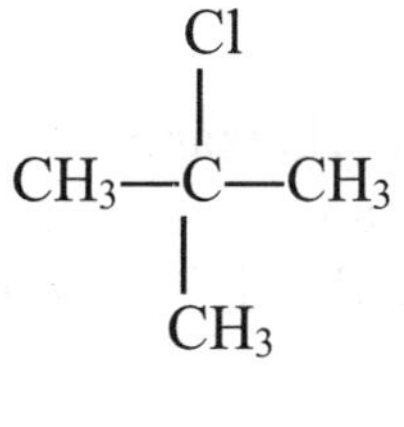

$$\begin{array}{c} Cl \\ | \\ CH_3-C-CH_3 \\ | \\ CH_3 \end{array}$$

c. 2,2-Dimethylhexane:

$$\begin{array}{c} CH_3 \\ | \\ CH_3-C-CH_2CH_2CH_2CH_3 \\ | \\ CH_3 \end{array}$$

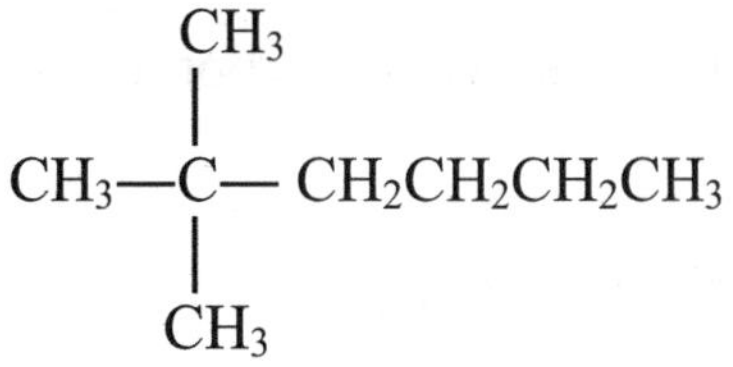

10.57

a. 2,2-Dibromobutane:

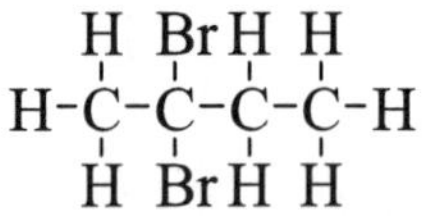

```
      H Br H  H
      |  |  |  |
   H-C--C--C--C-H
      |  |  |  |
      H Br H  H
```

b. 2-Iododecane:

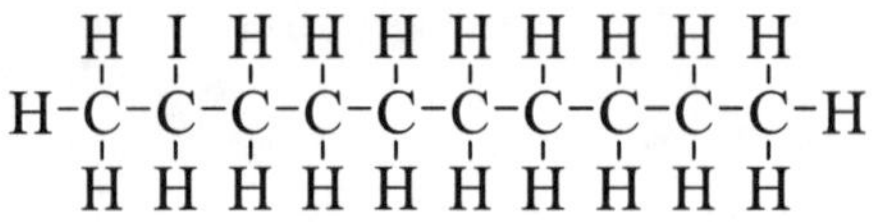

```
      H  I  H  H  H  H  H  H  H  H
      |  |  |  |  |  |  |  |  |  |
   H-C--C--C--C--C--C--C--C--C--C-H
      |  |  |  |  |  |  |  |  |  |
      H  H  H  H  H  H  H  H  H  H
```

c. 1,2-Dichloropentane:

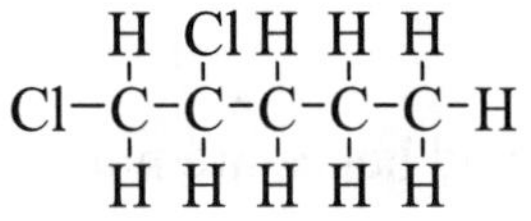

```
       H  Cl H  H  H
       |  |  |  |  |
   Cl--C--C--C--C--C-H
       |  |  |  |  |
       H  H  H  H  H
```

d. 1-Bromo-2-methylpentane:

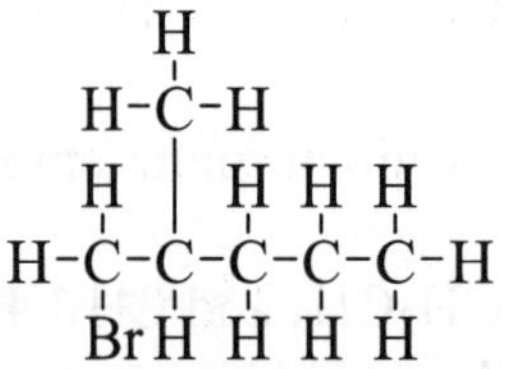

```
          H
          |
        H-C-H
        H |   H  H  H
        |  |  |  |  |
      H-C--C--C--C--C-H
        |  |  |  |  |
        Br H  H  H  H
```

10.59 a. 3-Methylpentane   c. 1-Bromoheptane
       b. 2,5-Dimethylhexane   d. 1-Chloro-3-methylbutane

10.61 a. 2-Chloropropane   d. 1-Chloro-2-methylpropane
       b. 2-Iodobutane   e. 2-Iodo-2-methylpropane
       c. 2,2-Dibromopropane

10.63 a. Identical – both are 2-Bromobutane
       b. Identical – both are 3-Bromo-5-methylhexane
       c. Identical – both are 2,2-Dibromobutane
       d. Isomers of molecular formula $C_6H_{12}Br$: 1,3-Dibromo-3-methylpentane and 1,4-
          Dibromo-2-ethylbutane

10.65 Cycloalkanes are a family of molecules having carbon-to-carbon bonds in a ring
       structure.

10.67 The general formula for a cycloalkane is C_nH_{2n}.

10.69 a. Chlorocyclopropane
       b. 1,2-Dichlorocyclopropane
       c. 1,2-Dichlorocyclopropane

10.71 a. 1-Bromo-2-methylcyclobutane: b. Iodocyclopropane:

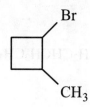

c. 1-Bromo-3-chlorocyclopentane: d. 1,2-Dibromo-3-methylcyclohexane:

10.73 a. $C_3H_8 + 5O_2 \rightarrow 4H_2O + 3CO_2$

b. $C_7H_{16} + 11O_2 \rightarrow 8H_2O + 7CO_2$

c. $C_9H_{20} + 14O_2 \rightarrow 10H_2O + 9CO_2$

d. $2C_{10}H_{22} + 31O_2 \rightarrow 22H_2O + 20CO_2$

10.75 a. $8\ CO_2 + 10\ H_2O$

b.

$$Br{-}\underset{\underset{CH_3}{|}}{\overset{\overset{CH_3}{|}}{C}}{-}CH_3 \qquad + \qquad CH_3CHCH_2Br + 2HBr$$

2-Bromo-2-methylpropane   1-Bromo-2-methylpropane

c. Cl_2 + light

10.77 The following molecules are all the constitutional isomers of C_6H_{14}:

$$CH_3CH_2CH_2CH_2CH_2CH_3$$

Hexane

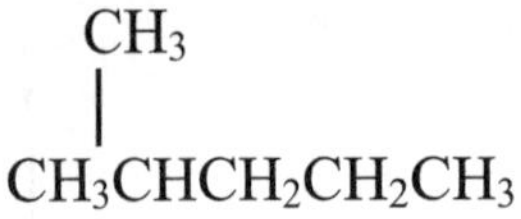

$$\underset{\underset{CH_3}{|}}{CH_3CHCH_2CH_2CH_3}$$

2-Methylpentane

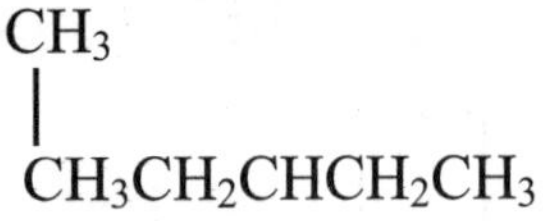

$$\underset{\underset{CH_3}{|}}{CH_3CH_2CHCH_2CH_3}$$

3-Methylpentane

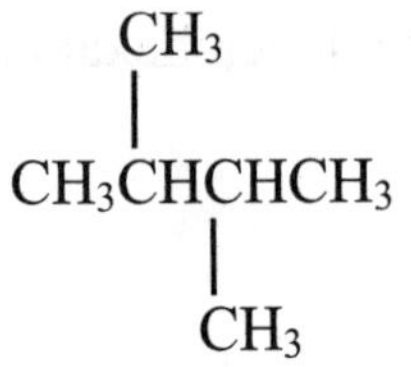

$$\underset{\underset{CH_3}{|}}{\overset{\overset{CH_3}{|}}{CH_3CHCHCH_3}}$$

2,3-Dimethylbutane

$$\underset{\underset{CH_3}{|}}{\overset{\overset{CH_3}{|}}{CH_3CCH_2CH_3}}$$

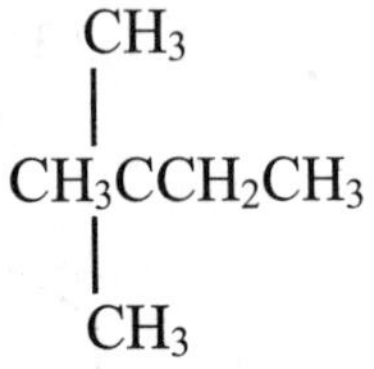

2,2-Dimethylbutane

a. 2,3-Dimethylbutane produces only two monobrominated derivatives: 1-bromo-2,3-dimethylbutane and 2-bromo-2,3-dimethylbutane.

b. Hexane produces three monobrominated products: 1-bromohexane, 2-bromohexane, and 3-bromohexane. 2,2-Dimethylbutane also produces three monobrominated products: 1-bromo-2,2-dimethylbutane, 2-bromo-3,3-dimethylbutane, and 1-bromo-3,3-dimethylbutane.

c. 3-Methylpentane produces four monobrominated products: 1-bromo-3-methylpentane, 2-bromo-3-methylpentane, 3-bromo-3-methylpentane, and 1-bromo-2-ethylbutane.

10.79 The hydrocarbon is cyclooctane having a molecular formula of C_8H_{16}.

Chapter 11
The Unsaturated Hydrocarbons: Alkenes, Alkynes, and Aromatics
Solutions to the Odd-Numbered Problems

In-Chapter Questions and Problems

11.1 a.

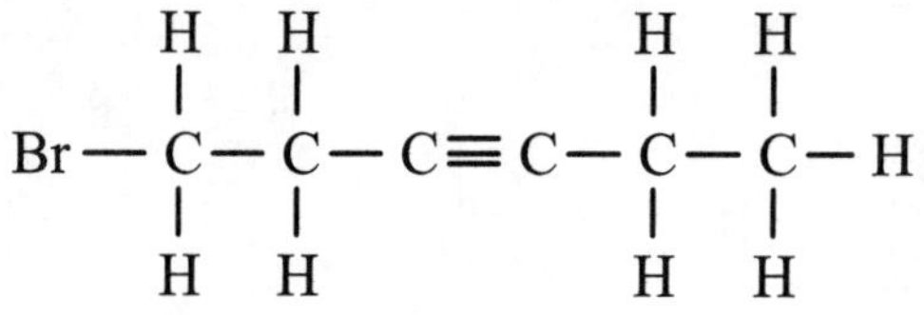

b.

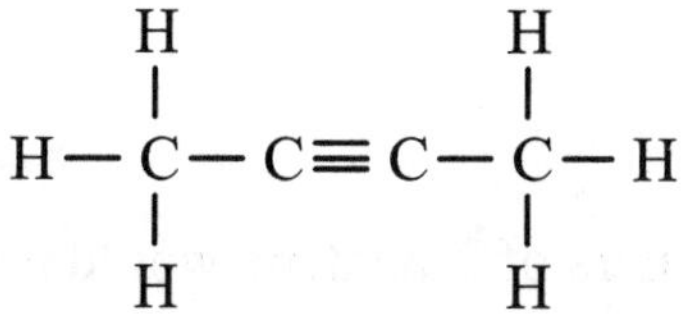

c. $Cl-C\equiv C-Cl$

d.

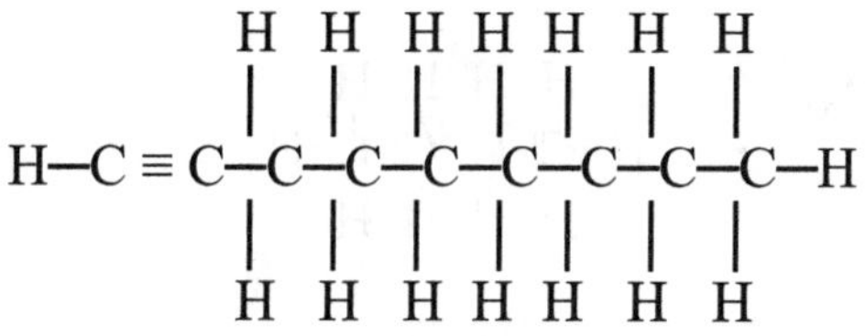

11.3 a.

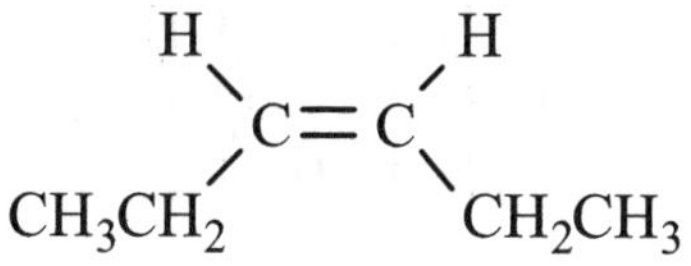

cis-3-Hexene

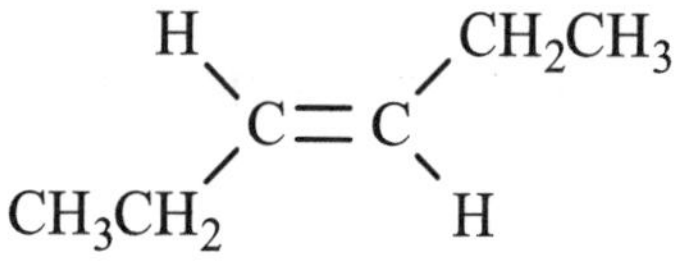

trans-3-Hexene

b.

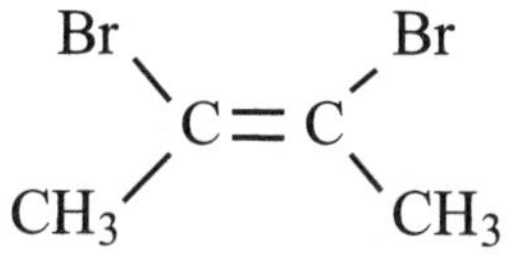

cis-2,3-Dibromo-2-butene

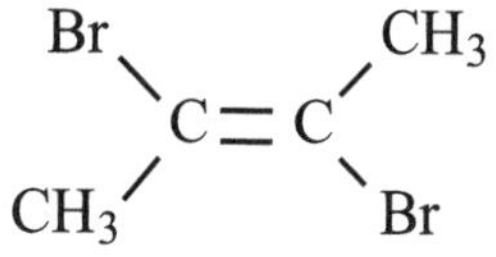

trans-2,3-Dibromo-2-butene

11.5   Molecule c can exist as *cis-* and *trans-*isomers because there are two different groups on each of the carbon atoms attached by the double bond.

11.7   a.

$$\underset{CH_3CH_2}{\overset{H}{\diagdown}} C = C \underset{CH_2CH_2CH_2CH_3}{\overset{H}{\diagup}}$$

b.

$$\underset{H}{\overset{CH_3}{\diagdown}} C = C \underset{CH_2CHCH_3}{\overset{H}{\diagup}}$$
                                |
                               Cl

c.

$$\underset{Cl}{\overset{CH_3}{\diagdown}} C = C \underset{CH_3}{\overset{Cl}{\diagup}}$$

11.9   The hydrogenation of the *cis-* and *trans-* isomers of 2-pentene would produce the same product, pentane.

11.11

a. $H_3C-C{\equiv}C-CH_3 \ + \ 2 \ H_2 \xrightarrow{\text{Ni}}$

$$\underset{\underset{H\ H\ H\ H}{|\ |\ |\ |}}{\overset{\overset{H\ H\ H\ H}{|\ |\ |\ |}}{H-C-C-C-C-H}}$$

      2-Butyne   Butane

b. $H_3C-C{\equiv}C-CH_2CH_3 \ + \ 2 \ H_2 \xrightarrow{\text{Ni}}$

$$\underset{\underset{H\ H\ H\ H\ H}{|\ |\ |\ |\ |}}{\overset{\overset{H\ H\ H\ H\ H}{|\ |\ |\ |\ |}}{H-C-C-C-C-C-H}}$$

      2-Pentyne   Pentane

11.13

a. $CH_3CH=CH_2$ $+ Br_2$ $\longrightarrow$

$$\begin{array}{c}
\text{H Br H} \\
\text{H-C-C-CH} \\
\text{H H Br}
\end{array}$$

b. $CH_3CH=CHCH_3 + Br_2$ $\longrightarrow$

$$\begin{array}{c}
\text{H H Br H} \\
\text{H-C-C-C-C-H} \\
\text{H Br H H}
\end{array}$$

11.15

a. $CH_3C \equiv CCH_3$ $+$ $2Cl_2$ $\rightarrow$ $CH_3CCl_2CCl_2CH_3$

b. $CH_3C \equiv CCH_2CH_3$ $+$ $2Cl_2$ $\rightarrow$ $CH_3CCl_2CCl_2CH_2CH_3$

11.17

a.

$$CH_3CH=CHCH_3 \;+\; H_2O \;\xrightarrow{H^+}\; \underset{\displaystyle CH_3CHCH_2CH_3}{\overset{\displaystyle OH}{\;|\;}}$$

b.

$$\underset{\displaystyle H_2C=CHCH_2CH_2CHCH_3}{\overset{\displaystyle CH_3}{|}} + H_2O \xrightarrow{H^+} \underset{\displaystyle CH_3CHCH_2CH_2CHCH_3}{\overset{\displaystyle OH \qquad CH_3}{\;|\qquad\quad|\;}}$$

(major product)

$$\underset{\displaystyle H_2C=CHCH_2CH_2CHCH_3}{\overset{\displaystyle CH_3}{|}} + H_2O \xrightarrow{H^+} \underset{\displaystyle CH_2CH_2CH_2CH_2CHCH_3}{\overset{\displaystyle OH \qquad CH_3}{\;|\qquad\quad|\;}}$$

(minor product)

c.   These products will be formed in approximately equal amounts:

$$CH_3CH_2CH_2CH=CHCH_2CH_3 + H_2O \xrightarrow{H^+} \underset{\displaystyle CH_3CH_2CH_2CHCH_2CH_2CH_3}{\overset{\displaystyle OH}{\;|\;}}$$

$$CH_3CH_2CH_2CH=CHCH_2CH_3 + H_2O \xrightarrow{H^+} \underset{\displaystyle CH_3CH_2CH_2CH_2CHCH_2CH_3}{\overset{\displaystyle OH}{\;|\;}}$$

d.

$$CH_3\underset{|}{\overset{Cl}{C}}HCH{=}CH\overset{Cl}{\underset{|}{C}}HCH_3 + H_2O \xrightarrow{H^+} CH_3\overset{Cl}{C}H\overset{OH}{C}HCH_2\overset{Cl}{C}HCH_3$$

(only product)

11.19 a.

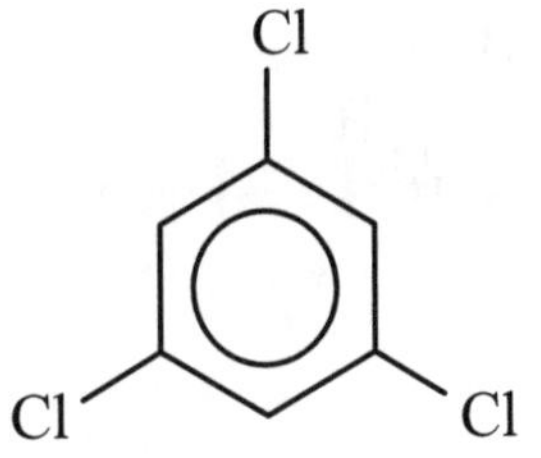

b.

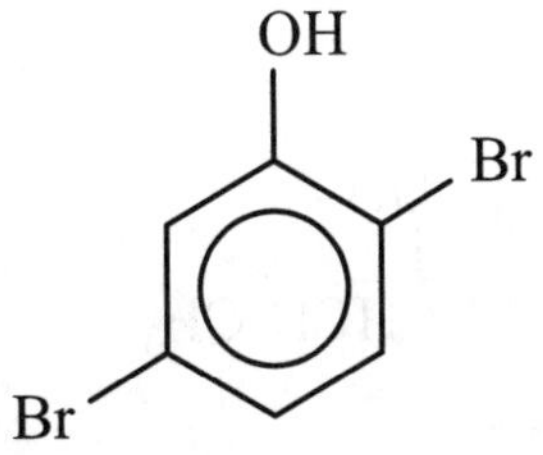

c.

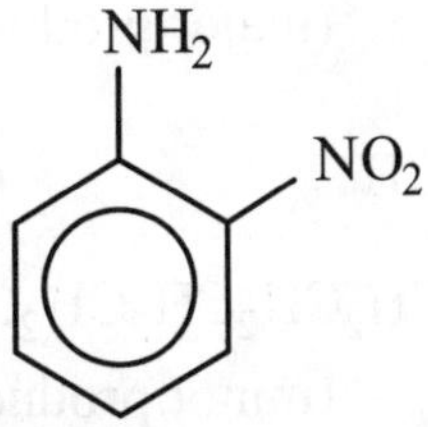

End-of-Chapter Questions and Problems

11.21 The longer the carbon chain of an alkene, the higher the boiling point.

11.23 The general formula for an alkane is C_nH_{2n+2}.
The general formula for an alkene is C_nH_{2n}.
The general formula for an alkyne is C_nH_{2n-2}.

11.25 Ethene is a planar molecule. All of the bond angles are 120°.

11.27 a. 2-Pentyne > Propyne > Ethyne
b. 3-Decene > 2-Butene > Ethene

11.29 Identify the longest carbon chain containing the carbon-to-carbon double or triple bond. Replace the –*ane* suffix of the alkane name with –*ene* for an alkene or -*yne* for an alkyne. Number the chain to give the lowest number to the first of the two carbons involved in the double or triple bond. Determine the name and carbon number of each substituent group and place that information as a prefix in front of the name of the parent compound.

11.31 Geometric isomers of alkenes differ from one another in the placement of substituents attached to each of the carbon atoms of the double bond. Of the pair of geometric isomers, the *cis*-isomer is the one in which the groups are on the same side of the double bond.

11.33 a. 2-Methyl-2-hexene:

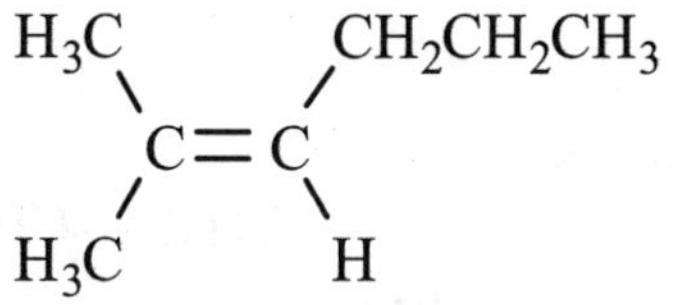

$$
\begin{array}{ccc}
H_3C & & CH_2CH_2CH_3 \\
\diagdown & & \diagup \\
 & C=C & \\
\diagup & & \diagdown \\
H_3C & & H
\end{array}
$$

b. *trans*-3-Heptene:

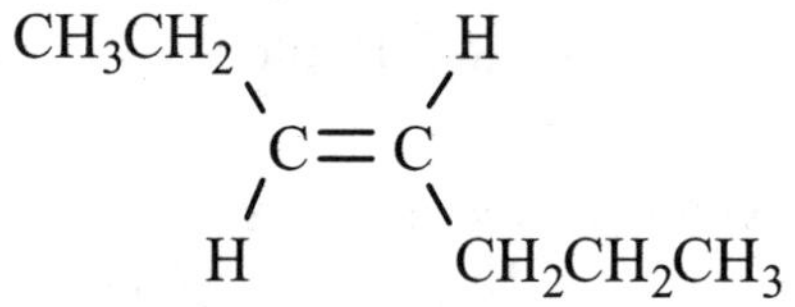

$$
\begin{array}{ccc}
CH_3CH_2 & & H \\
\diagdown & & \diagup \\
 & C=C & \\
\diagup & & \diagdown \\
H & & CH_2CH_2CH_3
\end{array}
$$

c. *cis*-1-Chloro-2-pentene:

$$
\begin{array}{ccc}
ClCH_2 & & CH_2CH_3 \\
\diagdown & & \diagup \\
 & C=C & \\
\diagup & & \diagdown \\
H & & H
\end{array}
$$

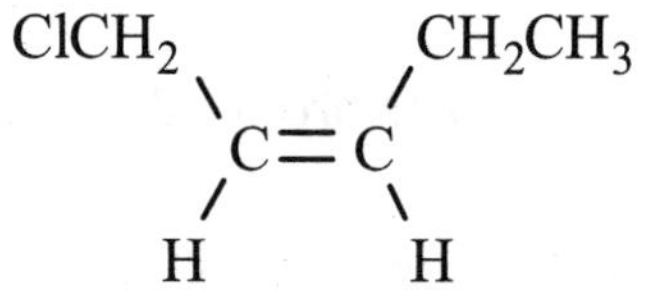

d. *cis*-2-Chloro-2-methyl-3-heptene:

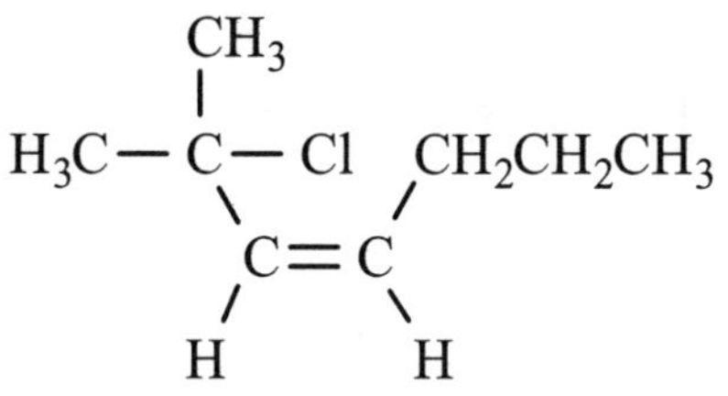

$$
\begin{array}{c}
CH_3 \\
| \\
H_3C-C-Cl \quad CH_2CH_2CH_3 \\
\diagdown \qquad \diagup \\
C=C \\
\diagup \quad \diagdown \\
H \qquad H
\end{array}
$$

e. *trans*-5-Bromo-2,6-dimethyl-3-octene:

$$
\begin{array}{c}
\text{CH}_3 \\
|\\
\text{H}_3\text{C}-\text{CH} \qquad \text{H} \\
\diagdown \qquad \diagup \\
\text{C}=\text{C} \\
\diagup \qquad \diagdown \\
\text{H} \qquad \text{CH}-\text{CH}-\text{CH}_2\text{CH}_3 \\
\qquad |\qquad\;\; | \\
\qquad \text{Br}\quad\;\; \text{CH}_3
\end{array}
$$

11.35 a. 3-Methyl-1-pentene
       b. 7-Bromo-1-heptene
       c. 5-Bromo-3-heptene
       d. 1-*t*-Butyl-4-methylcyclohexene

11.37 a. 2,3-Dibromobutane cannot exist as *cis*- and *trans*- isomers because it is an alkane.

       b. 2-Heptene can exist as *cis*- and *trans*- isomers:

$$
\begin{array}{cc}
\begin{array}{c}
\text{H}\qquad\qquad\text{H}\\
\diagdown\qquad\diagup\\
\text{C}=\text{C}\\
\diagup\qquad\diagdown\\
\text{CH}_3\qquad\text{CH}_2\text{CH}_2\text{CH}_2\text{CH}_3
\end{array}
&
\begin{array}{c}
\text{H}\qquad\qquad\text{CH}_2\text{CH}_2\text{CH}_2\text{CH}_3\\
\diagdown\qquad\diagup\\
\text{C}=\text{C}\\
\diagup\qquad\diagdown\\
\text{CH}_3\qquad\text{H}
\end{array}
\\
\\
\textit{cis}\text{-2-Heptene} & \textit{trans}\text{-2-Heptene}
\end{array}
$$

       c. 2,3-Dibromo-2-butene can exist as *cis*- and *trans*- isomers:

$$
\begin{array}{cc}
\begin{array}{c}
\text{Br}\qquad\qquad\text{Br}\\
\diagdown\qquad\diagup\\
\text{C}=\text{C}\\
\diagup\qquad\diagdown\\
\text{CH}_3\qquad\text{CH}_3
\end{array}
&
\begin{array}{c}
\text{Br}\qquad\qquad\text{CH}_3\\
\diagdown\qquad\diagup\\
\text{C}=\text{C}\\
\diagup\qquad\diagdown\\
\text{CH}_3\qquad\text{Br}
\end{array}
\\
\\
\textit{cis}\text{-2,3-Dibromo-2-butene} & \textit{trans}\text{-2,3-Dibromo-2-butene}
\end{array}
$$

       d. Propene cannot exist as *cis*- and *trans*- isomers.

11.39 Alkenes b and c would not exhibit *cis-trans* isomerism.

11.41 Alkenes b and d can exist as both *cis*- and *trans*- isomers.

11.43

$$
\begin{array}{c}
\text{R}\qquad\qquad\text{R}\\
\diagdown\qquad\diagup\\
\text{C}=\text{C} \qquad +\text{H}_2 \quad\xrightarrow[\text{heat or pressure}]{\text{Pt, Pd, or Ni}}\quad
\begin{array}{c}
\text{H}\quad\;\text{R}\\
|\qquad|\\
\text{R}-\text{C}-\text{C}-\text{R}\\
|\qquad|\\
\text{R}\quad\;\text{H}
\end{array}\\
\diagup\qquad\diagdown\\
\text{R}\qquad\qquad\text{R}
\end{array}
$$

11.45

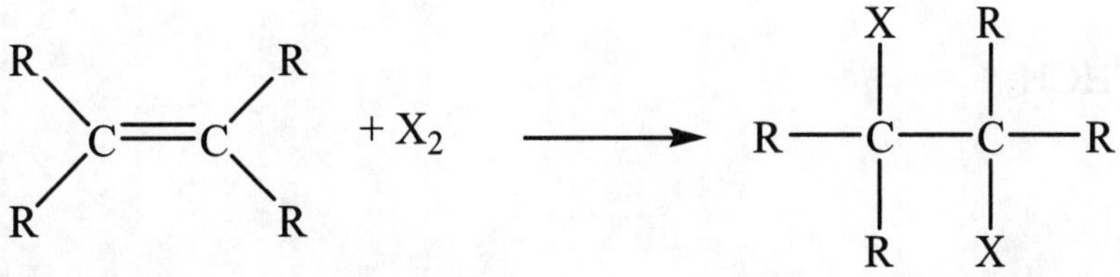

11.47

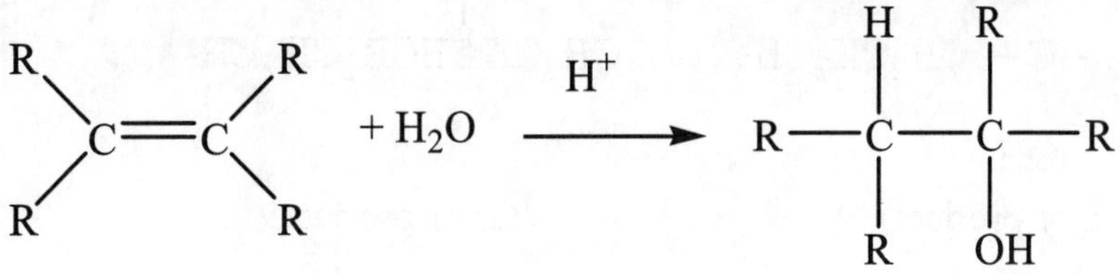

11.49 Addition of bromine (Br_2) to an alkene results in a color change from red to colorless. If equimolar quantities of Br_2 are added to hexene, the reaction mixture will change from red to colorless. This color change will not occur if cyclohexane is used.

11.51 a. H_2 d. $19\,O_2 \rightarrow 12\,CO_2\ +\ 14\,H_2O$

        b. H_2O e. Cl_2

        c. Br_2 f.

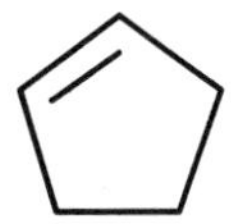

11.53

a.

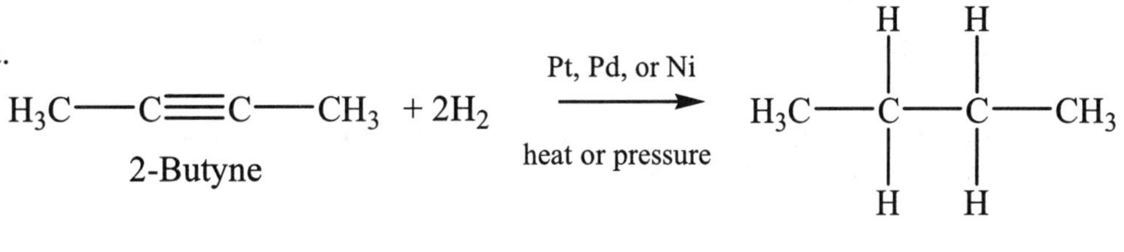

2-Butyne

b.

2-Pentyne

375

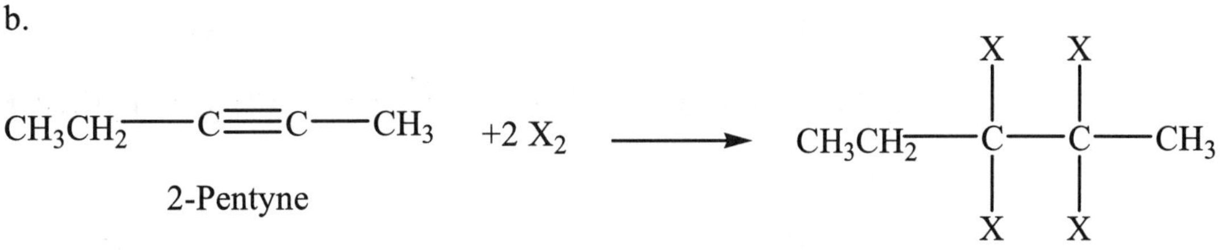

11.55 a. 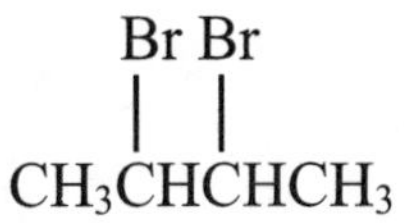 CH$_3$CHCHCH$_3$ with Br Br groups:

Br Br
| |
CH$_3$CHCHCH$_3$

b.

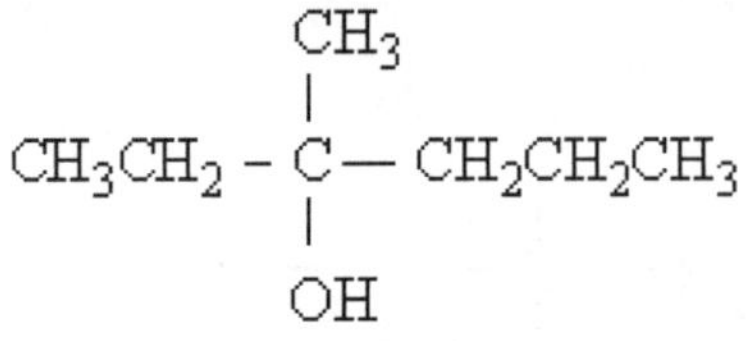

$$CH_3CH_2 - \underset{\underset{OH}{|}}{\overset{\overset{CH_3}{|}}{C}} - CH_2CH_2CH_3$$

(major product)

$$CH_3\underset{\underset{OH}{|}}{CH}\underset{\underset{CH_3}{|}}{CH}CH_2CH_2CH_3$$

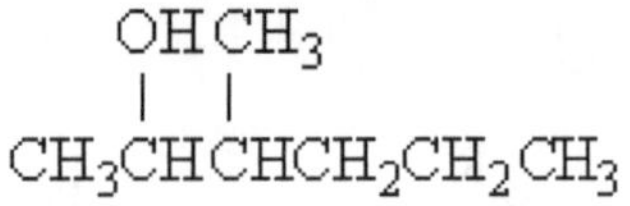

(minor product)

c.

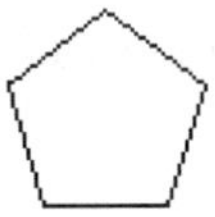

11.57 A polymer is a macromolecule composed of repeating structural units called *monomers*.

11.59

$$n \left[\begin{array}{c} F \\ C = C \\ F \end{array} \begin{array}{c} F \\ \\ F \end{array} \right] \longrightarrow \left[\begin{array}{cc} F & F \\ | & | \\ C - C \\ | & | \\ F & F \end{array} \right]_n$$

Tetrafluoroethane   Teflon

11.61 The I.U.P.A.C. name for (a) is 2-pentene, for (b) is 3-bromo-1-propene, and for (c) is 2,3-dimethylcyclohexene.

a. These products will be formed in approximately equal amounts.

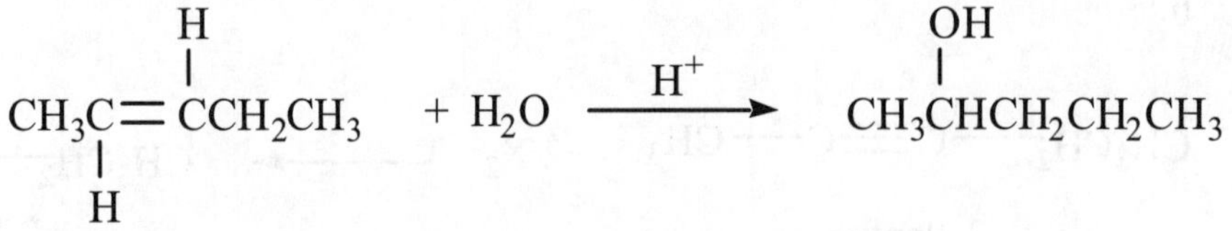

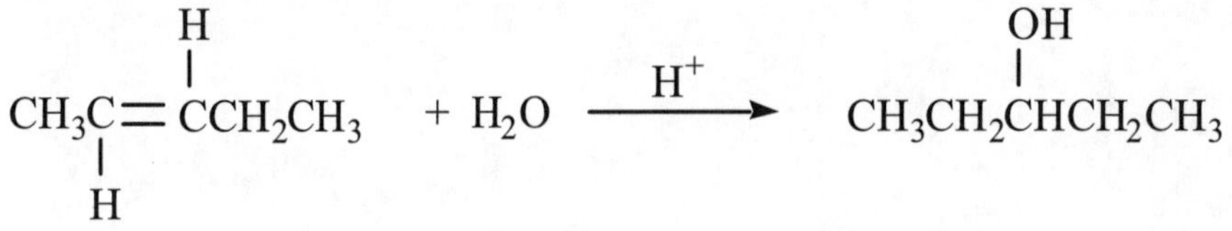

$$CH_3\underset{\underset{H}{|}}{\overset{\overset{H}{|}}{C}}=CCH_2CH_3 \ + \ H_2O \ \xrightarrow{H^+} \ CH_3\underset{\underset{OH}{|}}{CH}CH_2CH_2CH_3$$

$$CH_3\underset{\underset{H}{|}}{\overset{\overset{H}{|}}{C}}=CCH_2CH_3 \ + \ H_2O \ \xrightarrow{H^+} \ CH_3CH_2\underset{\underset{OH}{|}}{CH}CH_2CH_3$$

b.

$$\underset{\underset{\displaystyle CH_2}{|}}{\overset{\overset{\displaystyle Br}{|}}{CH_2}}\overset{\overset{\displaystyle H}{|}}{C}=CH_2 \;+\; H_2O \;\xrightarrow{\;H^+\;}\; \underset{\underset{\displaystyle CH_2}{}}{CH_2}\overset{\overset{\displaystyle Br\;\;OH}{|\;\;\;|}}{CHCH_3}$$

(major product)

$$\underset{}{\overset{\overset{\displaystyle Br\;\;H}{|\;\;\;|}}{CH_2C}}=CH_2 \;+\; H_2O \;\xrightarrow{\;H^+\;}\; \overset{\overset{\displaystyle Br}{|}}{CH_2}CH_2CH_2OH$$

(minor product)

c. These products will be formed in approximately equal amounts.

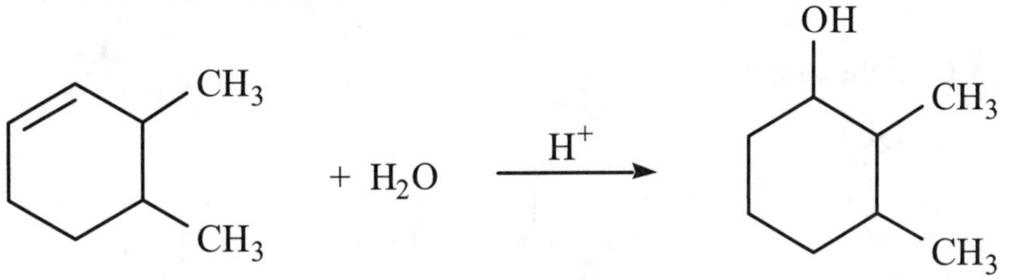

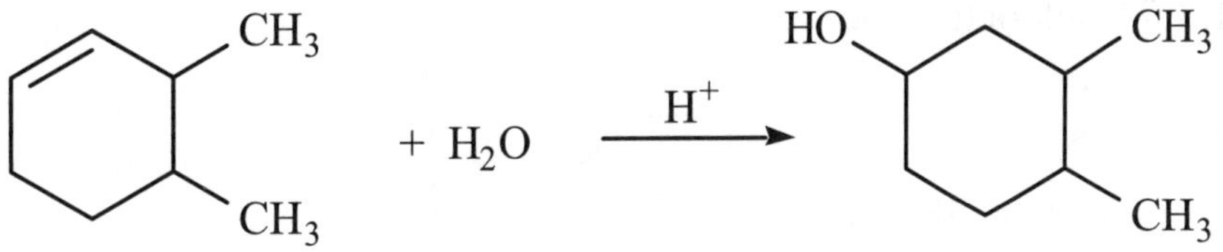

11.63 a. (This is the minor product of this reaction.)

$$H_2C=CHCH_2\overset{\overset{\displaystyle CH_3}{|}}{C}HCH_3 \;+\; H_2O \;\xrightarrow{\;H^+\;}\; CH_2CH_2CH_2\overset{\overset{\displaystyle OH\;\;\;\;\;\;\;CH_3}{|\;\;\;\;\;\;\;\;|}}{CHCH_3}$$

b.

$$CH_3CH=CHCH_2CH_2CH_3 \;+\; H_2O \;\rightarrow\; CH_3CH_2\underset{\underset{\displaystyle OH}{|}}{CH}CH_2CH_2CH_3$$

OR

$$CH_3CH_2CH=CHCH_2CH_3 \;+\; H_2O \;\rightarrow\; CH_3CH_2\underset{\underset{\displaystyle OH}{|}}{CH}CH_2CH_2CH_3$$

c.

11.65 a. 1,4-Hexadiene:

$$CH_2=CHCH_2CH=CHCH_3 \ + \ 2H_2 \ \rightarrow \ CH_3CH_2CH_2CH_2CH_2CH_3$$

b. 2,4,6-Octatriene:

$$CH_3CH=CHCH=CHCH=CHCH_3 \ + \ 3H_2 \rightarrow CH_3CH_2CH_2CH_2CH_2CH_2CH_2CH_3$$

c. 1,3-Cyclohexadiene:

d. 1,3,5-Cyclooctatriene:

11.67 The term aromatic hydrocarbon was first used as a term to describe the pleasant-smelling resins of tropical trees.

11.69 a. 2,4-Dibromotoluene: b. 1,2,4-Triethylbenzene:

c. Isopropylbenzene:

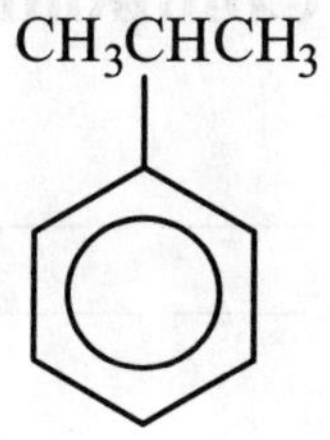

CH₃CHCH₃

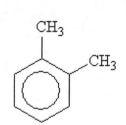

d. **2-Bromo-5-chlorotoluene:**

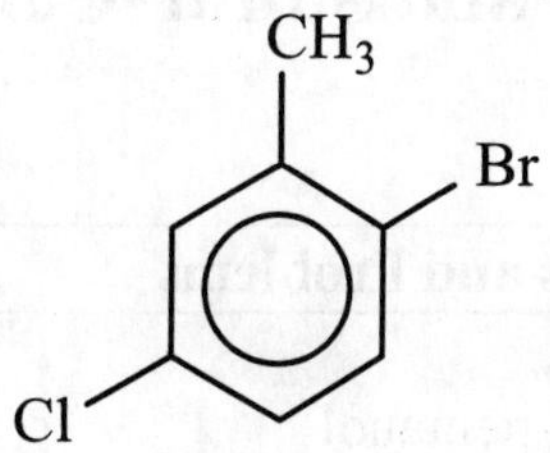

CH₃

Br

Cl

11.71 a.

CH₃

CH₃

b.

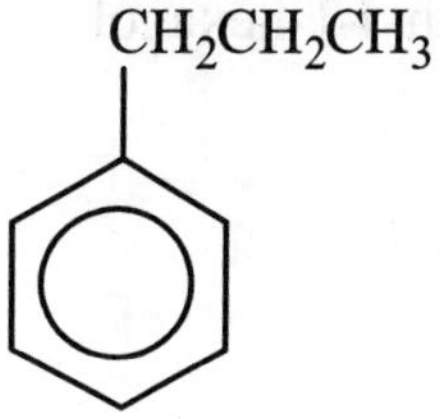

CH₂CH₂CH₃

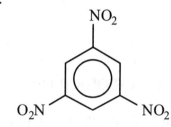

c.

NO₂

O₂N   NO₂

d.

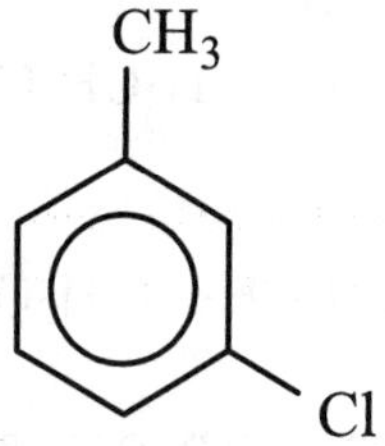

CH₃

Cl

11.73

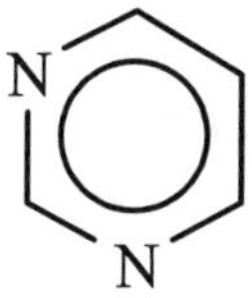

N

N

Pyrimidine

11.75

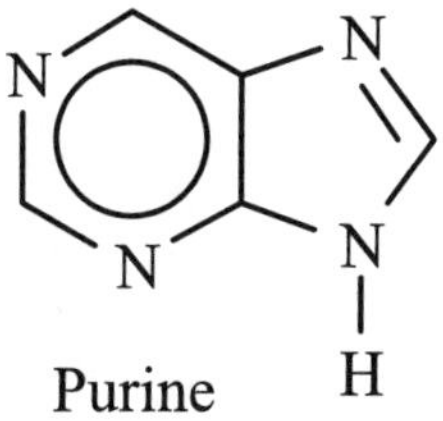

N

N

N

N

H

Purine   H

Chapter 12
Oxygen- and Sulfur-Containing Organic Compounds

In-Chapter Questions and Problems

12.1 a. 4-Methyl-1-pentanol
      b. 4-Methyl-2-hexanol
      c. 1,2,3-Propanetriol
      d. 4-Chloro-3-methyl-1-hexanol

12.3 a. Primary alcohol
      b. Secondary alcohol
      c. Tertiary alcohol

12.5

a. $CH_3CH{=}CH_2 + H_2O \xrightarrow{H^+} CH_3CH(OH)CH_3 + CH_3CH_2CH_2OH$
                                         major product   minor product

b. $CH_2{=}CH_2 + H_2O \xrightarrow{H^+} CH_3CH_2OH$

c. $CH_3CH_2CH{=}CHCH_2CH_3 + H_2O \xrightarrow{H^+} CH_3CH_2CH_2CH(OH)CH_2CH_3$

12.7 a. The major product is a secondary alcohol (2-propanol) and the minor product is a
          primary alcohol (1-propanol).
      b. The product, ethanol, is a primary alcohol.
      c. The product, 3-hexanol, is a secondary alcohol.

12.9 a. Ethanol
      b. 2-Propanol is the major product. 1-Propanol is the minor product.
      c. 2-Butanol

12.11 a. 2-Butanol is the major product. 1-Butanol is the minor product.
       b. 2-Methyl-2-propanol is the major product. 2-Methyl-1-propanol is the minor product.

12.13

               CH$_3$ CH$_3$ O
                | | ||

a. $CH_3CCH_2CH_2OH \rightarrow CH_3CCH_2C{-}H$

               | |
               CH$_3$ CH$_3$

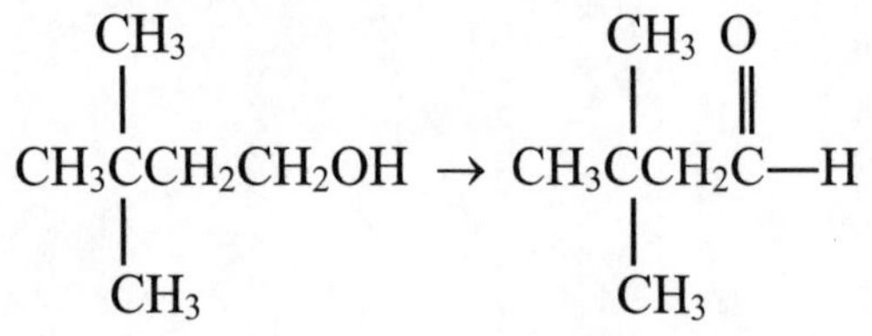

b. $CH_3CH_2OH \rightarrow CH_3\overset{\overset{\displaystyle O}{\|}}{C}\!-\!H$

12.15

a. $CH_3\overset{\overset{\displaystyle OH}{|}}{C}HCH_2CH_3 \rightarrow CH_3\overset{\overset{\displaystyle O}{\|}}{C}CH_2CH_3$

b. $CH_3\overset{\overset{\displaystyle OH}{|}}{C}HCH_2CH_2CH_3 \rightarrow CH_3\overset{\overset{\displaystyle O}{\|}}{C}CH_2CH_2CH_3$

12.17 a. I.U.P.A.C. name: 1-Ethoxypropane
       Common name: Ethylpropyl ether
       b. I.U.P.A.C. name: 1-Methoxypropane
       Common name: Methylpropyl ether

12.19

$$CH_3CH_2OH + CH_3CH_2OH \xrightarrow{H^+} CH_3CH_2\text{-}O\!-\!CH_2CH_3 + H_2O$$

Ethanol   Diethyl ether   Water

12.21 a.

$CH_3\!-\!\overset{\overset{\displaystyle O}{\|}}{C}\!-\!CH_3$

b.

$CH_3\overset{\overset{\displaystyle OH}{|}}{C}HCH_2CHCH_3$

12.23 a. I.U.P.A.C: 3,4-Dimethylpentanal
           Common: β,γ-Dimethylvaleraldehyde
       b. I.U.P.A.C: 2-Ethylpentanal
           Common: α-Ethylvaleraldehyde

12.25 a. 3-Iodobutanone
       b. 4-Methyl-2-octanone

12.27 Ethanal (acetaldehyde) is the aldehyde synthesized from ethanol in the liver.

$CH_3\!-\!\overset{\overset{\displaystyle O}{\|}}{C}\!-\!H$

12.29 The following equation represents the oxidation of 1-propanol to form propanal:

$$CH_3CH_2CH_2 - OH \xrightarrow{H_2Cr_2O_7} CH_3CH_2 - \overset{\overset{\displaystyle O}{\|}}{C} - H$$

Note that propanal may be further oxidized to form propanoic acid (a carboxylic acid).

12.31 The following equation represents the reaction between ethanal and Tollens' reagent:

$$\underset{\text{Ethanal}}{CH_3\overset{\overset{\displaystyle O}{\|}}{C} - H} + \underset{\substack{\text{Silver ammonia} \\ \text{complex}}}{Ag(NH_3)_2^+} \longrightarrow \underset{\substack{\text{Ethanoate} \\ \text{anion}}}{CH_3\overset{\overset{\displaystyle O}{\|}}{C} - O^-} + \underset{\substack{\text{Silver} \\ \text{metal}}}{Ag^0}$$

12.33 a. Reduction
b. Reduction
c. Reduction
d. Oxidation
e. Reduction

End-of-Chapter Questions and Problems

12.35 The longer the hydrocarbon tail of an alcohol becomes, the less water soluble it will be.

12.37 $a < d < c < b$

12.39 The I.U.P.A.C. rules for the nomenclature of alcohols require you to name the parent compound, that is the longest continuous carbon chain bonded to the –OH group. Replace the –*e* ending of the parent alkane with –*ol* of the alcohol. Number the parent chain so that the carbon bearing the hydroxyl group has the lowest possible number. Name and number all other substituents. If there is more than one hydroxyl group, the –*ol* ending will be modified to reflect the number. If there are two –OH groups the suffix –*diol* is used; if it has three –OH groups, the suffix –*triol* is used, etc.

12.41 a. 1-Heptanol
b. 2-Propanol
c. 2,2-Dimethylpropanol

12.43 a. 3-Hexanol:

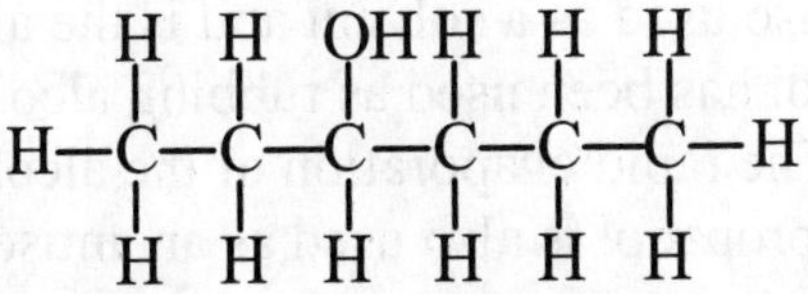

$$
\begin{array}{ccccccc}
& H & H & OH & H & H & H \\
& | & | & | & | & | & | \\
H- & C- & C- & C- & C- & C- & C-H \\
& | & | & | & | & | & | \\
& H & H & H & H & H & H \\
\end{array}
$$

b. 1,2,3-Pentanetriol:

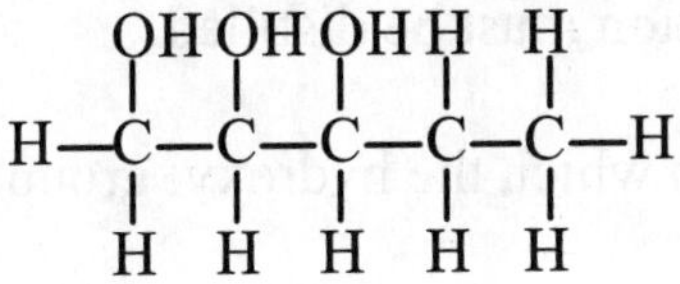

$$
\begin{array}{ccccc}
OH & OH & OH & H & H \\
| & | & | & | & | \\
H-C- & C- & C- & C- & C-H \\
| & | & | & | & | \\
H & H & H & H & H \\
\end{array}
$$

c. 2-Methyl-2-pentanol:

$$
\begin{array}{ccccc}
H & OH & H & H & H \\
| & | & | & | & | \\
H-C- & C- & C- & C- & C-H \\
| & | & | & | & | \\
H & & H & H & H \\
\end{array}
$$
$$
\begin{array}{c}
H-C-H \\
| \\
H
\end{array}
$$

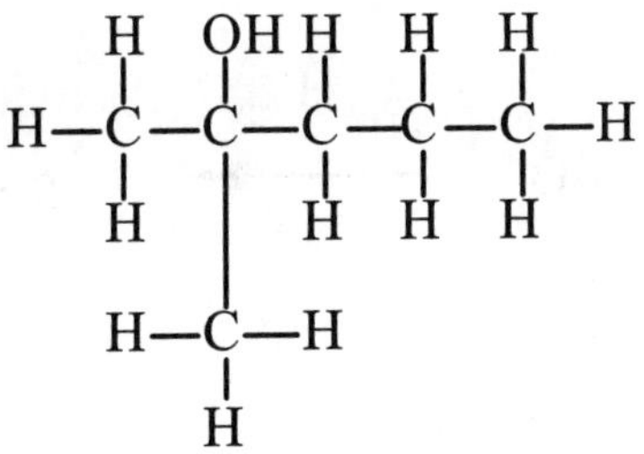

d. Cyclohexanol:

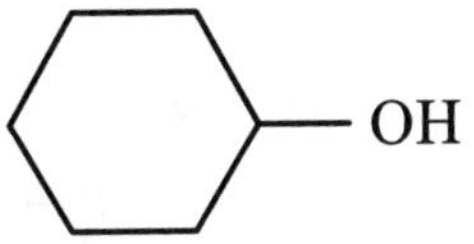

e. 3,4-Dimethyl-3-heptanol:

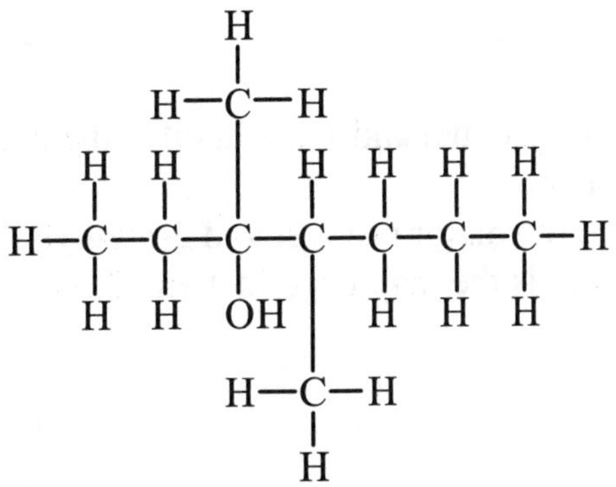

383

12.45 Methanol is commonly used as a solvent and as a starting material for the synthesis of formaldehyde. Ethanol is also used as a solvent and is the alcohol found in alcoholic beverages. Isopropyl alcohol has been used as rubbing alcohol. Patients with a high fever were given alcohol baths. The rapid evaporation of the alcohol causes cooling of the skin and helps reduce fever. Isopropanol is also used as an antiseptic.

12.47 When the ethanol concentration in a fermentation reaches 12–13%, the yeast producing the ethanol are killed by it. To produce a liquor of higher alcohol concentration, the product of the original fermentation must be distilled.

12.49 The carbinol carbon is the one to which the hydroxyl group is bonded.

12.51 a. Primary alcohol
b. Secondary alcohol
c. Tertiary alcohol
d. Tertiary alcohol
e. Tertiary alcohol

12.53 a.

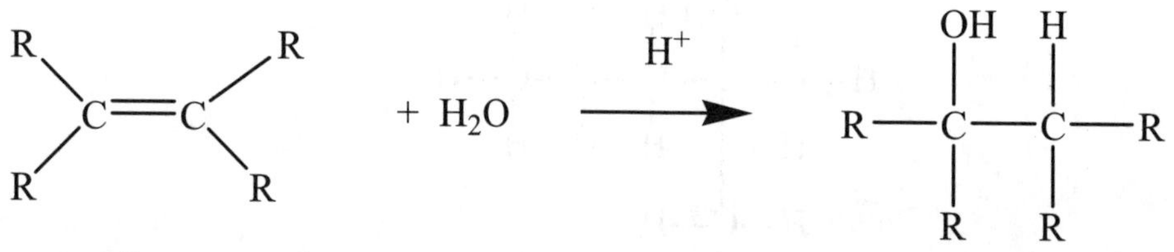

Alkene + H_2O $\xrightarrow{H^+}$ Alcohol

b.

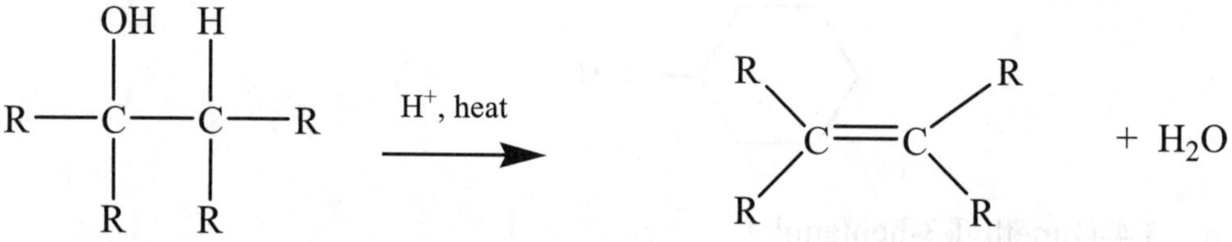

Alcohol $\xrightarrow{H^+, \text{heat}}$ Alkene + H_2O

12.55 a. 2-Pentanol is the major product and 1-pentanol is the minor product.
b. 2-Pentanol and 3-pentanol
c. 3-Methyl-2-butanol is the major product and 3-methyl-1-butanol is the minor product.
d. 3,3-Dimethyl-2-butanol is the major product and 3,3-dimethyl-1-butanol is the minor product.

12.57 a. Butanone
b. N.R.

c. Cyclohexanone

d. N.R.

12.59 Phenols are compounds in which the hydroxyl group is attached to a benzene ring. Like alcohols, phenols are polar compounds because of the polar hydroxyl group. Thus, the simpler phenols are somewhat soluble in water.

12.61 Alcohols of molecular formula $C_4H_{10}O$

$CH_3CH_2CH_2CH_2OH$ $CH_3CH(OH)CH_2CH_3$ $CH_3CH(CH_3)CH_2OH$ $CH_3C(CH_3)_2CH_3$

Ethers of molecular formula $C_4H_{10}O$

$CH_3-O-CH_2CH_2CH_3$ $CH_3CH_2-O-CH_2CH_3$ $CH_3-O-\underset{\underset{CH_3}{|}}{CH}CH_3$

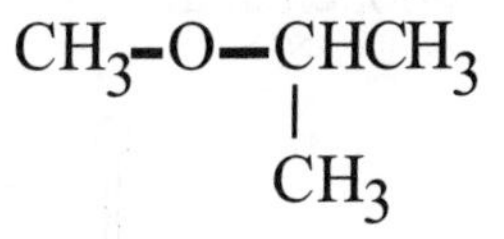

12.63 a. $CH_3CH_2-O-CH_2CH_3 \; + \; H_2O$

b. $CH_3CH_2-O-CH_2CH_3 \; + \; CH_3-O-CH_3$

    $+ \; CH_3-O-CH_2CH_3 \; + \; H_2O$

12.65 Cystine:

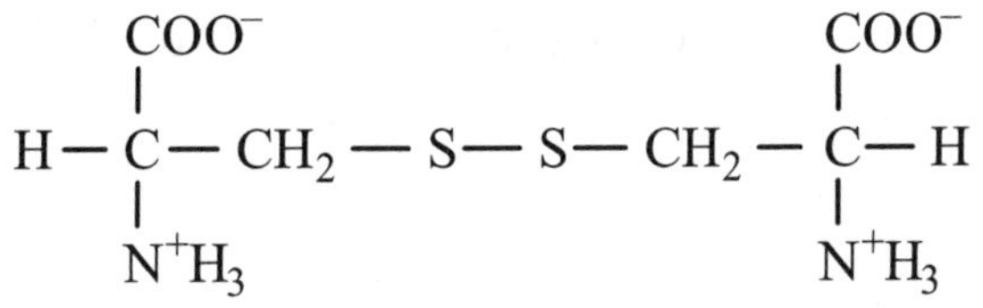

$$H-\underset{\underset{N^+H_3}{|}}{\overset{\overset{COO^-}{|}}{C}}-CH_2-S-S-CH_2-\underset{\underset{N^+H_3}{|}}{\overset{\overset{COO^-}{|}}{C}}-H$$

12.67 As the carbon chain length increases, the compounds become less polar and more hydrocarbonlike. As a result, their solubility in water decreases.

12.69

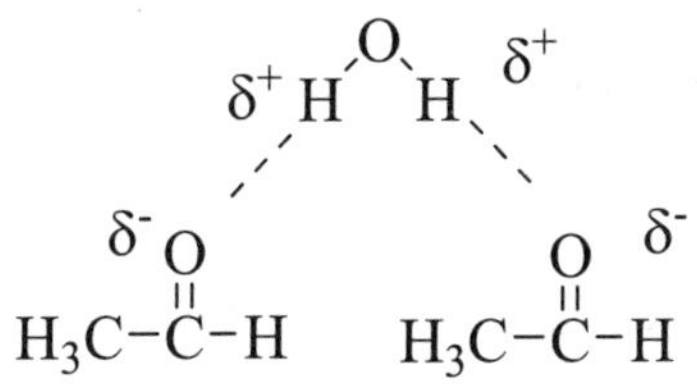

12.71 To name an aldehyde using the I.U.P.A.C. nomenclature system, identify and name the longest carbon chain containing the carbonyl group. Replace the final –e of the alkane name with – al. Number and name all substituents as usual. Remember that the carbonyl carbon is always carbon-1 and does not need to be numbered in the name of the compound.

12.73 a. 2-Butanone
 b. 2-Ethylhexanal
 c. Butanal
 d. 4-Bromo-4-methylpentanal

12.75

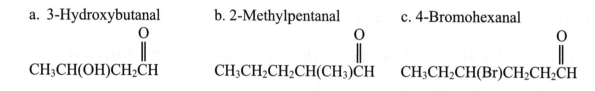

a. 3-Hydroxybutanal

$$CH_3CH(OH)CH_2\overset{\overset{\displaystyle O}{\|}}{C}H$$

b. 2-Methylpentanal

$$CH_3CH_2CH_2CH(CH_3)\overset{\overset{\displaystyle O}{\|}}{C}H$$

c. 4-Bromohexanal

$$CH_3CH_2CH(Br)CH_2CH_2\overset{\overset{\displaystyle O}{\|}}{C}H$$

d. 3-Iodopentanal

$$CH_3CH_2CH(I)CH_2\overset{\overset{\displaystyle O}{\|}}{C}H$$

e. 2-Hydroxy-3-methylheptanal

$$CH_3CH_2CH_2CH_2CH(CH_3)CH(OH)\overset{\overset{\displaystyle O}{\|}}{C}H$$

12.77 Acetone is a good solvent because it can dissolve a wide range of compounds. It has both a polar carbonyl group and nonpolar side chains. As a result, it dissolves organic compounds and is also miscible in water.

12.79   a. False   b. True   c. False   d. False

12.81 a.

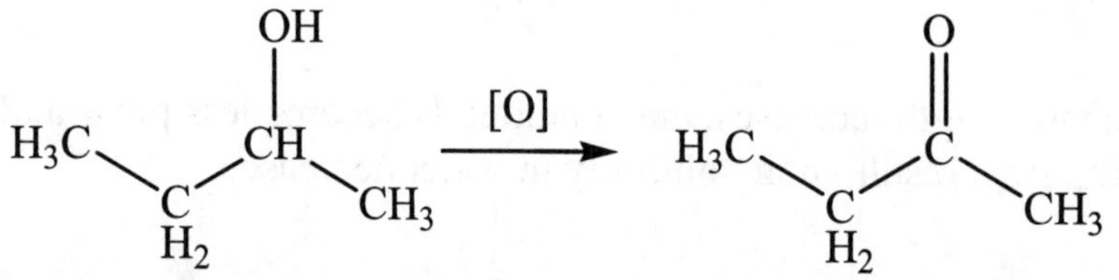

H₃C—C(H₂)—CH(OH)—CH₃   →[O]→   H₃C—C(H₂)—C(=O)—CH₃

2-Butanol   Butanone

b.

HO—C(H₂)—CH(CH₃)—CH₃   →[O]→   O=C(H)—CH(CH₃)—CH₃

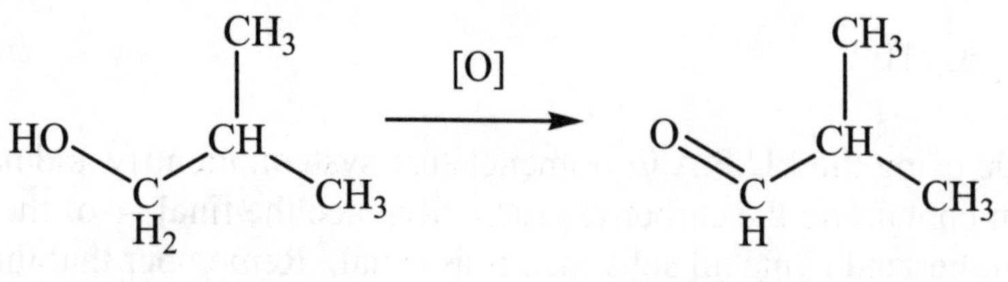

2-Methyl-1-propanol   Methylpropanal

Note that methylpropanal can be further oxidized to methylpropanoic acid.

c.

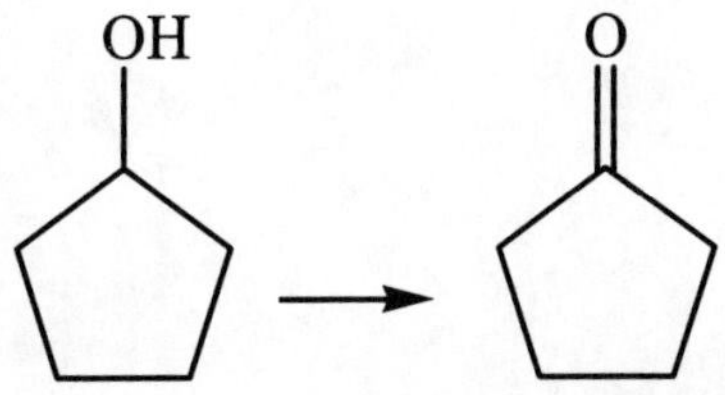

Cyclopentanol   Cyclopentanone

d.

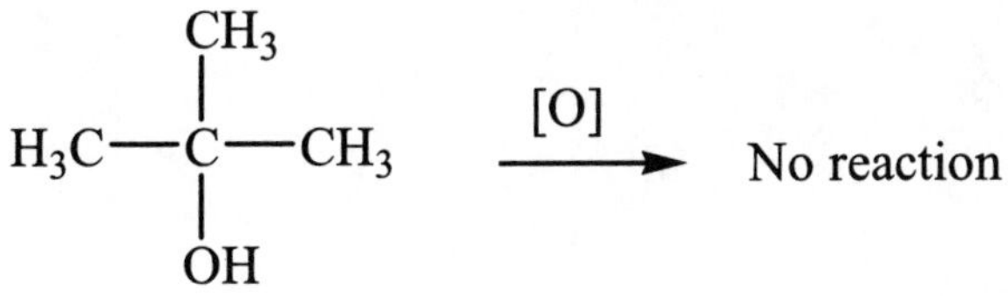

$$H_3C-\underset{\underset{OH}{|}}{\overset{\overset{CH_3}{|}}{C}}-CH_3 \xrightarrow{[O]} \text{No reaction}$$

2-Methyl-2-propanol

e.

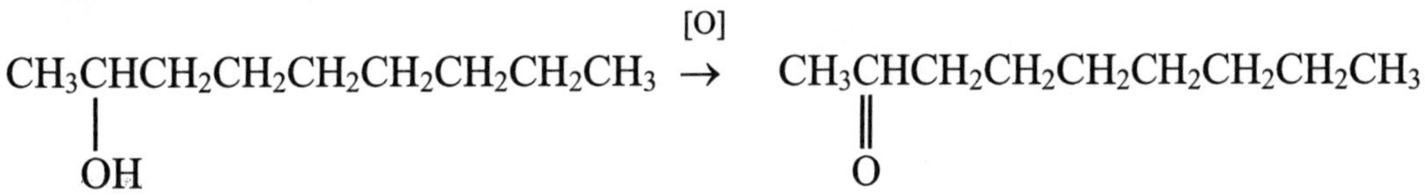

$$CH_3\underset{\underset{OH}{|}}{C}HCH_2CH_2CH_2CH_2CH_2CH_2CH_3 \xrightarrow{[O]} CH_3\underset{\underset{O}{\parallel}}{C}HCH_2CH_2CH_2CH_2CH_2CH_2CH_3$$

2-Nonanol   2-Nonanone

f.

$$\underset{\underset{OH}{|}}{C}H_2CH_2CH_2CH_2CH_2CH_2CH_2CH_2CH_2CH_3 \xrightarrow{[O]} \underset{\underset{O}{\parallel}}{C}HCH_2CH_2CH_2CH_2CH_2CH_2CH_2CH_2CH_3$$

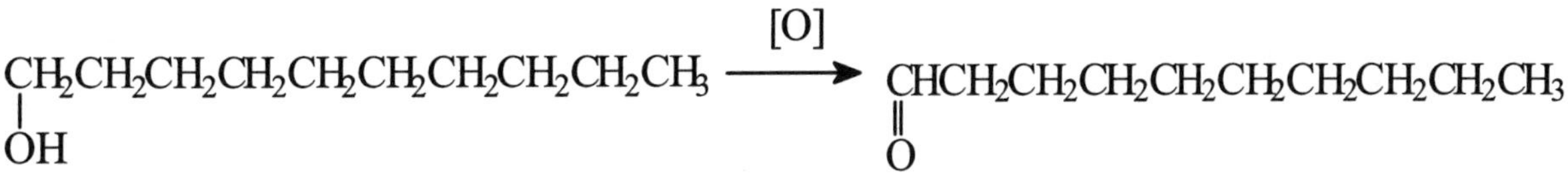

1-Decanol   Decanal

Note that decanal can be further oxidized to decanoic acid.

12.83

a. 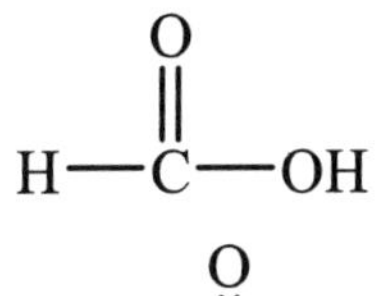 $H-\overset{\overset{O}{\parallel}}{C}-OH$

b. 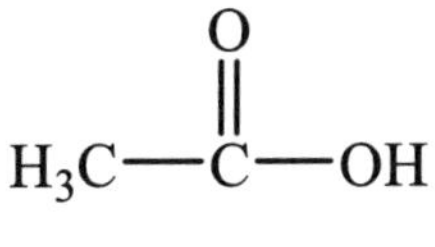 $H_3C-\overset{\overset{O}{\parallel}}{C}-OH$

387

c. $CH_3CH_2 \overset{\displaystyle O}{\underset{\displaystyle \|}{-C}} -OH$

d. $CH_3CH_2CH_2 \overset{\displaystyle O}{\underset{\displaystyle \|}{-C}} -OH$

Chapter 13
Carboxylic Acids, Esters, Amines, and Amides
Solutions to the Odd-Numbered Problems

In-Chapter Questions and Problems

13.1 a. Ketone
b. Ketone
c. Alkane

13.3 The carboxyl group consists of two very polar groups, the carbonyl group and the hydroxyl group. Thus, carboxylic acids are very polar, in addition to which, they can hydrogen bond to one another. Aldehydes are polar, as a result of the carbonyl group, but cannot hydrogen bond to one another. As a result, carboxylic acids have higher boiling points than aldehydes of the same carbon chain length.

13.5 a. 2,4-Dimethylpentanoic acid
b. 2,4-Dichlorobutanoic acid

13.7 a. α,γ-Dimethylvaleric acid
b. α,γ-Dichlorobutyric acid

13.9 a. 2,4,6-Tribromobenzoic acid:

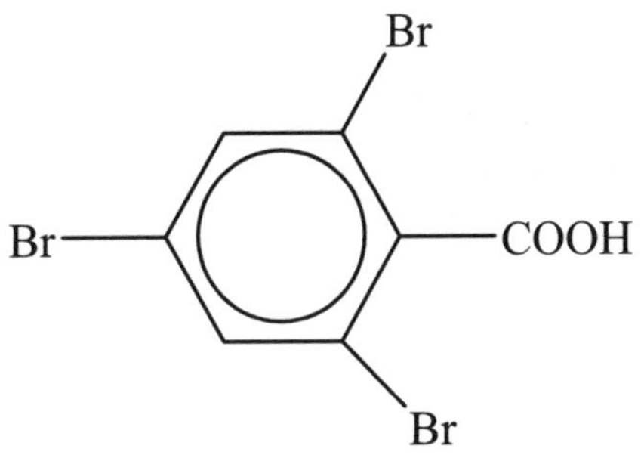

b. 2,3-Dibromobenzoic acid:

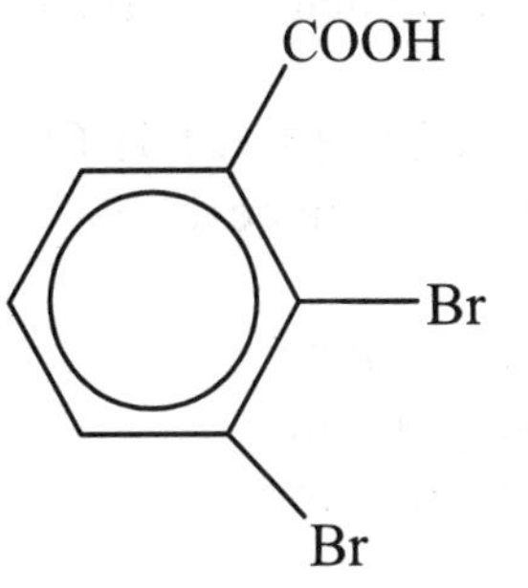

13.11 a. Propanol would be oxidized to propanal, which would quickly be oxidized to propanoic acid:

$$CH_3CH_2\overset{\overset{\displaystyle O}{\|}}{C}{-}H \rightarrow CH_3CH_2\overset{\overset{\displaystyle O}{\|}}{C}{-}OH$$

Propanal Propanoic acid

b. Pentanal would be oxidized to pentanoic acid:

$$CH_3CH_2CH_2CH_2\overset{\displaystyle O}{\overset{\|}{C}}-OH$$

13.13 a. I.U.P.A.C. name: Propyl butanoate
Common name: propyl butyrate
b. I.U.P.A.C. name: Ethyl butanoate
Common name ethyl butyrate

13.15 The following reaction between butanol and ethanoic acid produces butyl ethanoate. It requires a trace of acid and heat. It is also reversible

$$CH_3CH_2CH_2CH_2OH + CH_3COOH \leftrightarrow CH_3\overset{\displaystyle O}{\overset{\|}{C}}-OCH_2CH_2CH_2CH_3$$

b. The following reaction between ethanol and propanoic acid produces ethyl propanoate. It requires a trace of acid and heat. It is also reversible.

$$CH_3CH_2OH + CH_3CH_2COOH \leftrightarrow CH_3CH_2\overset{\displaystyle O}{\overset{\|}{C}}-OCH_2CH_3$$

13.17 a. $CH_3COOH + CH_3CH_2CH_2OH$
Ethanoic acid   1-Propanol

b. $CH_3CH_2CH_2CH_2CH_2COO^-K^+ + CH_3CH_2CH_2OH$
Potassium hexanoate   1-Propanol

13.19 a. Tertiary amine
b. Primary amine
c. Secondary amine

13.21

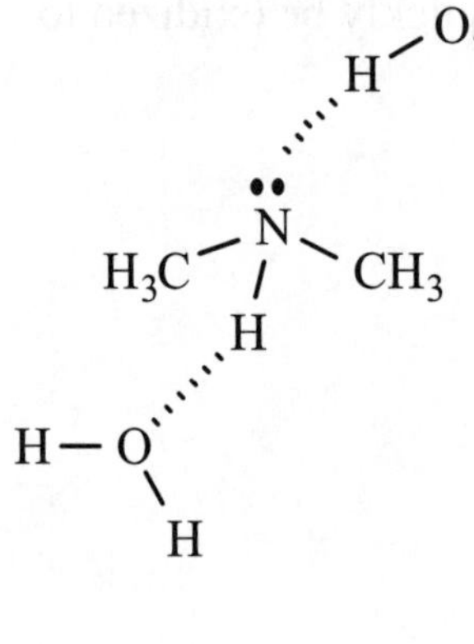

390

13.23 a. Methanol would have the higher boiling point than methylamine. The intermolecular hydrogen bonds between alcohol molecules will be stronger than those between two amines because oxygen is more electronegative than nitrogen.

b. Water would have a higher boiling point than dimethylamine. The intermolecular hydrogen bonds between water molecules will be stronger than those between two amines because oxygen is more electronegative than nitrogen.

c. Ethylamine will have a higher boiling point that methylamine because it has a higher molecular weight.

d. Propylamine will have a higher boiling point that butane because propylamine molecules can form intermolecular hydrogen bonds while the nonpolar butane cannot do so.

13.25 a. 2-Propanamine:

```
        H    H    H
        |    |    |
   H —  C —  C —  C — H
        |    |    |
        H    N    H
            / \
           H   H
```

b. 3-Octanamine:

```
                 H     H
                  \   /
    H   H    N    H   H   H   H   H
    |   |    |    |   |   |   |   |
H — C — C — C — C — C — C — C — C — H
    |   |    |    |   |   |   |   |
    H   H    H    H   H   H   H   H
```

c. N-Ethyl-2-heptanamine:

```
    H   H   H   H   H   H   H
    |   |   |   |   |   |   |
H — C — C — C — C — C — C — C — H
    |   |   |   |   |   |   |
    H   |   H   H   H   H   H
        |
        N — H
        |
    H — C — H
        |
    H — C — H
        |
        H
```

d. 2-Methyl-2-pentanamine:

```
        H   H
         \ /
   H   N   H   H   H
   |   |   |   |   |
H— C — C — C — C — C —H
   |   |   |   |   |
   H   |   H   H   H
       H — C — H
           |
           H
```

e. 4-Chloro-5-iodo-1-nonanamine:

```
   H   H
    \ /
     N   H   H   H   H   H   H   H   H
     |   |   |   |   |   |   |   |   |
H— C — C — C — C — C — C — C — C — C —H
     |   |   |   |   |   |   |   |   |
     H   H   H   Cl  I   H   H   H   H
```

f. *N,N*-Diethyl-1-pentanamine:

```
           H
           |
       H — C — H
           |
       H — C — H
           |
   H   H   |   H   H   H   H   H
   |   |   |   |   |   |   |   |
H— C — C — N — C — C — C — C — C —H
   |   |       |   |   |   |   |
   H   H       H   H   H   H   H
```

End-of-Chapter Questions and Problems

13.27 Aldehydes are polar, as a result of the carbonyl group, but cannot hydrogen bond to one another. Alcohols are polar and can hydrogen bond as a result of the polar hydroxyl group. The carboxyl group of the carboxylic acids consists of both of these polar groups: the carbonyl group and the hydroxyl group. Thus, carboxylic acids are more polar than either aldehydes or alcohols, in addition to which, they can hydrogen bond to one another. As a result, carboxylic acids have higher boiling points than aldehydes or alcohols of the same carbon chain length.

13.29 a. 3-Hexanone   d. Dipropyl ether
       b. 3-Hexanone   e. Hexanal
       c. Hexane   f. Ethanol

13.31 Determine the name of the parent compound, that is the longest carbon chain containing the carboxyl group. Change the –e ending of the alkane name to –oic acid. Number the chain so that the carboxyl carbon is carbon-1. Name and number substituents in the usual way.

13.33 a. 2-Bromopentanoic acid:

H H H H O
| | | | ‖
H—C—C—C—C—C—OH
| | | |
H H H Br

b. 2-Bromo-3-methylbutanoic acid:

H
|
H—C—H
|
H | H O
| | ‖
H—C———C———C—C—OH
| | |
H H Br

c. 2,4,6-Trimethylstearic acid:

H   H
| |
H—C—H   H—C—H
| |
H H H H H H H H H H H H | H H H | O
| | | | | | | | | | | | | | | | ‖
H—C—C—C—C—C—C—C—C—C—C—C—C—C—C—C—C—C—C—OH
| | | | | | | | | | | | | | | | |
H H H H H H H H H H H H H H | H H
                           H—C—H
                           |
                           H

d. Propenoic acid:

H   H
\ /
C=C
/ \
H   C=O
|
OH

13.35 a. I.U.P.A.C. name: 2-Hydroxypropanoic acid
Common name: α-Hydroxypropionic acid

b. I.U.P.A.C. name: 3-Hydroxybutanoic acid
Common name: β-Hydroxybutyric acid

393

c. I.U.P.A.C. name: 4,4-Dimethylpentanoic acid
Common name: γ,γ-Dimethylvaleric acid

d. I.U.P.A.C. name: 3,3-Dichloropentanoic acid
Common name: β,β-Dichlorovaleric acid

13.37 Soaps are made from water, a strong base, and natural fats or oils. The fats and oils are triesters of glycerol. In the presence of the strong base, the ester bonds are hydrolyzed and the salts of the long chain fatty acids are formed. The salts of fatty acids are soaps.

13.39 a. (1) H_2CrO_4 (3) HCl

(2) $CH_3CH(OH)CH_3$ (4) $CH_3CH_2-C(=O)-O^-\ Na^+$

b. $CH_3-C(=O)-O^-\ Na^+$

c. $CH_3CH_2CH_2CH_2CH_2-C(=O)-O^-\ Na^+$

d. CH_3COOH

13.41 Esters are mildly polar as a result of the polar carbonyl group within the structure.

13.43 a. Methyl benzoate:

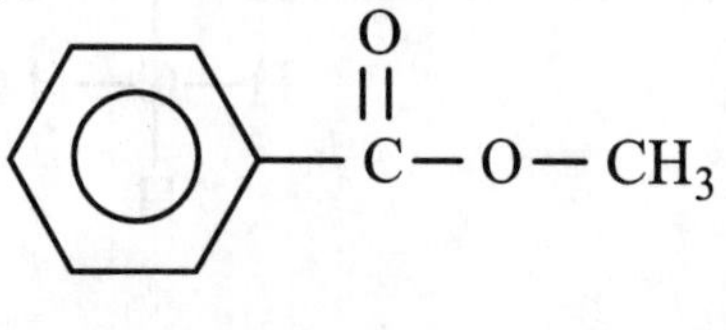

b. Butyl decanoate:

$CH_3CH_2CH_2CH_2CH_2CH_2CH_2CH_2CH_2-C(=O)-O-CH_2CH_2CH_2CH_3$

c. Methyl propionate:

$$CH_3CH_2-\overset{\overset{\displaystyle O}{\|}}{C}-O-CH_3$$

d. Ethyl propionate:

$$CH_3CH_2-\overset{\overset{\displaystyle O}{\|}}{C}-O-CH_2CH_3$$

13.45 The synthesis of an ester is referred to as a dehydration reaction because a water molecule is eliminated during the course of the reaction.

13.47

13.49 a.

Propyl propanoate   Propanoic acid   1-Propanol

b.

Butyl methanoate   Methanoic acid   1-Butanol

c.

$$H-\overset{\overset{\displaystyle O}{\|}}{C}-OCH_2CH_3 \quad \underset{\longrightarrow}{\overset{H^+,\ heat}{\longleftarrow}} \quad H-\overset{\overset{\displaystyle O}{\|}}{C}-OH \quad + \quad CH_3CH_2OH$$

Ethyl methanoate   Methanoic   Ethanol
acid

d.

Methyl pentanoate   Pentanoic acid   Ethanol

13.51 In systematic nomenclature, primary amines are named by determining the name of the parent compound, the longest continuous carbon chain containing the amine group. The –e ending of the alkane chain is replaced with –amine. Thus, an alkane becomes an alkanamine. The parent chain is then numbered to give the carbon bearing the amine group the lowest possible number. Finally, all substituents are named and numbered and added as prefixes to the "alkanamine" name.

13.53 Amphetamines elevate blood pressure and pulse rate. They also decrease the appetite.

13.55 a. Ethanol has a higher boiling point than ethanamine because the hydroxyl group oxygen is more electronegative than the amine nitrogen atom. As a result, the intermolecular hydrogen bonds between alcohol molecules are stronger than those between primary amines.

b. 1-Propanamine has a higher boiling point than butane due to the polar –NH_2 of 1-propanamine that can form intermolecular hydrogen bonds with other 1-propanamine molecules. Butane is unable to form intermolecular hydrogen bonds.

c. Water has a higher boiling point because the H····OH hydrogen bonds between water molecules are stronger than the H····NH hydrogen bonds between methanamine molecules. This is because oxygen is more electronegative than nitrogen.

d. Ethylmethylamine has a higher boiling point than butane because the polar -NH of ethylmethylamine can form intermolecular hydrogen bonds with other ethylmethylamine molecules. Butane is unable to form intermolecular hydrogen bonds.

13.57 a. Systematic name: 2-Butanamine
Common name: 2-Butylamine
b. Systematic name: 3-Hexanamine
Common name: 3-Hexylamine

c. Systematic name: 2-Methyl-2-propanamine
   Common name: Tertiary butylamine
d. Systematic name: 1-Octanamine
   Common name: 1-Octylamine
e. Systematic name: *N,N*-dimethylethanamine
   Common name: Ethyldimethylamine

13.59 a. Cyclohexanamine:

b. 2-Bromocyclopentanamine:

c. Tetraethylammonium iodide:

d. 3-Bromobenzenamine:

13.61 a. Cyclohexanamine is a primary amine.
       b. Dibutylamine is a secondary amine.
       c. 2-Methyl-2-heptanamine is a primary amine.
       d. Tripentylamine is a tertiary amine.

13.63 a. H_2O
       b. HBr

c. $CH_3CH_2CH_2-N^+H_3$

d.

$$CH_3CH_2 - \overset{\overset{\displaystyle H}{|}}{\underset{\underset{\displaystyle H}{|}}{\overset{+}{N}}} - CH_2CH_3 \qquad Cl^-$$

13.65 Drugs containing amine groups are generally administered as ammonium salts because the salt is more soluble in water and, hence, in body fluids.

13.67 Amides have very high boiling points because the amide group consists of two very polar functional groups, the carbonyl group and the amino group. Strong intermolecular hydrogen bonding between the N-H bond of one amide and the C=O group of a second amide results in very high boiling points.

13.69 a. I.U.P.A.C. name: Propanamide
   Common name: Propionamide

   b. I.U.P.A.C. name: Pentanamide
   Common name: Valeramide

   c. I.U.P.A.C. name: *N,N*-Dimethylethanamide
   Common name: *N,N*-Dimethylacetamide

13.71 a. Ethanamide:

$$H_3C - \overset{\overset{\displaystyle O}{||}}{C} - NH_2$$

   b. *N*-Methylpropanamide:

$$CH_3CH_2 - \overset{\overset{\displaystyle O}{||}}{C} - \overset{\overset{\displaystyle H}{|}}{N} - CH_3$$

   c. *N,N*-Diethylbenzamide:

$$C_6H_5 - \overset{\overset{\displaystyle O}{||}}{C} - \underset{\underset{\displaystyle CH_2CH_3}{|}}{N} - CH_2CH_3$$

   d. 3-Bromo-4-methylhexanamide:

$$CH_3CH_2\underset{\underset{\displaystyle Br}{|}}{C}H\overset{\overset{\displaystyle CH_3}{|}}{C}HCH_2 - \overset{\overset{\displaystyle O}{||}}{C} - NH_2$$

e. *N,N*-Dimethylacetamide:

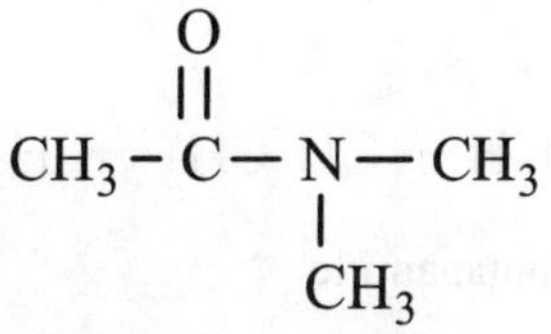

$$CH_3 - \overset{\overset{\displaystyle O}{\|}}{C} - \underset{\underset{\displaystyle CH_3}{|}}{N} - CH_3$$

13.73 Amides are not proton acceptors (bases) because the highly electronegative carbonyl oxygen has a strong attraction for the nitrogen lone pair of electrons. As a result they cannot "hold" a proton. They are not proton donors because there are no hydrogen atoms bonded to the oxygen.

13.75 Penicillin BT

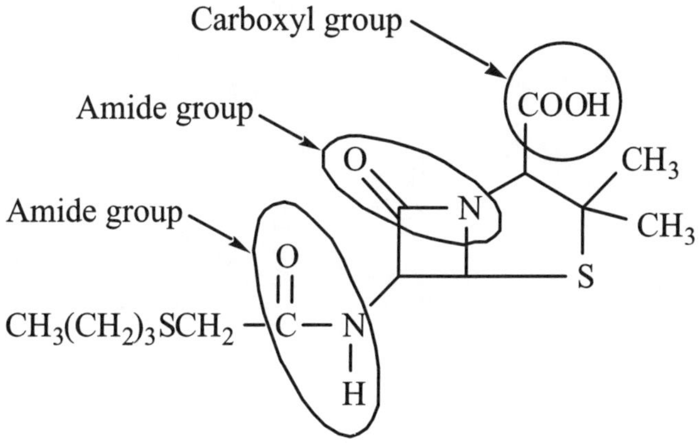

Carboxyl group

Amide group

Amide group

COOH

CH$_3$

CH$_3$

O

N

S

O

$CH_3(CH_2)_3SCH_2 - C - N$

H

13.77 a.

$$CH_3 - \overset{\overset{\displaystyle O}{\|}}{C} - NH - CH_3 \ + \ H_3O^+ \ \longrightarrow \ CH_3COOH \ + \ CH_3NH_3{}^+$$

N-Methylethanamide   Ethanoic acid   Methanamine

b.

$$CH_3CH_2CH_2 - \overset{\overset{\displaystyle O}{\|}}{C} - NH - CH_3 \ + \ H_3O^+ \ \longrightarrow$$

N-Methylbutanamide

$$CH_3CH_2CH_2COOH \ + \ CH_3NH_3{}^+$$

Butanoic acid   Methanamine

c.

$$\underset{\text{N-Ethyl-3-methylbutanamide}}{\overset{\displaystyle CH_3 \qquad O}{CH_3CHCH_2 - \overset{||}{C} - NH - CH_2CH_3}} + H_3O^+ \longrightarrow$$

$$\underset{\text{3-Methylbutanoic acid}}{\overset{\displaystyle CH_3}{CH_3CHCH_2COOH}} + \underset{\text{Methanamine}}{CH_3NH_3^+}$$

In-Chapter Questions and Problems

14.1 It is currently recommended that 45–65% of the calories in the diet should be carbohydrates. Of that amount, no more than 10% should be simple sugars.

14.3 An aldose is a sugar with an aldehyde functional group. A ketose is a sugar with a ketone functional group.

14.5
a. ketose
b. aldose
c. ketose
d. aldose
e. ketose
f. aldose

14.7

a.
```
      CH3
       |
       C=O
       |
  H ---*--- OH
       |
      CH2OH
```

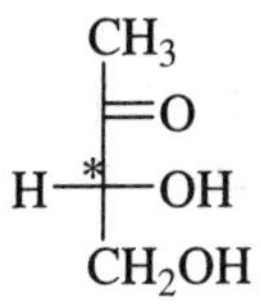

b.
```
      CHO
       |
  H ---*--- OH
       |
  H ---*--- OH
       |
 HO ---*--- H
       |
      CH2OH
```

c.
```
     CH2OH
       |
       C=O
       |
 HO ---*--- H
       |
  H ---*--- OH
       |
  H ---*--- OH
       |
      CH2OH
```

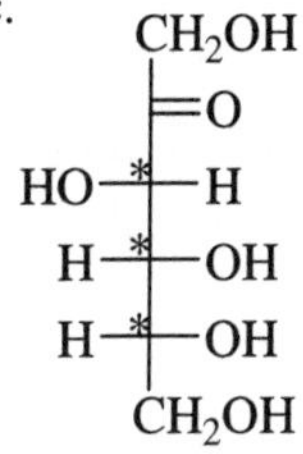

d.
```
      CHO
       |
  H ---*--- OH
       |
      CH2OH
```

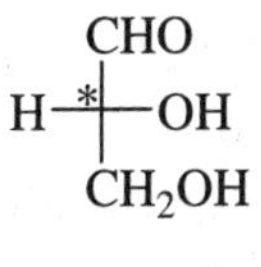

e,
```
      CH3
       |
       C=O
       |
  H ---*--- OH
       |
  H ---*--- OH
       |
  H ---*--- OH
       |
      CH2OH
```

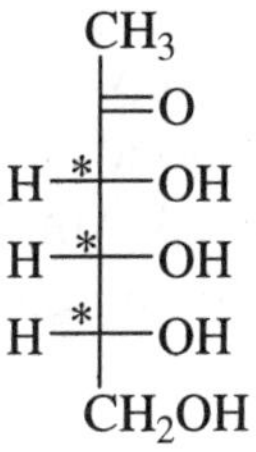

f.
```
      CHO
       |
 HO ---*--- H
       |
  H ---*--- OH
       |
 HO ---*--- H
       |
      CH2OH
```

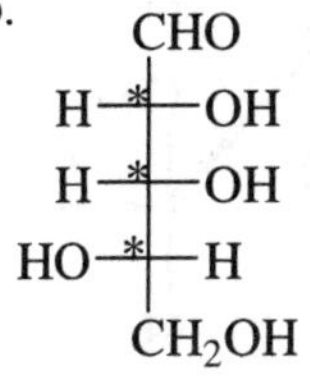

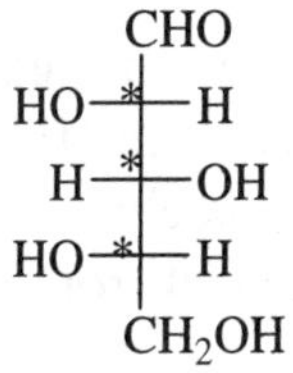

14.9
a. D-
b. L-
c. D-
d. D-
e. D-
f. L-

14.11

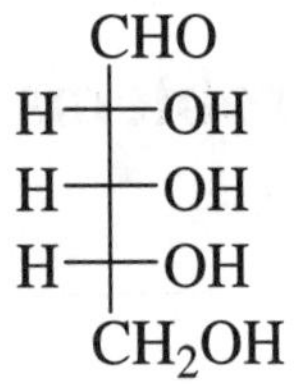

$$\begin{array}{c} \text{CHO} \\ \text{H}\!\!-\!\!\text{OH} \\ \text{H}\!\!-\!\!\text{OH} \\ \text{H}\!\!-\!\!\text{OH} \\ \text{CH}_2\text{OH} \end{array}$$

D-Ribose

14.13

$$\begin{array}{c} \text{CHO} \\ \text{HO}\!\!-\!\!\text{H} \\ \text{HO}\!\!-\!\!\text{H} \\ \text{HO}\!\!-\!\!\text{H} \\ \text{CH}_2\text{OH} \end{array}$$

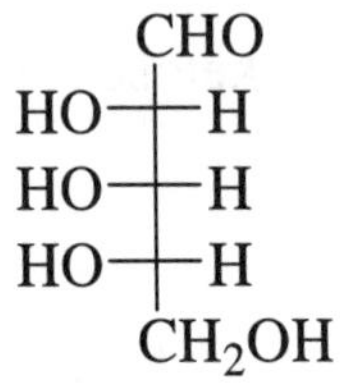

L-Ribose

14.15

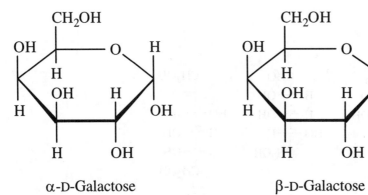

α-D-Galactose   β-D-Galactose

14.17 α-Amylase and β-amylase are digestive enzymes that break down the starch amylose. α-Amylase cleaves glycosidic bonds of the amylose chain at random, producing shorter polysaccharide chains. β-Amylase sequentially cleaves maltose (a disaccharide of glucose) from the reducing end of the polysaccharide chain.

End-of-Chapter Questions and Problems

14.19 A monosaccharide is the simplest sugar and consists of a single saccharide unit. A monosaccharide cannot be broken down into a simpler molecule by hydrolysis. A disaccharide is made up of two monosaccharides joined covalently by a glycosidic bond. Hydrolysis of a disaccharide yields two monosaccharides.

14.21 The molecular formula for a simple sugar is $(CH_2O)_n$. Typically n is an integer from 3 to 7.

14.23 Mashed potato flakes, rice, and corn starch would contain amylose and amylopectin, both of which are polysaccharides. A candy bar contains sucrose, a disaccharide. Orange juice contains fructose, a monosaccharide. It may also contain sucrose if the label indicates that sugar has been added.

14.25 Four kilocalories of energy are released for each gram of carbohydrate "burned" or oxidized.

14.27

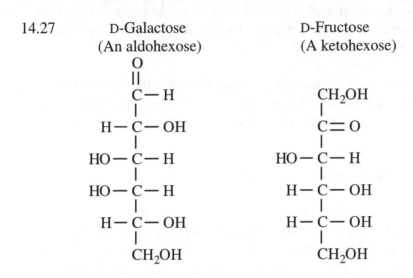

D-Galactose
(An aldohexose)

D-Fructose
(A ketohexose)

14.29 An *aldose* is a sugar that contains an aldehyde (carbonyl) group.

14.31 A ketopentose is a sugar with a five-carbon backbone and containing a ketone (carbonyl) group.

14.33   a.   β-D-Glucose
        b.   β-D-Fructose
        c.   α-D-Galactose

14.35 The two aldotrioses of molecular formula C_3H_6O:

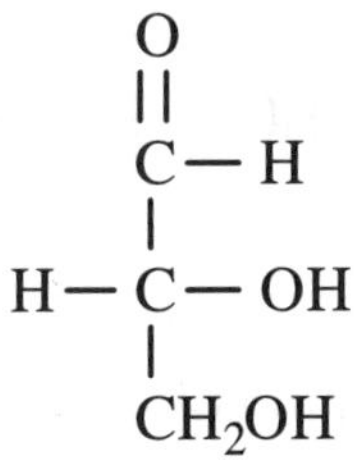

$$
\begin{array}{cc}
\begin{array}{c}
O \\
\parallel \\
C-H \\
| \\
H-C-OH \\
| \\
CH_2OH
\end{array}
&
\begin{array}{c}
O \\
\parallel \\
C-H \\
| \\
HO-C-H \\
| \\
CH_2OH
\end{array}
\end{array}
$$

D-Glyceraldehyde 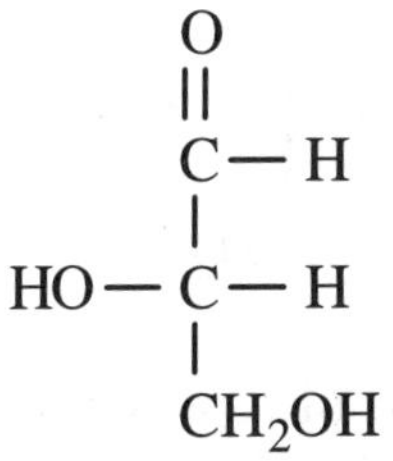   L-Glyceraldehyde

14.37 *Stereoisomers* are pairs of molecules having the same structural formula and bonding pattern but differing in the arrangement of the atoms in space.

14.39 A chiral carbon is a carbon atom that is bonded to four different groups.

14.41 A polarimeter is an instrument that is used to measure the optical activity of molecules. It has a monochromatic light source. Light waves from the source are directed through a polarizer. The light that emerges is plane-polarized. This passes through the sample and into the analyzer, which will measure the degree to which the plane of light has been rotated by the sample. If the plane of light is not altered by the sample, the compound is optically inactive. However, if the plane of light is rotated either clockwise or counterclockwise, the sample is optically active.

14.43 A Fischer Projection is a two-dimensional drawing of a molecule that shows a chiral carbon at the intersection of two lines. Horizontal lines at the intersection represent bonds projecting out of the page and vertical lines represent bonds that project into the page.

14.45 Dextrose is a common name used for D-glucose.

14.47 D- and L-Glyceraldehyde are a pair of enantiomers; that is, they are nonsuperimposable mirror images of one another. In D-glyceraldehyde, the hydroxyl group on the chiral carbon farthest from the aldehyde group (C-2) is on the right of the structure. In L-glyceraldehyde, the hydroxyl group on C-2 is on the left of the structure.

14.49 a.   b.   c.

14.51 A Haworth projection is a means of representing the orientation of substituent groups around a cyclic sugar molecule.

14.53

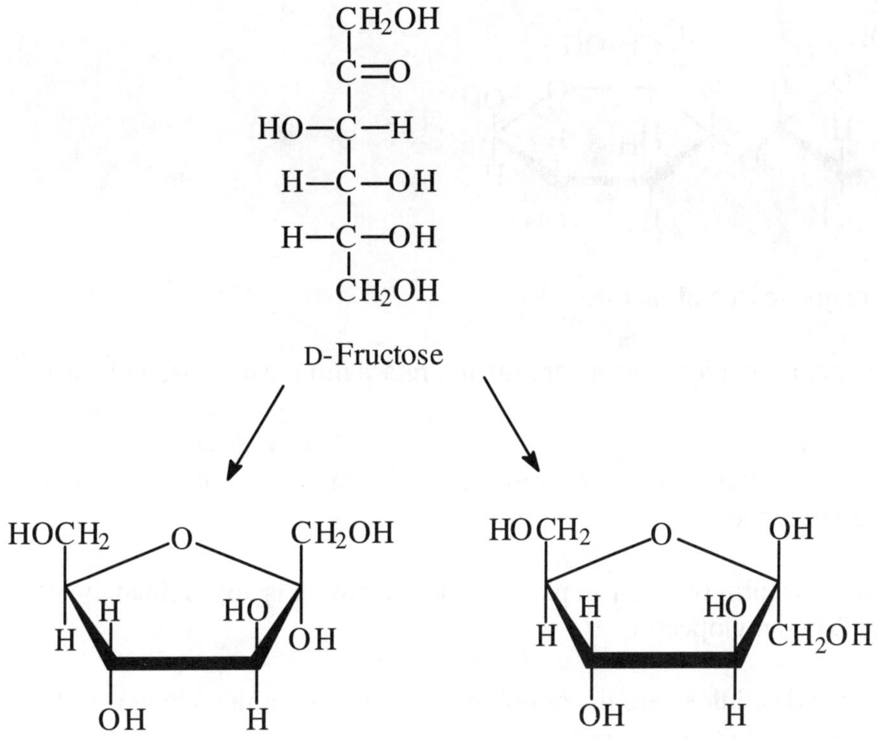

$$CH_2OH$$
$$C=O$$
$$HO-C-H$$
$$H-C-OH$$
$$H-C-OH$$
$$CH_2OH$$

D-Fructose

α-D-Fructose

β-D-Fructose

14.55 β-maltose and α-lactose would give positive Benedict's tests. Glycogen would give only a weak reaction because it is a long polymer and thus there are fewer reducing ends for a given mass of the carbohydrate.

14.57 Enantiomers are stereoisomers that are nonsuperimposable mirror images of one another. For instance:

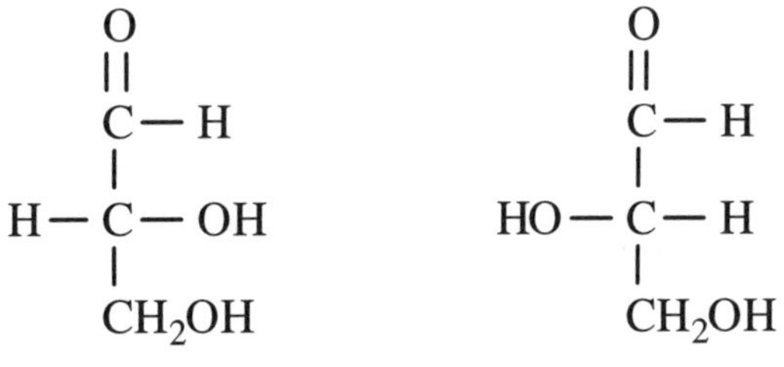

D-Glyceraldehyde

L-Glyceraldehyde

14.59 A glycosidic bond is the bond formed between the hydroxyl group of the C-1 carbon of one sugar and a hydroxyl group of another sugar.

14.61 The structure of β-maltose is:

14.63 Milk is the major source of lactose.

14.65 Untreated galactosemia leads to severe mental retardation, cataracts, and early death.

14.67 A polymer is a very large molecule formed by the combination of many small molecules called monomers.

14.69 The major storage form of sugar in plants is starch, which is approximately 80% amylase and 20% amylopectin.

14.71 Both amylose and cellulose are linear polymers of glucose units. However, the glucose units of amylose are joined by $\alpha(1 \rightarrow 4)$ glycosidic bonds, and those of cellulose are bonded together by $\beta(1 \rightarrow 4)$ glycosidic bonds.

14.73 The major physiological purpose of glycogen is to serve as a storage molecule for glucose. This represents an energy reservoir for the body. Glycogen synthesis and degradation in the liver are involved in regulation of blood glucose levels. Glycogen is stored in the liver and skeletal muscles.

Chapter 15
Lipids and Their Functions in Biochemical Systems
Solutions to the Odd-Numbered Problems

In-Chapter Questions and Problems

15.1   a. Oleic acid: $CH_3(CH_2)_7CH=CH(CH_2)_7COOH$

        b. Lauric acid: $CH_3(CH_2)_{10}COOH$

        c. Linoleic acid: $CH_3(CH_2)_4CH=CH–CH_2–CH=CH(CH_2)_7COOH$

        d. Stearic acid: $CH_3(CH_2)_{16}COOH$

15.3

$$CH_3(CH_2)_{10}COOH + CH_3CH_2OH \xrightarrow{H^+, heat} CH_3(CH_2)\overline{_{10}}\overset{\overset{\displaystyle O}{\|}}{C}—OCH_2CH_3 + H_2O$$

Lauric acid   Ethanol   Ethyl dodecanoate
Dodecanoic acid

15.5

$$CH_3CH_2\overset{\overset{\displaystyle O}{\|}}{—C—}OCH_2CH_2CH_2CH_3 + H_2O \xrightarrow{H^+, heat} CH_3CH_2CH_2CH_2OH + CH_3CH_2COOH$$

              Butyl propionate   Butyl alcohol   Propionic acid

15.7

$$CH_3CH_2\overset{\overset{\displaystyle O}{\|}}{—C—}OCH_2CH_2CH_2CH_3 + KOH \longrightarrow CH_3CH_2COO^- K^+ + CH_3CH_2CH_2CH_2OH$$

              Butyl propionate   Potassium propionoate   Butyl alcohol

15.9

$$CH_3(CH_2)_5CH==CH(CH_2)_7COOH + H_2 \xrightarrow{Ni} CH_3(CH_2)_{14}COOH$$

          cis-9-Hexadecenoic acid   Hexadecanoic acid

15.11

$$
\begin{array}{c}
\text{H} \\
| \\
\text{H}-\text{C}-\text{OH} \\
| \\
\text{H}-\text{C}-\text{OH} \quad + \; 2 \; CH_3(CH_2)_{16}COOH \quad \longrightarrow \\
| \\
\text{H}-\text{C}-\text{OH} \\
| \\
\text{H}
\end{array}
\qquad
\begin{array}{c}
\text{H} \quad\quad\quad\; O \\
| \quad\quad\quad\quad \| \\
\text{H}-\text{C}-\text{O}-\text{C}-(CH_2)_{16}CH_3 \\
| \\
\text{H}-\text{C}-\text{O}-\text{C}-(CH_2)_{16}CH_3 \\
| \quad\quad\quad\quad\| \\
\text{H}-\text{C}-\text{OH} \quad O \\
| \\
\text{H}
\end{array}
$$

15.13 a. Mono-, di-, and triglycerides of oleic acid:

$$
CH_3(CH_2)_7CH = CH(CH_2)_7 - \overset{\overset{\displaystyle O}{\|}}{C} - O - \overset{\displaystyle CH_2}{\underset{\underset{\displaystyle CH_2 - OH}{|}}{\underset{\displaystyle CH - OH}{|}}}
$$

$$
\begin{array}{l}
CH_3(CH_2)_7CH = CH(CH_2)_7 - \overset{\overset{\displaystyle O}{\|}}{C} - O - CH_2 \\
\qquad\qquad\qquad\qquad\qquad\qquad\qquad\quad | \\
CH_3(CH_2)_7CH = CH(CH_2)_7 - \overset{\overset{\displaystyle O}{\|}}{C} - O - CH \\
\qquad\qquad\qquad\qquad\qquad\qquad\qquad\quad | \\
\qquad\qquad\qquad\qquad\qquad\qquad\qquad CH_2 - OH
\end{array}
$$

$$
\begin{array}{l}
CH_3(CH_2)_7CH = CH(CH_2)_7 - \overset{\overset{\displaystyle O}{\|}}{C} - O - CH_2 \\
\qquad\qquad\qquad\qquad\qquad\qquad\qquad\quad | \\
CH_3(CH_2)_7CH = CH(CH_2)_7 - \overset{\overset{\displaystyle O}{\|}}{C} - O - CH \\
\qquad\qquad\qquad\qquad\qquad\qquad\qquad\quad | \\
CH_3(CH_2)_7CH = CH(CH_2)_7 - \overset{\overset{\displaystyle O}{\|}}{C} - O - CH_2
\end{array}
$$

b.   Mono-, di-, and triglycerides of capric acid:

$$CH_3(CH_2)_8 - \overset{\displaystyle O}{\overset{\|}{C}} - O - CH_2$$
$$| $$
$$CH - OH$$
$$| $$
$$CH_2 - OH$$

$$CH_3(CH_2)_8 - \overset{\displaystyle O}{\overset{\|}{C}} - O - CH_2$$
$$CH_3(CH_2)_8 - \overset{\displaystyle O}{\overset{\|}{C}} - O - CH$$
$$| $$
$$CH_2 - OH$$

$$CH_3(CH_2)_8 - \overset{\displaystyle O}{\overset{\|}{C}} - O - CH_2$$
$$CH_3(CH_2)_8 - \overset{\displaystyle O}{\overset{\|}{C}} - O - CH$$
$$CH_3(CH_2)_8 - \overset{\displaystyle O}{\overset{\|}{C}} - O - CH_2$$

15.15 Structure of the steroid nucleus:

15.17 Diffusion is the net movement of a solute from an area of higher concentration to an area of lower concentration.

409

15.19 The four main groups of lipids are fatty acids, glycerides, nonglyceride lipids, and complex lipids.

15.21 Lipid-soluble vitamins are transported into cells of the small intestine in association with dietary fat molecules. Thus, a diet low in fat reduces the amount of vitamins A, D, E, and K that enters the body.

15.23 A saturated fatty acid is one in which the hydrocarbon tail has only carbon-to-carbon single bonds. Thus, each carbon atom is bonded to the maximum number of hydrogen atoms. An unsaturated fatty acid has at least one carbon-to-carbon double bond. Palmitic acid (hexadecanoic acid) is a saturated fatty acid that has the following structure:

$$CH_3(CH_2)_{14}-\overset{\overset{\textstyle O}{\|}}{C}-OH$$

Palmitoleic acid (*cis*-9-hexadecenoic acid) is a monounsaturated fatty acid that has the following structure:

$$CH_3(CH_2)_5CH=CH(CH_2)_7\overset{\overset{\textstyle O}{\|}}{C}-OH$$

15.25 The melting points of fatty acids increase as the length of the hydrocarbon chains increase. This is because the intermolecular attractive forces, including van der Waals forces, increase as the length of the hydrocarbon chain increases.

15.27 a. Decanoic acid:

$$CH_3(CH_2)_8-\overset{\overset{\textstyle O}{\|}}{C}-OH$$

b. Stearic acid:

$$CH_3(CH_2)_{16}-\overset{\overset{\textstyle O}{\|}}{C}-OH$$

c. *trans*-5-Decenoic acid:

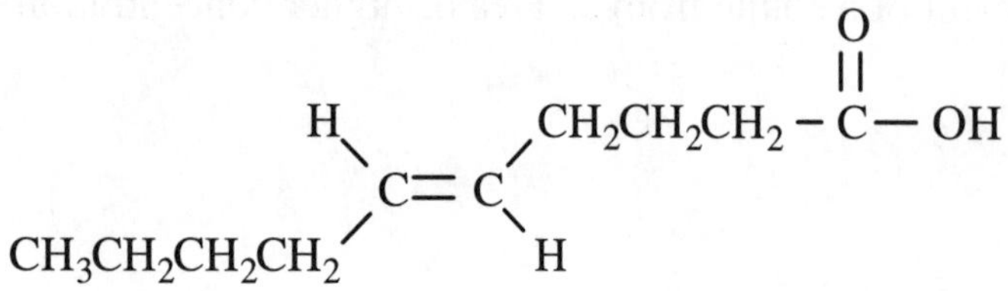

d. *cis*-5-Decenoic acid:

$$CH_3CH_2CH_2CH_2 \diagdown \qquad \diagup CH_2CH_2CH_2 - \overset{\overset{\displaystyle O}{\|}}{C} - OH$$
$$C = C$$
$$\diagup \qquad \diagdown$$
$$H \qquad H$$

15.29 a. Esterification of glycerol with three molecules of myristic acid:

$$\begin{array}{l} CH_2OH \\ | \\ CHOH \\ | \\ CH_2OH \end{array} \quad + \quad 3 \ CH_3(CH_2)_{12} - \overset{\overset{\displaystyle O}{\|}}{C} - OH \quad \xrightarrow{\text{H}^+, \text{ heat}}$$

$$CH_3(CH_2)_{12} - \overset{\overset{\displaystyle O}{\|}}{C} - O - CH_2$$
$$CH_3(CH_2)_{12} - \overset{\overset{\displaystyle O}{\|}}{C} - O - CH \qquad + \ 3 \ H_2O$$
$$CH_3(CH_2)_{12} - \overset{\overset{\displaystyle O}{\|}}{C} - O - CH_2$$

b. Acid hydrolysis of a triglyceride containing three stearic acid molecules:

$$CH_3(CH_2)_{16} - \overset{\overset{\displaystyle O}{\|}}{C} - O - CH_2$$
$$CH_3(CH_2)_{16} - \overset{\overset{\displaystyle O}{\|}}{C} - O - CH \qquad + \ 3 \ H_2O \quad \xrightarrow{\text{H}^+, \text{ heat}}$$
$$CH_3(CH_2)_{16} - \overset{\overset{\displaystyle O}{\|}}{C} - O - CH_2$$

$$3 \ CH_3(CH_2)_{16} - \overset{\overset{\displaystyle O}{\|}}{C} - OH \quad + \quad \begin{array}{l} CH_2OH \\ | \\ CHOH \\ | \\ CH_2OH \end{array}$$

411

c. Reaction of decanoic acid with KOH:

$$CH_3(CH_2)_8-\overset{\displaystyle O}{\overset{\|}{C}}-OH \ + \ KOH \ \longrightarrow \ CH_3(CH_2)_8-\overset{\displaystyle O}{\overset{\|}{C}}-O^-K^+ \ + \ H_2O$$

d. Hydrogenation of linoleic acid:

$$CH_3(CH_2)_4CH=CHCH_2CH=CH(CH_2)_7-\overset{\displaystyle O}{\overset{\|}{C}}-OH$$

$$+ \ 2 \ H_2$$

$$\Big\downarrow Ni$$

$$CH_3(CH_2)_{16}-\overset{\displaystyle O}{\overset{\|}{C}}-OH$$

15.31 An essential fatty acid is one that the body cannot synthesize and thus must be supplied in the diet. The essential fatty acid linoleic acid is required for the synthesis of arachidonic acid, a precursor for the synthesis of the prostaglandins, a group of hormonelike molecules. Arachidonic acid is a starting material for the biosynthesis of prostaglandins. The prostaglandins are hormonelike molecules that exert a variety of effects on the body.

15.33 Prostaglandins stimulate smooth muscle contraction, especially uterine contractions during labor. They enhance fever and swelling associated with the inflammatory response. Some prostaglandins cause bronchial dilation. Others inhibit secretion of acid into the stomach and stimulate the secretion of a mucous layer that protects the stomach lining.

15.35 An emulsifying agent is a molecule that aids in the suspension of triglycerides in water. They are amphipathic molecules, such as lecithin, that serve as bridges holding together the highly polar water molecules and the nonpolar triglycerides.

15.37 A triglyceride with three saturated fatty acid tails would be a solid at room temperature. The long, straight fatty acid tails would stack with one another because of strong intermolecular and intramolecular attractions.

15.39

$$CH_3(CH_2)_{14}-\overset{\displaystyle O}{\overset{\displaystyle ||}{C}}-O-CH_2$$

$$CH_2(CH_2)_6-\overset{\displaystyle O}{\overset{\displaystyle ||}{C}}-O-CH$$

$$CH_2(CH_2)_6-\overset{\displaystyle O}{\overset{\displaystyle ||}{C}}-O-CH_2$$

Left structures:

$$\underset{CH_3(CH_2)_4CH_2}{}\overset{H}{\diagdown}C=C\overset{CH_2(CH_2)_6}{\diagup}\overset{}{\diagdown}H$$

$$\underset{CH_3(CH_2)_4CH_2}{}\diagdown C=C \diagup \overset{CH_2(CH_2)_6}{}$$

15.41   A sphingolipid is one that is not derived from glycerol, but rather from sphingosine, a long-chain, nitrogen-containing (amino) alcohol. Like phospholipids, sphingolipids are amphipathic. The two major types of sphingolipids are the sphingomyelins and the glycosphingolipids.

15.43   Sphingomyelins are important structural lipid components of nerve cell membranes. They are found in the myelin sheath that surrounds and insulates cells of the central nervous system.

15.45   Cholesterol is readily soluble in the hydrophobic region of biological membranes. It is involved in regulating the fluidity of the membrane.

15.47   Progesterone is the most important hormone associated with pregnancy. It is needed for the successful initiation and completion of the pregnancy. It prepares the lining of the uterus to accept the fertilized egg, facilitates development of the fetus, and suppresses ovulation during pregnancy. Testosterone is needed for development of male secondary sexual characteristics. Estrone is required for proper development of female secondary sexual characteristics.

15.49   Cortisone is used to treat rheumatoid arthritis, asthma, gastrointestinal disorders, and many skin conditions.

15.51   Myricyl palmitate (beeswax) is made up of the fatty acid palmitic acid and the alcohol myricyl alcohol:

$$CH_3(CH_2)_{28}CH_2OH$$

15.53   The four major types of plasma lipoproteins are chylomicrons, high-density lipoproteins, low-density lipoproteins, and very low-density lipoproteins.

15.55   *Chylomicrons* carry dietary lipids from the intestine to other tissues. They are approximately 4% phospholipids, 90% triglycerides, 5% cholesterol, and 1% protein.

Very low-density lipoproteins carry triglycerides synthesized in the liver to adipose tissue for storage. They are approximately 18% phospholipids, 60% triglycerides, 14% cholesterol, and 8% protein.

Low-density lipoproteins carry cholesterol to peripheral tissues and help regulate cholesterol levels in those tissues. They are approximately 20% phospholipids, 10% triglycerides, 45% cholesterol, and 25% protein.

High-density lipoproteins transport cholesterol from peripheral tissues to the liver. They are approximately 30% phospholipids, 5% triglycerides, 20% cholesterol, and 45% protein.

15.57 The basic structure of a biological membrane is a bilayer of phospholipid molecules arranged so that the hydrophobic hydrocarbon tails are packed in the center and the hydrophilic head groups are exposed on the inner and outer surfaces.

15.59 A peripheral membrane protein is bound to only one surface of the membrane, either inside or outside the cell.

15.61 Cholesterol is freely soluble in the hydrophobic layer of a biological membrane. It moderates the fluidity of the membrane by disrupting the stacking of the fatty acid tails of membrane phospholipids.

15.63 If the fatty acid tails of membrane phospholipids are converted from saturated to unsaturated, the fluidity of the membrane will increase. Each carbon-to-carbon double bond that is added will introduce a "kink" into the fatty acid tail. As a result, the tails cannot pack together as they would if they were saturated. The result is increased fluidity.

15.65 Both simple diffusion and facilitated diffusion are means of passive transport. Simple diffusion involves the net movement of a solute directly across a membrane from a region of higher concentration to a region of lower concentration. Facilitated diffusion also involves movement of a substance from a region of high concentration to an area of low concentration. However, facilitated diffusion requires a channel protein or permease through which the solute must pass.

15.67 Both active transport and facilitated diffusion require a protein channel or permease through which solutes pass into or out of the cell. Active transport requires an energy input to transport molecules or ions against the gradient (from an area of lower concentration to an area of higher concentration). Facilitated diffusion is a means of passive transport in which molecules or ions pass from regions of higher concentration to regions of lower concentration through a permease protein. No energy is expended by the cell in facilitated diffusion.

15.69 An antiport transport mechanism is one in which one molecule or ion is transported into the cell while a different molecule or ion is transported out of the cell.

15.71 One ATP is hydrolyzed to transport 3 Na^+ out of the cell and 2 K^+ into the cell.

In-Chapter Questions and Problems

16.1 a. Glycine (gly)

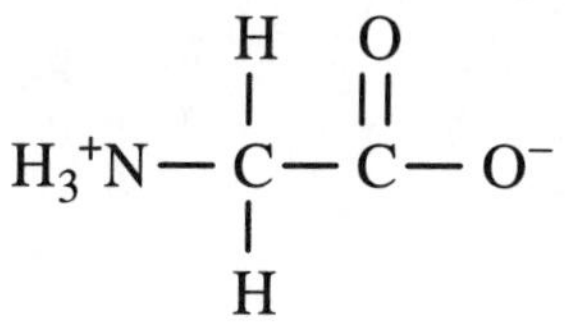

b. Proline (pro)

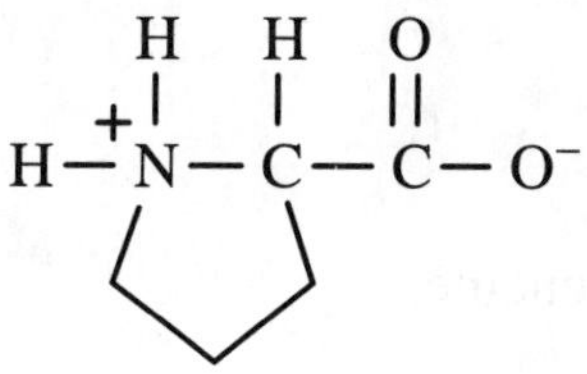

c. Threonine (thr)

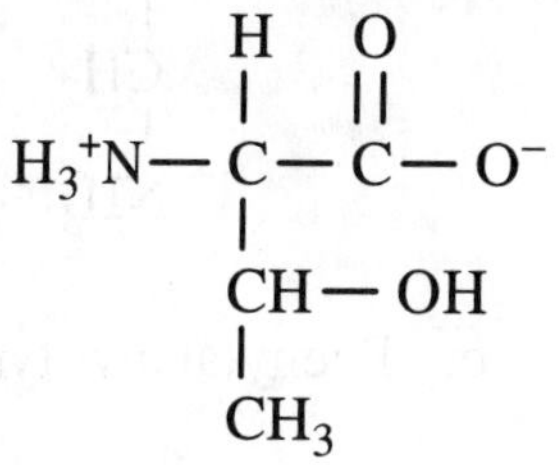

$$H_3^+N-\overset{\overset{\displaystyle H}{|}}{\underset{\underset{\displaystyle H}{|}}{C}}-\overset{\overset{\displaystyle O}{\|}}{C}-O^-$$

$$H-\overset{+}{N}\overset{\overset{\displaystyle H}{|}}{-}\overset{\overset{\displaystyle H}{|}}{C}-\overset{\overset{\displaystyle O}{\|}}{C}-O^-$$

$$H_3^+N-\overset{\overset{\displaystyle H}{|}}{\underset{\underset{\displaystyle CH-OH}{|}}{C}}-\overset{\overset{\displaystyle O}{\|}}{C}-O^-$$
$$CH_3$$

d. Aspartate (asp)

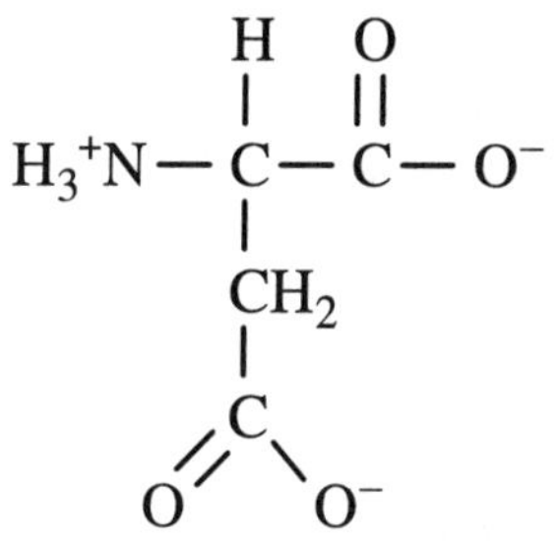

e. Lysine (lys)

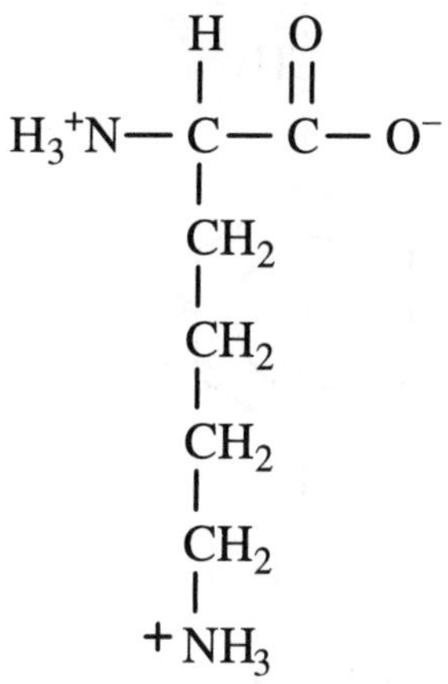

$$H_3^+N-\overset{\overset{\displaystyle H}{|}}{\underset{\underset{\displaystyle CH_2}{|}}{C}}-\overset{\overset{\displaystyle O}{\|}}{C}-O^-$$
$$\underset{O \diagup \diagdown O^-}{C}$$

$$H_3^+N-\overset{\overset{\displaystyle H}{|}}{\underset{\underset{\displaystyle CH_2}{|}}{C}}-\overset{\overset{\displaystyle O}{\|}}{C}-O^-$$
$$CH_2$$
$$CH_2$$
$$CH_2$$
$$^+NH_3$$

16.3 a. Alanyl-phenylalanine:

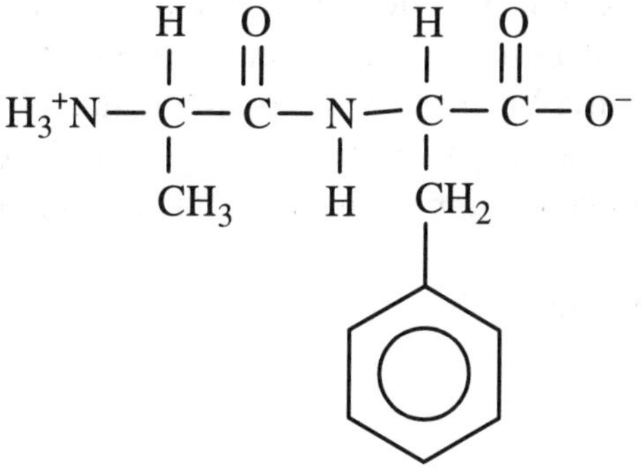

$$H_3^+N-\overset{\overset{\displaystyle H}{|}}{\underset{\underset{\displaystyle CH_3}{|}}{C}}-\overset{\overset{\displaystyle O}{\|}}{C}-\overset{\overset{\displaystyle H}{|}}{N}-\overset{\overset{\displaystyle H}{|}}{\underset{\underset{\displaystyle CH_2}{|}}{C}}-\overset{\overset{\displaystyle O}{\|}}{C}-O^-$$

b. Lysyl-alanine:

c. Phenylalanyl-tyrosyl-leucine:

16.5   The primary structure of a protein is the amino acid sequence of the protein chain.
Regular, repeating folding of the peptide chain caused by hydrogen bonding between the
amide hydrogens and carbonyl oxygens of the peptide bond is the secondary structure of
a protein. The two most common types of secondary structure are the α-helix and the β-
pleated sheet. Tertiary structure is the further folding of the regions of α-helix and β-
pleated sheet into a compact, globular structure. Formation and maintenance of the
tertiary structure result from weak attractions between amino acid R groups. The binding
of two or more peptides to produce a functional protein defines the quaternary structure.

16.7   a. transferase   d. oxidoreductase
        b. transferase   e. hydrolase
        c. isomerase

16.9   The induced-fit model assumes that the enzyme is flexible. Both the enzyme and the
substrate are able to change shape to form the enzyme-substrate complex.

The lock-and-key model assumes that the enzyme is inflexible (the lock) and the substrate (the key) fits into a specific rigid site (the active site) on the enzyme to form the enzyme-substrate complex.

16.11 An enzyme might distort a bond, thereby catalyzing bond breakage. An enzyme could bring two reactants into close proximity and in the proper orientation for the reaction to occur. Finally, an enzyme could alter the pH of the microenvironment of the active site, thereby serving as a transient donor or acceptor of H^+.

16.13 Water-soluble vitamins are required by the body for the synthesis of coenzymes that are required for the function of a variety of enzymes.

16.15 A decrease in pH will change the degree of ionization of the R groups within a peptide chain. This disturbs the weak interactions that maintain the structure of an enzyme, which may denature the enzyme. Less drastic alterations in the charge of R groups in the active site of the enzyme can inhibit enzyme-substrate binding or destroy the catalytic ability of the active site.

End-of-Chapter Questions and Problems

16.17 An enzyme is a protein that serves as a biological catalyst, speeding up biological reactions.

16.19 Enzymes speed up reactions that might take days or weeks to occur on their own. They also catalyze reactions that might require very high temperatures or harsh conditions if carried out in the laboratory. In the body, these reactions occur quickly under physiological conditions.

16.21 The general structure of an L-α-amino acid:

$$H_3^+N - \underset{\underset{R}{|}}{\overset{\overset{COO^-}{|}}{C}} - H$$

16.23 A chiral carbon is one that has four different atoms or groups of atoms attached to it.

16.25 Interactions between the R groups of the amino acids in a polypeptide chain are important for the formation and maintenance of the tertiary and quaternary structures of proteins.

16.27 A peptide bond is an amide bond between two amino acids in a peptide chain.

16.29 a. His-trp-cys:

$$H_3{}^+N-CH-\underset{\underset{CH_2}{|}}{\overset{\overset{O}{\|}}{C}}-\underset{\underset{H}{|}}{N}-CH-\overset{\overset{O}{\|}}{C}-\underset{\underset{H}{|}}{N}-CH-\overset{\overset{O}{\|}}{C}-O^-$$

with side chains: histidine imidazole (H^+N, NH), tryptophan indole (HN), cysteine CH_2-SH

b. Gly-leu-ser:

$$H_3{}^+N-\underset{\underset{H}{|}}{\overset{\overset{H}{|}}{C}}-\overset{\overset{O}{\|}}{C}-\underset{\underset{H}{|}}{N}-\underset{\underset{CH_2}{|}}{\overset{\overset{H}{|}}{C}}-\overset{\overset{O}{\|}}{C}-\underset{\underset{H}{|}}{N}-\underset{\underset{CH_2}{|}}{\overset{\overset{H}{|}}{C}}-\overset{\overset{O}{\|}}{C}-O^-$$

leucine: $CH_2-CHCH_3-CH_3$; serine: CH_2-OH

c. Arg-ile-val:

$$H_3{}^+N-\underset{\underset{CH_2}{|}}{\overset{\overset{H}{|}}{C}}-\overset{\overset{O}{\|}}{C}-\underset{\underset{H}{|}}{N}-\underset{\underset{CHCH_3}{|}}{\overset{\overset{H}{|}}{C}}-\overset{\overset{O}{\|}}{C}-\underset{\underset{CH}{|}}{\overset{\overset{H}{|}}{C}}-\overset{\overset{O}{\|}}{C}-O^-$$

arginine side chain: $CH_2-CH_2-CH_2-NH-C{=}NH_2{}^+-NH_2$

isoleucine side chain: $CHCH_3-CH_2-CH_3$

valine side chain: CH with H_3C and CH_3

16.31 The primary structure of a protein is the sequence of amino acids bonded to one another by peptide bonds.

16.33　The primary structure of a protein determines its three-dimensional shape because the location of R groups along the protein chain is determined by the primary structure. The interactions among the R groups, based on their location in the chain, will govern how the protein folds. This, in turn, dictates its three-dimensional structure and biological function.

16.35　The secondary structure of a protein is the folding of the primary structure into an α-helix or β-pleated sheet.

16.37　a.　α-Helix
　　　　b.　β-Pleated sheet

16.39　The tertiary structure of a protein is the globular, three-dimensional structure of a protein that results from folding the regions of secondary structure.

16.41　Cystine is formed in an oxidation reaction between two cysteine amino acids which may be in the same peptide chain or in different peptide chains.　Thus, cystine forms a covalent bridge between two peptide chains or between two regions of the same peptide chain.　In the first case, cystine formation helps maintain the quaternary structure of a protein composed of more than one peptide chain.　In the second case, cystine formation helps stabilize the tertiary structure of the protein.

16.43　Quaternary protein structure is the aggregation of two or more folded peptide chains to produce a functional protein.

16.45　A glycoprotein is a protein with sugars as prosthetic groups. Glycoproteins are often receptors on the cell surface.

16.47　Hemoglobin is the protein in red blood cells that is responsible for carrying oxygen to the cells of the body.　It is composed of four subunits: two α subunits and two β subunits. Each of the subunits contains a heme group with an Fe^{2+} which serves as the binding site for the oxygen.

16.49　Because carbon monoxide binds tightly to the heme groups of hemoglobin, it is not easily removed or replaced by oxygen. As a result, the effects of oxygen deprivation (suffocation) occur.

16.51　When sickle cell hemoglobin (Hb S) is deoxygenated, the amino acid valine fits into a hydrophobic pocket on the surface of another Hb S molecule. Many such sickle cell hemoglobin molecules polymerize into long rods that cause the red blood cell to sickle. In normal hemoglobin, glutamic acid is found in place of the valine. This negatively charged amino acid will not "fit" into the hydrophobic pocket.

16.53　*Denaturation* is the process by which the organized structure of a protein is disrupted, resulting in a completely disorganized, nonfunctional form of the protein.

16.55 Heat is an effective means of sterilization because it destroys the proteins of microbial life-forms, including fungi, bacteria, and viruses.

16.57 The common name of an enzyme is often derived from the name of the substrate and/or the type of reaction that it catalyzes.

16.59 A substrate is the chemical reactant in a chemical reaction that binds to an enzyme active site and is converted to product.

16.61 An enzyme-substrate complex is a molecular aggregate formed when the substrate binds to the active site of the enzyme.

16.63 A cofactor is an organic group, often containing a metal ion, that must be bound to an apoenzyme to maintain the correct configuration of the active site.

16.65 a. Citrate decarboxylase catalyzes the cleavage of a carboxyl group from citrate.
b. Adenosine diphosphate phosphorylase catalyzes the addition of a phosphate group to ADP.
c. Oxalate reductase catalyzes the reduction of oxalate.
d. Nitrite oxidase catalyzes the oxidation of nitrite.
e. cis-trans Isomerase catalyzes the interconversion of cis and trans isomers.

16.67 The activation energy of a reaction is the energy required for the reaction to occur.

16.69 The lock-and-key model of enzyme-substrate binding was proposed by Emil Fischer in 1894. He thought that the active site was a rigid region of the enzyme into which the substrate fit perfectly. Thus, the model purports that the substrate simply snaps into place within the active site, like two pieces of a jigsaw puzzle fitting together.

16.71 The first step of an enzyme-catalyzed reaction is the formation of the enzyme-substrate complex. In the second step, the transition state is formed. This is the state in which the substrate assumes a form intermediate between the original substrate and the product. In step 3, the substrate is converted to product and the enzyme-product complex is formed. Step 4 involves the release of the product and regeneration of the enzyme in its original form.

16.73 NAD^+/NADH serves as an acceptor/donor of hydride anions in biochemical reactions. NAD^+/NADH serves as a coenzyme for oxidoreductases.

16.75 Each of the following answers assumes that the enzyme was purified from an organism with optimal conditions for life near 37°C, pH 7.

a. Decreasing the temperature from 37°C to 10°C will cause the rate of an enzyme-catalyzed reaction to decrease because the frequency of collisions between enzyme and substrate will decrease as the rate of molecular movement decreases.

b. Increasing the pH from 7 to 11 will generally cause a decrease in the rate of an enzyme-catalyzed reaction. In fact, most enzymes would be denatured by a pH of 11 and enzyme activity would cease.

c. Heating an enzyme from 37°C to 100°C will destroy enzyme activity because the enzyme would be denatured by the extreme heat.

16.77 The sulfa drugs are structural analogs of *para*-aminobenzoic acid (PABA). PABA is the substrate of an enzyme involved in the pathway for the biosynthesis of folic acid. The sulfa drugs act as competitive inhibitors of this enzyme.

Folic acid is a vitamin required for the synthesis of a coenzyme needed to make the amino acid methionine and the purine and pyrimidine nitrogenous bases for DNA and RNA. When a sulfa drug binds to the enzyme, no product is formed, folic acid is not made, and the biosynthesis of methionine and the nitrogenous bases ceases. This eventually kills the microorganism. Humans are not harmed by the sulfa drugs because we do not synthesize our own folic acid. It is obtained in the diet.

16.79 The compound would be a competitive inhibitor of the enzyme.

16.81 Creatine phosphokinase (CPK), lactate dehydrogenase (LDH), and aspartate aminotransferase (AST/SGOT).

Chapter 17
Introduction to Molecular Genetics
Solutions to the Odd-Numbered Problems

In-Chapter Questions and Problems

17.1   a. Adenosine diphosphate:

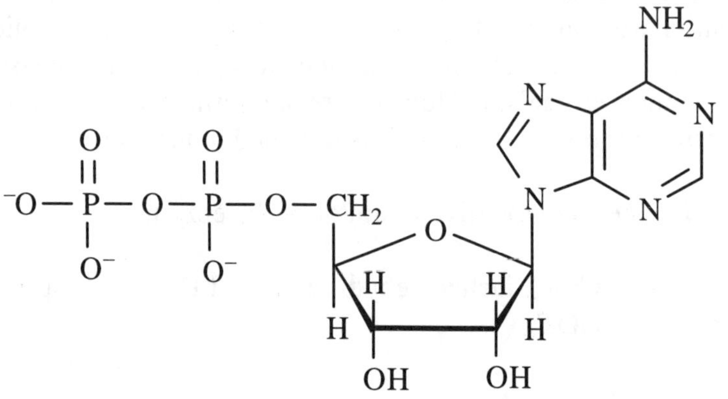

b. Deoxyguanosine triphosphate:

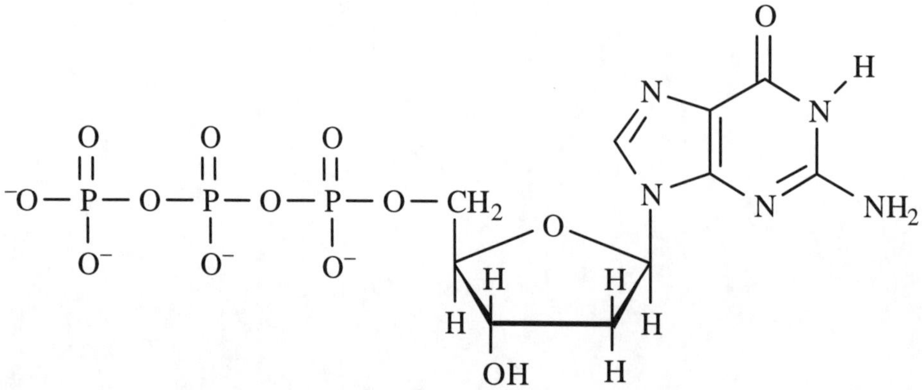

17.3   The RNA polymerase recognizes the promoter site for a gene, separates the strands of DNA, and catalyzes the polymerization of an RNA strand complementary to the DNA strand that carries the genetic code for a protein. It recognizes a termination site at the end of the gene and releases the RNA molecule.

17.5   The genetic code is said to be degenerate because several different triplet codons may serve as code words for a single amino acid.

422

17.7 The nitrogenous bases of the codons are complementary to those of the anticodons. As a result they are able to hydrogen bond to one another according to the base pairing rules.

17.9 The ribosomal P-site holds the peptidyl tRNA during protein synthesis. The peptidyl tRNA is the tRNA carrying the growing peptide chain. The only exception to this is during initiation of translation when the P-site holds the initiator tRNA.

17.11 The normal mRNA sequence, AUG-CCC-GAC-UUU, would encode the peptide sequence, methionine-proline-aspartate-phenylalanine. The mutant mRNA sequence, AUG-CGC-GAC-UUU, would encode the mutant peptide sequence, methionine-arginine-aspartate-phenylalanine. This would not be a silent mutation because a hydrophobic amino acid (proline) has been replaced by a positively charged amino acid (arginine).

End-of-Chapter Questions and Problems

17.13 It is the N-9 of the purine that forms the *N*-glycosidic bond with C-1 of the five-carbon sugar. The general structure of the purine ring is shown below:

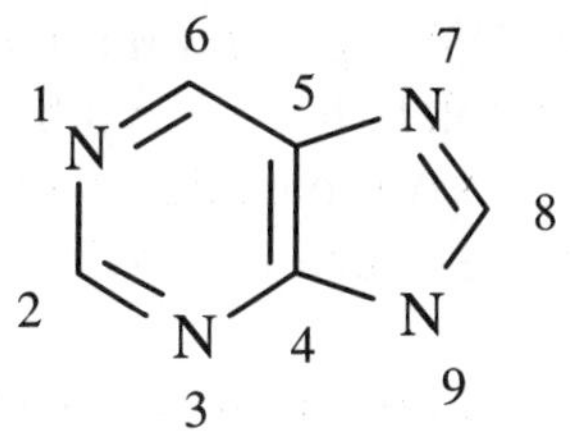

17.15 The ATP nucleotide is composed of the five-carbon sugar ribose, the purine adenine, and a triphosphate group.

17.17 The two strands of DNA in the double helix are said to be *antiparallel* because they run in opposite directions. One strand progresses in the $5' \rightarrow 3'$ direction, and the opposite strand progresses in the $3' \rightarrow 5'$ direction.

17.19 Two hydrogen bonds link the adenine-thymine base pair.

17.21

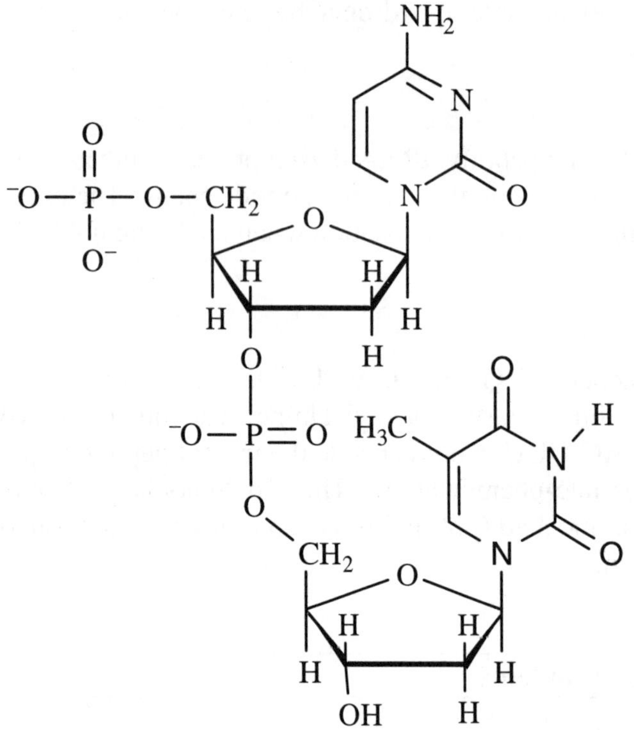

NH_2

$^-O-P-O-CH_2$ (O double bond O, O^-)

N, O

H H H H

O

$^-O-P=O$ H_3C H, N, O, O

O

CH_2

O

H H H H

OH H

17.23 The bacterial chromosome is a circular DNA molecule that is supercoiled, that is, the helix is coiled on itself.

17.25 The term *semiconservative DNA replication* refers to the fact that each parental DNA strand serves as the template for the synthesis of a daughter strand. As a result, each of the daughter DNA molecules is made up of one strand of the original parental DNA and one strand of newly synthesized DNA.

17.27 If the parental DNA strand had the following nucleotide sequence: 5'-ATGCGGCTAGAATATTCCA-3', the sequence of the complementary daughter strand would be 3'-TACGCCGATCTTATAAGGT-5'.

17.29 The central dogma of molecular biology states that information flow in cellular biological systems is unidirectional: DNA → RNA → Protein. The DNA carries the genetic information; RNA molecules carry out the expression of the genetic information to produce proteins; the final products are proteins that carry out the work of the cell and serve as cellular structural components.

17.31 Anticodons are found on transfer RNA molecules.

17.33 If a gene had the sequence, 5'-TACCTAGCTCTGGTCATTAAGGCAGTA-3', the mRNA would have the sequence, 3'-AUGGAUCGAGACCAGUAAUUCCGUCAU-5'.

17.35   *RNA splicing* is the process by which the noncoding sequences (introns) of the primary transcript of a eukaryotic mRNA are removed and the protein coding sequences (exons) are spliced together.

17.37   The three classes of RNA molecules are messenger RNA (mRNA), transfer RNA (tRNA), and ribosomal RNA (rRNA).

17.39   Introns must be removed from a primary transcript because they do not code for protein sequences and would result in synthesis of a nonfunctional protein.

17.41   The *poly(A) tail* is a stretch of 100–200 adenosine nucleotides polymerized onto the 3' end of a mRNA by the enzyme poly(A) polymerase.

17.43   The *cap structure* is made up of the nucleotide 7-methylguanosine attached to the 5' end of a mRNA by a 5'-5' triphosphate bridge. Generally, the first two nucleotides of the mRNA are also methylated. The cap structure is required for efficient translation of the mRNA.

17.45   There are 64 codons in the genetic code.

17.47   The reading frame of a gene is the sequential set of triplet codons that carries the genetic code for the primary structure of a protein. Each triplet specifies the addition of a particular amino acid to the growing peptide chain.

17.49   Methionine (AUG) and tryptophan (UGG) are encoded by only one codon.

17.51   The codon 5'-UUU-3' encodes the amino acid phenylalanine. The mutant codon 5'-UUA-3' encodes the amino acid leucine. Both leucine and phenylalanine are hydrophobic amino acids, however, leucine has a smaller R group. It is possible that the smaller R group would disrupt the structure of the protein.

17.53   The ribosomes serve as a platform on which protein synthesis can occur. They also carry the enzymatic activity that forms peptide bonds.

17.55   The sequence of DNA nucleotides in a gene is transcribed to produce a complementary sequence of RNA nucleotides in a messenger RNA (mRNA). In the process of translation, the sequence of the mRNA is read sequentially in words of three nucleotides (codons) to produce a protein. Each codon calls for the addition of a particular amino acid to the growing peptide chain. Through these processes, the sequence of nucleotides in a gene determines the sequence of amino acids in the primary structure of a protein.

17.57   In the initiation of translation, initiation factors, methionyl tRNA (the initiator tRNA), the mRNA, and the small and large ribosomal subunits form the initiation complex. During the elongation stage of translation, an aminoacyl tRNA binds to the A-site of the ribosome. Peptidyl transferase catalyzes the formation of a peptide bond and the peptide chain is transferred to the tRNA in the A-site. Translocation shifts the peptidyl tRNA from the A-site into the P-site, leaving the A-site available for the next aminoacyl tRNA. In the termination stage of translation, a termination codon is encountered. A release factor binds to the empty A-site and peptidyl transferase catalyzes the hydrolysis of the bond between the peptidyl tRNA and the completed peptide chain.

17.59   The bond between an amino acid and a tRNA is an ester bond formed between the carboxylate group of the amino acid and the 3'-OH of the sugar ribose in the tRNA molecule.

17.61   A point mutation is the substitution of one nucleotide pair for another in a gene.

17.63   Some mutations are silent because the change in the nucleotide sequence does not alter the amino acid sequence of the protein. This can happen because there are many amino acids encoded by multiple codons.

17.65   UV light causes the formation of pyrimidine dimers, the covalent bonding of two adjacent pyrimidine bases. Mutations occur when the UV damage repair system makes an error during the repair process. This causes a change in the nucleotide sequence of the DNA.

17.67   A *carcinogen* is a compound that causes cancer. Cancers are caused by mutations in the genes responsible for controlling cell division. Carcinogens cause DNA damage that results in changes in the nucleotide sequence of the gene. Thus, carcinogens are also mutagens.

Chapter 18
Carbohydrate Metabolism
Solutions to the Odd-Numbered Problems

In-Chapter Questions and Problems

18.1   ATP is called the universal energy currency because it is the major molecule used by all organisms to store energy. Hydrolysis of the high-energy bonds releases energy that is used for cellular work.

18.3   The first stage of catabolism is the digestion (hydrolysis) of dietary macromolecules in the stomach and intestine. Polysaccharides are hydrolyzed to monosaccharides; proteins are degraded to amino acids; and triglycerides are broken down into glycerol and fatty acids. The small molecules produced by digestion are taken into the cells lining the intestine by active or passive transport.

In the second stage of catabolism, monosaccharides, amino acids, fatty acids, and glycerol are converted by metabolic reactions into molecules that can be completely oxidized. Often they are converted into acetyl CoA.

In the third stage of catabolism, the two-carbon acetyl group of acetyl CoA is completely oxidized by the reactions of the citric acid cycle. The energy of the electrons harvested in these oxidation reactions is used to make ATP.

18.5   Substrate level phosphorylation is one way the cell can make ATP. In this reaction, a high-energy phosphoryl group of a substrate in the reaction is transferred to ADP to produce ATP. Substrate-level phosphorylation can be summarized as follows:

$$\text{Substrate} \sim P + ADP \rightarrow \text{Product} + ATP$$

18.7   Glycolysis is a pathway involving ten reactions. In reactions 1–3, energy is invested in the beginning substrate, glucose. This is done by transferring high-energy phosphoryl groups from ATP to the intermediates in the pathway. The product is fructose-1,6-bisphosphate. In the energy harvesting reactions of glycolysis, fructose-1,6-bisphosphate is split into two three-carbon molecules that begin a series of rearrangement, oxidation-reduction, and substrate-level phosphorylation reactions that produce 4 ATP, 2 NADH, and 2 pyruvate molecules. Because of the investment of two ATP in the early steps of glycolysis, the net yield of ATP is two.

18.9   Both the alcohol and lactate fermentations are anaerobic reactions that use the pyruvate and reoxidize the NADH produced in glycolysis. In the alcohol fermentation, pyruvate is first decarboxylated to produce acetaldehyde. The acetaldehyde is then reduced as NADH is oxidized. The products are CO_2, ethanol, and NAD^+. In the lactate fermentation, pyruvate is reduced to lactate and NADH is oxidized to NAD^+.

18.11

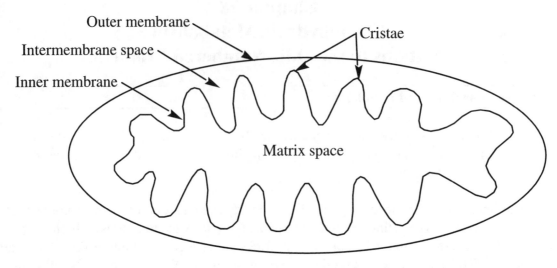

18.13 Pyruvate is converted to acetyl CoA by the pyruvate dehydrogenase complex. This huge enzyme complex requires four coenzymes, each of which is made from a different vitamin. The four coenzymes are thiamine pyrophosphate (made from thiamine), FAD (made from riboflavin), NAD^+ (made from niacin), and coenzyme A (made from the vitamin pantothenic acid). The coenzyme lipoamide is also involved in this reaction.

18.15 *Oxidative phosphorylation* is the process by which the energy of electrons harvested from oxidation of a fuel molecule is used to phosphorylate ADP to produce ATP.

18.17 $NAD^+ + H:^- \rightarrow NADH$

18.19 Glucagon indirectly stimulates glycogen phosphorylase, the first enzyme of glycogenolysis. This speeds up glycogen degradation. Glucagon also inhibits glycogen synthase, the first enzyme in glycogenesis. This inhibits glycogen synthesis.

End-of-Chapter Questions and Problems

18.21 ATP is the molecule that is primarily responsible for conserving the energy released in catabolism.

18.23 A high-energy bond is a weak bond that, on breaking, can form a much stronger bond and, in the process, release energy.

18.25 Carbohydrates are the most readily used energy source in the diet.

18.27 The following equation represents the hydrolysis of maltose:

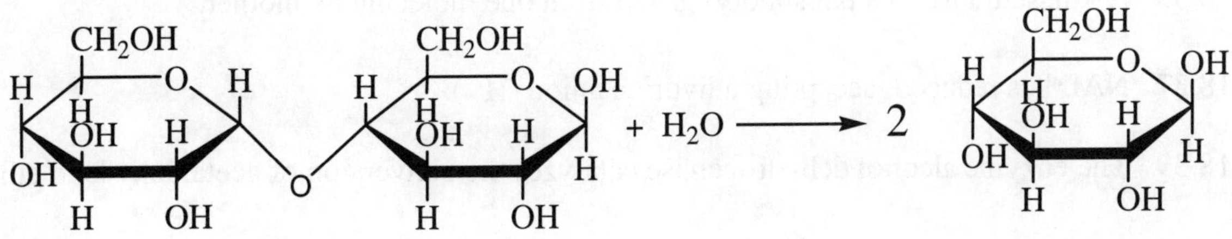

β-Maltose   + H_2O ⟶ 2   β-D-Glucose

18.29 The hydrolysis of a triglyceride containing oleic acid, stearic acid, and linoleic acid is represented in the following equations:

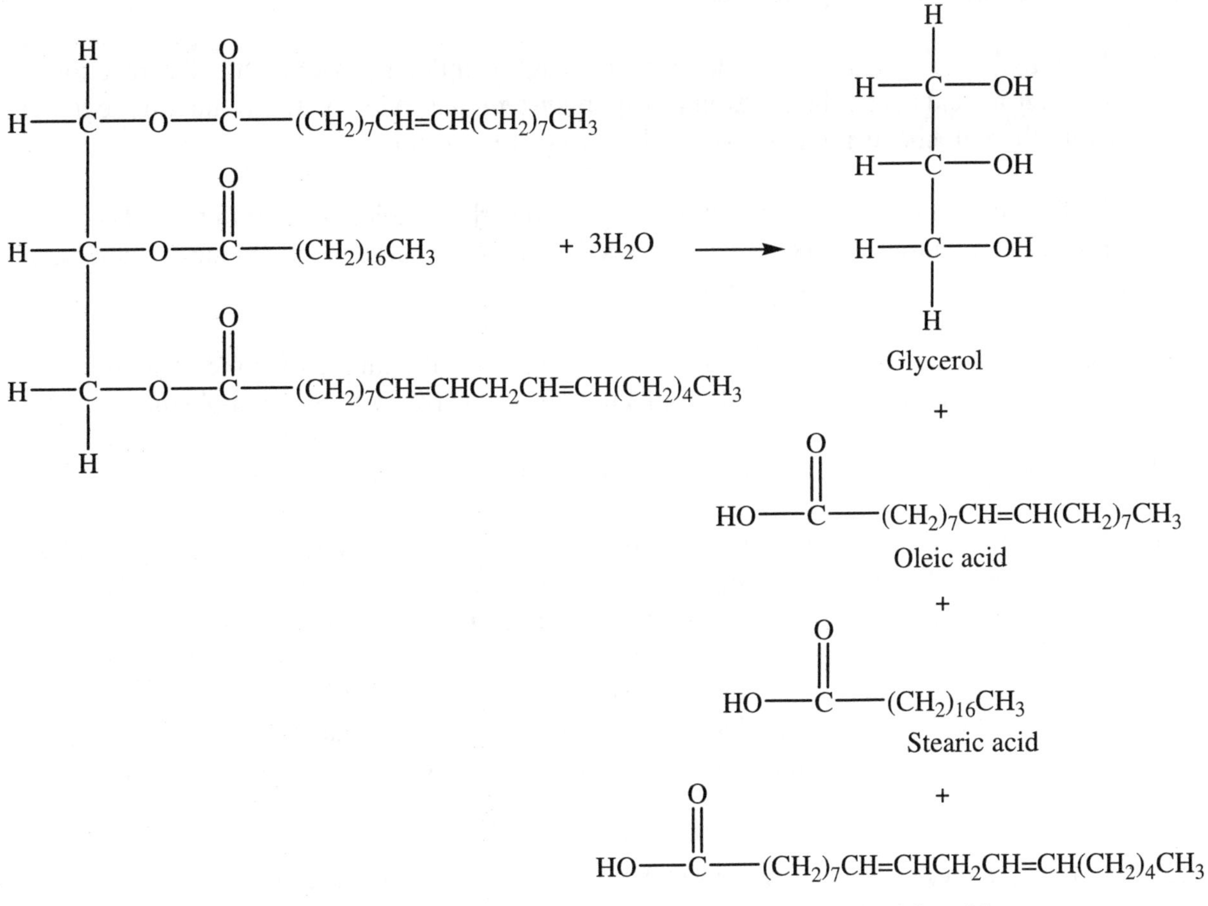

$$+ 3H_2O \longrightarrow$$

Glycerol

+

Oleic acid

+

Stearic acid

+

Linoleic acid

18.31 The end products of glycolysis are 2 molecules of pyruvate, 4 molecules of ATP (net production of 2 ATP), and 2 molecules of NADH.

429

18.33 Glycolysis occurs in the cytoplasm of the cell.

18.35 A kinase transfers a phosphoryl group from one molecule to another.

18.37 NAD^+ is reduced, accepting a hydride anion ($H:^-$).

18.39 The enzyme alcohol dehydrogenase catalyzes the conversion of acetaldehyde to ethanol.

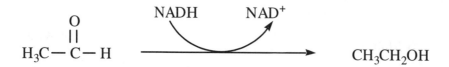

18.41 This child must have the enzymes to carry out the alcohol fermentation. When the child exercised hard, there was not enough oxygen in the cells to maintain aerobic respiration. As a result, glycolysis and the alcohol fermentation were responsible for the majority of the ATP production by the child. The accumulation of alcohol (ethanol) in the child caused the symptoms of drunkenness.

18.43 The mitochondrion is an organelle that serves as the cellular power plant. The reactions of the citric acid cycle, the electron transport system, and ATP synthase function together within the mitochondrion to harvest ATP energy for the cell.

18.45 The final oxidation of carbohydrates, amino acid carbon skeletons, and fatty acids occur in the mitochondrial matrix. The pathways that carry out these reactions are the citric acid cycle and β-oxidation of fatty acids.

18.47 Coenzyme A is a molecule derived from ATP and the vitamin pantothenic acid. It functions in the transfer of acetyl groups in lipid and carbohydrate metabolism.

18.49 Under aerobic conditions, pyruvate is converted to acetyl CoA.

18.51 The coenzymes NAD^+, FAD, thiamine pyrophosphate, and coenzyme A are required by the pyruvate dehydrogenase complex for the conversion of pyruvate to acetyl CoA. These coenzymes are synthesized from the vitamins niacin, riboflavin, thiamine, and pantothenic acid, respectively. If the vitamins are not available, the coenzymes will not be available and pyruvate cannot be converted to acetyl CoA. Since the complete oxidation of the acetyl group of acetyl CoA produces the vast majority of the ATP for the body, ATP production would be severely inhibited by a deficiency of any of these vitamins.

18.53 Isomers are molecules having the same molecular formula but different chemical structures.

18.55 a. false   c. true
       b. false   d. true

18.57 The acetyl group of acetyl CoA is converted into 2 CO_2 during oxidation in the reactions of the citric acid cycle. One ATP molecule, one $FADH_2$, and 3 NADH are produced in the process.

18.59 First, write an equation representing the conversion of pyruvate to acetyl CoA to determine which carbon in acetyl CoA is labeled:

$$H_3\overset{*}{C}-\overset{\overset{\displaystyle O}{\|}}{C}-COO^- \longrightarrow H_3\overset{*}{C}-\overset{\overset{\displaystyle O}{\|}}{C}{\sim}S-CoA$$

Pyruvate   Acetyl CoA

Next, write out the aldol condensation reaction of acetyl CoA and oxaloacetate to show the location of the radiolabeled carbon in citrate:

Oxaloacetate   Acetyl CoA   Citrate

Now draw out the intermediates of the citric acid cycle. Place an asterisk on the radio-labeled carbon and circle the $-COO^-$ groups that are released as CO_2.

Citrate   Isocitrate   α–Ketoglutarate   Succinyl CoA

431

The structures across the top represent a pathway:

$$\text{Succinate} \longrightarrow \text{Fumarate} \longrightarrow \text{Malate} \longrightarrow \text{Oxaloacetate}$$

Succinate:
COO^- — $*CH_2$ — CH_2 — COO^-

Fumarate:
COO^- — $*CH$ = CH — COO^-

Malate:
COO^- — $HO\text{—}*C\text{—}H$ — CH_2 — COO^-

Oxaloacetate:
COO^- — $*C$ = O — CH_2 — COO^-

18.61 It is a kinase because it transfers a phosphoryl group from one molecule to another. Kinases are a specific type of transferase.

18.63 The electron transport system is series of electron transport proteins embedded in the inner mitochondrial membrane that accept high-energy electrons from NADH and $FADH_2$ and transfer them in stepwise fashion to molecular oxygen (O_2).

18.65 The oxidation of NADH via oxidative phosphorylation yields 3 ATP.

18.67 The oxidation of a variety of fuel molecules, including carbohydrates, the carbon skeletons of amino acids, and fatty acids provides the electrons. The energy of these electrons is used to produce an H^+ reservoir. The energy of this proton reservoir is used for ATP synthesis.

18.69 a. Two ATP per glucose (net yield) are produced in glycolysis, while the complete oxidation of glucose in aerobic respiration (glycolysis, the citric acid cycle, and oxidative phosphorylation) results in the production of 36 ATP per glucose.

b. Aerobic respiration harvests nearly 40% of the potential energy of glucose, while anaerobic glycolysis harvests only about 2% of the potential energy of glucose.

18.71 Gluconeogenesis is production of glucose from noncarbohydrate starting materials. This pathway can provide glucose when starvation or strenuous exercise leads to a depletion of glucose from the body.

18.73 Gluconeogenesis produces glucose from noncarbohydrate molecules in times when blood glucose levels are low, as during strenuous exercise, a high protein/low carbohydrate diet, or starvation. This ensures proper function of brain and red blood cells, which only use glucose for fuel.

18.75 The liver is instrumental in the maintenance of blood glucose levels by serving as a reservoir for glycogen, a branched polymer of D-glucose. The pancreas is also critical to the control of blood glucose. When blood glucose levels are too high, the pancreas secretes the hormone insulin, which stimulates uptake and storage of glucose from the blood. When blood glucose levels are too low, the pancreas secretes the hormone glucagon, which stimulates the breakdown of liver glycogen and the release of glucose into the blood.

18.77 *Hypoglycemia* is the condition in which blood glucose levels are too low.

18.79 a. Insulin stimulates glycogen synthase, the first enzyme in glycogen synthesis. It also stimulates uptake of glucose from the bloodstream into cells and phosphorylation of glucose by the enzyme glucokinase.

b. This traps glucose within liver cells and increases the storage of glucose in the form of glycogen.

c. The ultimate result is a decrease of blood glucose levels.

Chapter 19
Fatty Acid and Amino Acid Metabolism
Solutions to the Odd-Numbered Problems

In-Chapter Questions and Problems

19.1   Because dietary lipids are hydrophobic, they arrive in the small intestine as large fat globules. The bile salts emulsify these fat globules into tiny fat droplets. This greatly increases the surface area of the lipids, allowing them to be more accessible to pancreatic lipases and thus more easily digested.

19.3   a. The products of the β-oxidation of hexanoic acid are 3 acetyl CoA, 2 NADH, and 2 FADH$_2$.

b. The products of the β-oxidation of octadecanoic acid are 9 acetyl CoA, 8 NADH, and 8 FADH$_2$.

19.5

$$CH_3CH_2CH_2\text{-}\overset{\overset{\displaystyle O}{\|}}{C}\text{~}S\text{-}CoA \ + \ FAD \longrightarrow CH_3CH{=}CH\text{-}\overset{\overset{\displaystyle O}{\|}}{C}\text{~}S\text{-}CoA \ + \ FADH_2$$

H$_2$O

$$NADH \ + \ CH_3\text{-}\overset{\overset{\displaystyle O}{\|}}{C}\text{-}CH_2\text{-}\overset{\overset{\displaystyle O}{\|}}{C}\text{~}S\text{-}CoA \longleftarrow CH_3\overset{\overset{\displaystyle OH}{|}}{C}H\text{-}CH_2\text{-}\overset{\overset{\displaystyle O}{\|}}{C}\text{~}S\text{-}CoA$$

Coenzyme A   NAD$^+$

$$2 \quad CH_3\text{-}\overset{\overset{\displaystyle O}{\|}}{C}\text{~}S\text{-}CoA$$

19.7   Starvation, a diet low in carbohydrates, and diabetes mellitus are conditions that lead to the production of ketone bodies. Lack of carbohydrates causes a decreased amount of oxaloacetate. This slows the citric acid cycle so that less acetyl CoA is oxidized. The excess acetyl CoA is converted to ketone bodies to recover coenzyme A.

19.9   The following are four differences between fatty acid biosynthesis and β-oxidation:

a. Fatty acid biosynthesis occurs in the cytoplasm, and β-oxidation occurs in the mitochondria.

b. The acyl group carrier in fatty acid biosynthesis is Acyl Carrier Protein (ACP), and the acyl group carrier in β-oxidation is coenzyme A.

c. The seven enzymes of fatty acid biosynthesis are associated as a multienzyme complex called fatty acid synthase. The enzymes involved in β-oxidation are not physically associated with one another.

d. NADPH is the reducing agent used in fatty acid biosynthesis. NADH and FADH$_2$ are produced by β-oxidation.

19.11 The purpose of the urea cycle is to convert toxic ammonium ions to urea, which is excreted in the urine of land animals. This keeps toxic ammonium ions out of the bloodstream.

19.13 Insulin stimulates uptake of glucose and amino acids by cells, glycogen and protein synthesis, and storage of lipids. It inhibits glycogenolysis, gluconeogenesis, breakdown of stored triglycerides, and ketogenesis.

End-of-Chapter Questions and Problems

19.15 Bile consists of micelles of lecithin, cholesterol, bile salts, proteins, inorganic ions, and bile pigments that aid in lipid digestion by emulsifying fat droplets. Bile salts are detergents because they have polar heads which make them soluble in aqueous solutions and hydrophobic tails that dissolve in nonpolar solvents and, in this case, bind triglycerides.

19.17 A chylomicron is a plasma lipoprotein that carries triglycerides from the intestine to all body tissues via the bloodstream. That function is reflected in the composition of the chylomicron, which is approximately 85% triglycerides, 9% phospholipids, 3% cholesterol esters, 2% protein, and 1% cholesterol.

19.19 Triglycerides in adipose tissue are the major storage form of lipids.

19.21 The major metabolic function of adipocytes is the storage of triglycerides.

19.23 Triglycerides represent a highly reduced source of fuel. The complete oxidation of fatty acids releases much more energy than the complete oxidation of the same amount of glycogen. Thus, more energy can be stored in a smaller space as triglycerides.

19.25 When dietary lipids in the form of fat globules reach the duodenum, they are emulsified by bile salts. The triglycerides in the resulting tiny fat droplets are hydrolyzed into monoglycerides and fatty acids by the action of pancreatic lipases, assisted by colipase. The monoglycerides and fatty acids are absorbed by cells lining the intestine. Within intestinal cells, triglycerides are reassembled and are packaged into chylomicrons (lipoprotein particles made up of protein and dietary triglycerides). Chylomicrons are secreted into the lymphatic vessels and eventually reach the bloodstream. In the bloodstream, the triglycerides are hydrolyzed once again and the products (glycerol and free fatty acids) are absorbed by the cells of the body.

19.27 The energy source for the activation of a fatty acid entering β-oxidation is the hydrolysis of ATP into AMP and PP_i (pyrophosphate group), an energy expense of two ATP high-energy bonds.

19.29 An alcohol is the product of the hydration of an alkene.

19.31 The following diagram summarizes the energy harvested by the β-oxidation of tetradecanoic acid:

Step 1 (Activation):   – 2 ATP

Steps 2–6 (repeated six times):

     6 $FADH_2$ x 2 ATP/$FADH_2$ 12 ATP
     6 NADH x 3 ATP/NADH   18 ATP

7 acetyl CoA to citric acid cycle:

     7 x 1 GTP x 1 ATP/GTP   7 ATP
     7 x 3 NADH x 3 ATP/NADH   63 ATP
     7 x 1 $FADH_2$ x 2 ATP/$FADH_2$ 14 ATP

                                  112 ATP

19.33 Two molecules of ATP are produced for each $FADH_2$ produced by β-oxidation.

19.35 The acetyl CoA produced by β-oxidation will enter the citric acid cycle unless there is too little oxaloacetate available. If that is the case, the acetyl CoA will be used in ketogenesis.

19.37 Ketone bodies include the compounds acetone, acetoacetone, and β-hydroxybutyrate, which are produced from fatty acids in the liver via acetyl CoA.

19.39 Ketoacidosis is a drop in the pH of the blood caused by elevated concentrations of ketone bodies.

19.41 If β-oxidation is going on at a rapid rate, it will produce large amounts of acetyl CoA. Normally the acetyl CoA would enter the citric acid cycle. However, if there is not enough oxaloacetate to allow all the acetyl CoA to enter the citric acid cycle, it will be used for the synthesis of ketone bodies.

19.43 Ketone bodies are relatively strong acids. If they are present in high concentrations in the blood, they will dissociate to release large amounts of H^+. This causes the blood to become more acidic, a condition called ketoacidosis.

19.45 Fatty acid biosynthesis occurs in the cytoplasm of the cell.

19.47   a. The phosphopantetheine group allows formation of a high-energy thioester bond with a fatty acid.
b. The phosphopantetheine group is derived from the vitamin pantothenic acid.

19.49   Fatty acid synthase is a huge multienzyme complex consisting of the seven enzymes involved in fatty acid synthesis. It is found in the cell cytoplasm. The enzymes involved in β-oxidation are not physically associated with one another. They are free in the mitochondrial matrix space.

19.51   Transaminases transfer amino groups from amino acids to ketoacids.

19.53   The glutamate family of transaminases is very important because the ketoacid corresponding to glutamate is α-ketoglutarate, one of the citric acid cycle intermediates. This provides a link between the citric acid cycle and amino acid metabolism. These transaminases provide amino groups for amino acid synthesis and collect amino groups during catabolism of amino acids.

19.55   a. Pyruvate
b. α-Ketoglutarate
c. Oxaloacetate
d. Acetyl CoA
e. Succinate
f. α-Ketoglutarate

19.57   The oxidative deamination of glutamate catalyzed by glutamate dehydrogenase can be represented by the following equation:

$$
\underset{\text{Glutamate}}{\overset{\overset{\textstyle N^+H_3}{|}}{\underset{\underset{\underset{\underset{COO^-}{|}}{CH_2}}{|}}{\underset{CH_2}{\overset{|}{HC}}\!-\!COO^-}}} + NAD^+ + H_2O \longrightarrow \underset{\alpha\text{-Ketoglutarate}}{\overset{\overset{\textstyle O}{\overset{\|}{}}}{\underset{\underset{\underset{\underset{COO^-}{|}}{CH_2}}{|}}{\underset{CH_2}{\overset{|}{C}}\!-\!COO^-}}} + NADH
$$

19.59   Hyperammonemia, an elevation of the concentration of ammonium ions in the body, results when the urea cycle is not functioning. A complete deficiency of a urea cycle enzyme results in death in early infancy. A partial deficiency causes retardation, convulsions, and vomiting and can be treated with a low protein diet.

19.61   a.   The source of one amino group of urea is the ammonium ion and the source of the other is the α-amino group of the amino acid aspartate.

b. The carbonyl group of urea is derived from CO_2.

19.63 In general, insulin stimulates anabolic processes, including glycogen synthesis, uptake of amino acids and protein synthesis, and triglyceride synthesis. At the same time, catabolic processes such as glycogenolysis are inhibited.

19.65 A target cell is one that has a receptor for a particular hormone.

19.67 Decreased blood glucose levels trigger the secretion of glucagon into the bloodstream.

19.69 Insulin is produced in the β-cells of the islets of Langerhans in the pancreas.

19.71 Insulin stimulates the uptake of glucose from the blood into cells. It enhances glucose storage by stimulating glycogenesis and inhibiting glycogen degradation and gluconeogenesis.

19.73 Insulin stimulates synthesis and storage of triglycerides.

19.75 Untreated diabetes mellitus is starvation in the midst of plenty because blood glucose levels are very high. However, in the absence of insulin, blood glucose can't be taken up into cells. The excess glucose is excreted into the urine while the cells of the body are starved for energy.

APPENDIX B

Answers to Chapter Self Tests

Chapter 1

1. 1.02 g/mL
2. 1.48×10^{-1}
3. 3.01×10^{-6}
4. 2.24×10^{3} cm/s
5. 3.95×10^{-22} g
6. 2.99×10^{10} cm/s
7. $-20.3°C$
8. 4.21×10^{3} g
9. $-40.0°F$
10. 0.545
11. energy
12. data
13. measurement
14. metric system
15. conversion factors
16. Kelvin
17. 310.0 K
18. 3.7×10^{-3}
19. b, c, d, f, and g
20. b, c, d, e, and f
21. gas, liquid, and solid
22. gas
23. gas
24. a. color
    b. odor
    c. taste
    d. compressibility
    e. melting point
    f. boiling point
25. heterogeneous

Chapter 2

1. 12 protons and 12 electrons
2. 17 protons and 18 electrons

3. a. Lithium atom (Li): 3 protons; 3 electrons
    b. Lithium ion (Li^{1+}): 3 protons; 2 electrons
    c. $Li \rightarrow 1e^- + Li^{1+}$
4. nucleus
5. electron
6. atomic number
7. neutron
8. positive, or cation, or +1
9. ^1_1H ^2_1H ^3_1H
    normal hydrogen   deuterium   tritium
10. valence electrons
11. six
12. helium
13. noble gas
14. a more stable energy state
15. isoelectric ion
16. representative elements
17. transition elcments
18. n = 1, 2, 3, etc.
19. two
20. orbital
21. *s*
22. opposite
23. electron configuration
24. $1s^2, 2s^2, 2p^2$
25. noble gases
26. Group IA
27. electronegativity

Chapter 3

1. ionic bond
2. covalent bond
3. positive and negative ions
4. two
5. left
6. right
7. metal or cation
8. sodium oxide
9. lithium sulfide
10. aluminum bromide
11. ferrous ion
12. iron (III) ion
13. higher positive charge
14. ferrous sulfate

15. copper (II) oxide
16. Na_2SO_4
17. four
18. eight
19. Cl
20. difference in electronegativities
21. H_2O
22. ICl
23. ammonia:

24. water:

25. valence electrons
26. covalent bond
27. $Na^+ + Cl^-$
28. $Ca^{2+} + S^{2-}$
29. electron transfer
30. two
31. four

Chapter 4

1.   1
2.   2.0 grams
3. 6.02×10^{23} atoms, or Avogadro's number
4.   4.81 grams
5. N_2
6.   covalent compound
7.   amu
8.   18.02 amu
9.   molecular weight
10. reactants
11. 2
12. 0.5
13. coefficients
14. 2
15. 2
16. 1

17. 80.09 grams
18. 169.9 grams
19. 7.149 grams
20. 3.01 moles
21. Avogadro's number
22. O_2

Chapter 5

1. 1.40×10^3 calories
2. high
3. nonspontaneous
4. gas
5. positive
6. calorimeter
7. fuel value
8. activated complex
9. equilibrium
10. LeChatelier
11. to the left
12. to the right

Chapter 6

1. evaporation
2. ideal
3. 759.0 mm of mercury pressure
4. 5.00 atmospheres
5. 2.09 liters
6. 1.96×10^{-2} moles of helium are present
7. δ^+ H–F δ^- ------- δ^+ H–F δ^-
8. nitrogen and oxygen
9. Dalton's
10. surface tension
11. surfactants
12. vapor pressure
13. melting point
14. very high
15. 20.0 liters
16. condensation
17. amorphous
18. barometer
19. calorimeter
20. covalent

Chapter 7

1.   1.10 % (W / V)
2.   800.0 mL
3.   1.25 liters
4. 5.52×10^{-2} moles
5.   0.466 gram
6.   0.19 molar HCl solution
7.   0.185 molar KNO_3 solution
8.   52.1 mL
9.   concentration
10. 13.5 grams
11. molarity
12. osmosis
13. hypertonic
14. isotonic

Chapter 8

1.   4.19
2.   0.0592 molar HCl solution
3. 5.00×10^{-14} moles/liter of H_3O^+ are present
4.   pH = 3.0
5.   neutralization
6.   Brønsted-Lowry
7.   dissociation
8.   b
9.   b

Chapter 9

1.   a. $_{0}^{1}n$ b. $_{83}^{210}Bi$
2.   3.125 mg remain
3.   9
4.   alpha particles, beta particles, and gamma rays
5.   gamma ray
6.   alpha particle
7.   radioactive
8.   half-life, or $t_{1/2}$
9.   fission
10. cancer or malignant cells

11. thyroid gland
12. background radiation
13. inversely
14. nuclear imaging device
15. film badge
16. gamma ray
17. natural radioactivity
18. alpha particle
19. radon

Chapter 10

1.   b, c, and d
2.   tetrahedral
3.   120°, or planar
4.   180°
5.   functional group
6.   hydroxyl
7.   carbonyl group
8.   aromatic
9.   a
10. c
11. a
12. b
13. a
14. c
15. e
16. inorganic chemistry
17. urea
18. hydrogen
19. carbon-to-carbon double bond
20. C_9H_{20}
21. a. 2-methylbutane
     b. 2,2-dimethylpentane
     c. 2,2,3-trimethylpentane
22. 1-bromo-4-chloro-2-methylpentane
23. 2-bromo-2-methylpropane
24. 2-bromo-4,4-dimethylhexane
25. $C_3H_8 + 5O_2 \rightarrow 3CO_2 + 4H_2O$
26. Two products are formed: 1-chloropropane and 2-chloropropane
27. none or little
28. lower
29. parent compound
30. trichloromethane
31. cycloalkane or cycloalkanes

32. ring
33. combustion
34. two
35. light, or UV, or high temperature
36. substitution

Chapter 11

1. a. alkane: C_4H_{10} could be butane

    b. alkene: C_4H_8 could be 1-butene or 2-butene

    c. alkyne: C_4H_6 could be 1-butyne or 2-butyne

2. 3-ethyl-2-pentene
3. 4,5-dibromo-2-hexyne
4. 2-methyl-2-pentene:

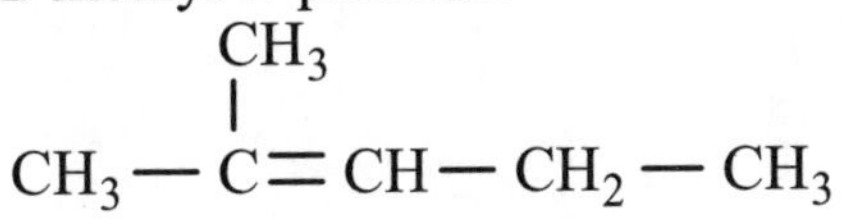

$$CH_3 - \underset{\underset{CH_3}{|}}{C} = CH - CH_2 - CH_3$$

5. *cis*-3-methyl-2-pentene:

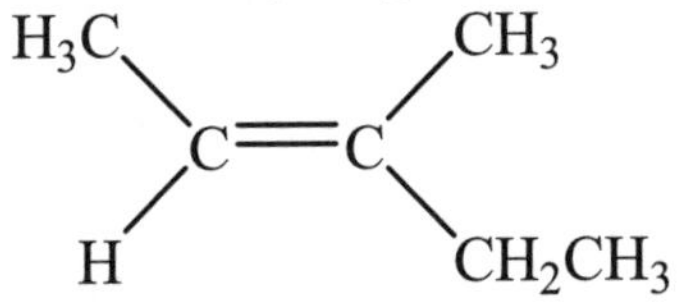

6. *trans*-2-bromo-2-butene
7.

$$H_2C{=}CHCH_2CH_3 \;+\; H_2 \;\xrightarrow{\text{Ni}}\; CH_3CH_2CH_2CH_3$$

   1-Butene   Butane

8.

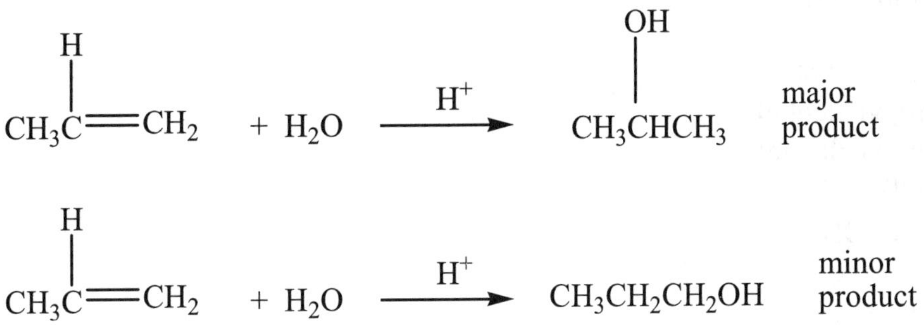

$$CH_3\underset{\underset{H}{|}}{C}{=}CH_2 \;+\; H_2O \;\xrightarrow{H^+}\; CH_3\underset{\underset{OH}{|}}{C}HCH_3 \quad \text{major product}$$

$$CH_3\underset{\underset{H}{|}}{C}{=}CH_2 \;+\; H_2O \;\xrightarrow{H^+}\; CH_3CH_2CH_2OH \quad \text{minor product}$$

9. a. benzene

    b. phenol or hydroxybenzene

    c. 1,2-dichlorobenzene

    d. 1,2-dichloro-3-bromobenzene

    e. 1,4-dibromobenzene

10. acetylene

11. *trans*-2-bromo-2-butene
12. *trans*-1,2-dibromopropene
13. hydrogenation
14. palladium and nickel
15. CH_3CH_2OH
16. benzene
17. chloroform
18. carbon-to-carbon double bond
19. addition
20. all carbon atoms have four single bonds
21. at least one carbon-to-carbon triple bond

22. $HC \equiv CHCH_2CH_3 \quad + \quad 2H_2 \quad \xrightarrow{Pt,\ heat} \quad CH_3CH_2\ CH_2CH_3$

23. $H_2C = CHCH_2CH_3 + H_2 \quad \xrightarrow{Pd,\ heat} \quad CH_3CH_2CH_2CH_3$
24. Vegetable oil

Chapter 12

1. a. 1-nonanol   d. dimethyl ether
   b. methanol   e. 1-propanol
   c. 1-propanol   f. propanal

2. a
3. a. 3-bromo-4-methyl-2-pentanol
   b. 5-chloro-2-pentanol
4. 1,2,3-propanetriol

5. a. propyl alcohol
   b. isopropyl alcohol
   c. *tert*-butyl alcohol
6. a. CH_3OH
   b. CH_3CH_2OH
7. a.

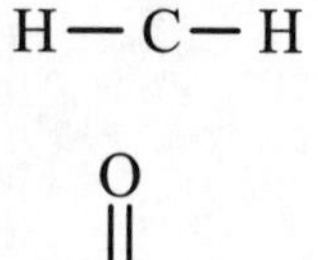

$$\begin{array}{c} O \\ \parallel \\ H-C-H \end{array}$$

   b.

$$\begin{array}{c} O \\ \parallel \\ CH_3C-H \end{array}$$

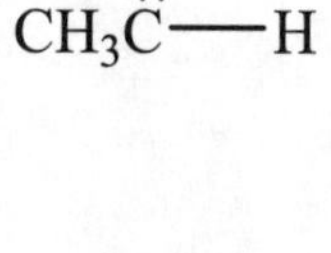

8.   a. ethene + H_2O

      b. propene + H_2O

      c. 1-butene + H_2O + 2-butene + H_2O

        (minor product)   (major product)

9.   a. dimethyl ether

      b. diethyl ether

      c. ethyl methyl ether (or methyl ethyl ether)

10.   d. acetic acid

11.   a. methane

12.   thiol

13.   They are polar and can form hydrogen bonds between alcohol molecules

14.   1

15.   tertiary

16.   oxidation

17.   thiol or thiols

18.   b. ethers

19.   b. ethers

20.   a. ethanal   d. 3-pentanone

      b. propanone   e. 4-bromohexanal

      c. propanal

21.   a.

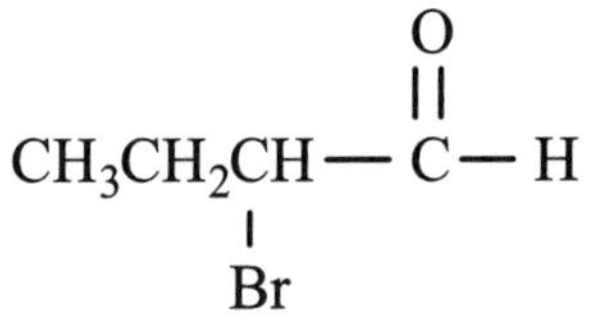

$$CH_3CH_2\underset{\underset{Br}{|}}{CH}-\overset{\overset{O}{\|}}{C}-H$$

      b.

$$CH_3-\underset{\underset{CH_3}{|}}{CH}-\overset{\overset{O}{\|}}{C}-CH_2CH_3$$

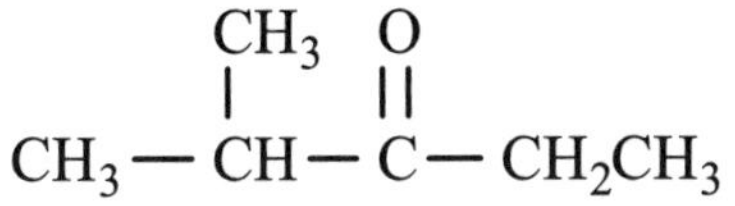

22.   a. HCOOH   d. CH_3CH_2COOH

      b. CH_3COOH e. no reaction

      c. no reaction

23.   a. ethanal (can be further oxidized to ethanoic acid)

      b. propanal (can be further oxidized to propanoic acid)

      c. propanone

      d. no reaction

24.   a. $CH_3COOH + Ag^0$

      b. No reaction: CH_3COCH_3 is a ketone.

      c. No reaction: $CH_3CH_2COCH_3$ is a ketone.

25.   ketone

26.   carboxylic acids

Chapter 13

1. a. ROH
    b. ROR
    c. RCOOR

    d. RCHO
    e. RCOOH
    f. RCOR

2. The general structure of an acyl group is:

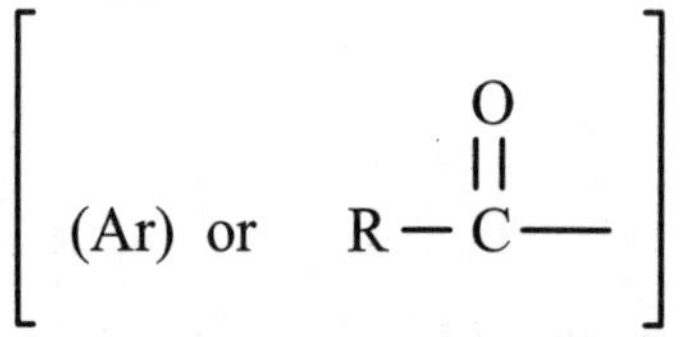

$$\left[\text{(Ar)} \quad \text{or} \quad R - \overset{\overset{\displaystyle O}{\overset{\displaystyle \|}{}}}{C} - \right]$$

3. a. methanoic acid
    b. hexanoic acid

    c. ethanoic acid
    d. 3-methylbutanoic acid

4. a. CH_3COOH
    c.

    b. $HCOOH$
    d. $CH_3(CH_2)_8COOH$

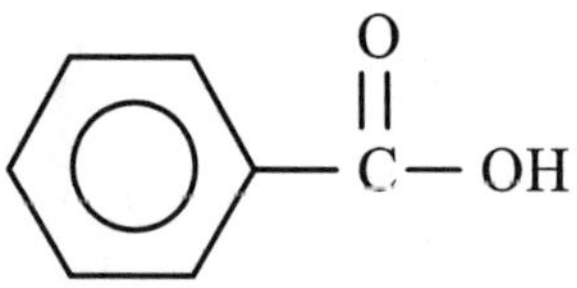

5. a. butyric acid
    b. propionic acid
    c. α-methylpropionic acid

6. a. ethyl formate
    b. methyl propionate
    c. pentyl acetate

7. a. methyl ethanoate
    b. ethyl propanoate

8. a. acetic acid

9. a. $CH_3COO^- + H_3O^+$
    b. $CH_3COO^- Na^+ + H_2O$
    c.

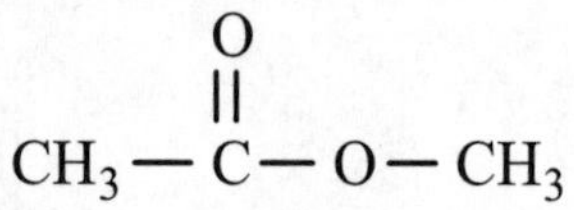

    $$CH_3 - \overset{\overset{\displaystyle O}{\overset{\displaystyle \|}{}}}{C} - O - CH_3$$

10. alcohol + carboxylic acid
11. carboxylate ion + hydronium ion
12. a. sodium butyrate + water
    b. calcium acetate + water
13. soaps
14. hydrophobic end
15. methanol + propanoic acid
16. carboxyl group
17. methyl ethanoate
18. oxidation reaction; $KMnO_4$ is an oxidizing agent.

19. hydrophilic
20. micelles
21. CH_3OH
22. a. tertiary amine   c. primary amine
    b. quaternary ammonium salt   d. secondary amine
23. a. ethanamine
    b. 2-propanamine
    c. *N,N*-dimethylmethanamine
24. a. $CH_3COOH + NH_4^+$
    b. $CH_3CH_2COOH + CH_3N^+H_3$
25. a. $(CH_3)_2NH_2^+I^-$
    b. $(CH_3)_3NH^+Cl^-$
26. a. RNH_2
    b. R_3N
27. a. ethanamide
    b. *N*-propylhexanamide
    c. *N*-methylbutanamide
28. ammonia
29. nitrogen
30. alkylammonium salt
31. methylammonium chloride
32. solid
33. a carboxylic acid and an amine or ammonia
34. because they are detergents

Chapter 14

1. The monosaccharides are a, b, c and f. Sucrose is a disaccharide, and starch is a polysaccharide.
2. a. aldohexose   d. aldopentose
   b. ketohexose   e. aldopentose
   c. aldohexose   f. aldotriose
3. enantiomers
4. cellulose
5. D-glucose
6. glyceraldehyde
7. α-amylase and β-amylase
8. glycosidic bond
9. An aldose has an aldehyde group and a ketose contains a ketone group.
10. A triose contains three carbon atoms and a pentose contains five carbon atoms.
11. a. D-glyceraldehyde   c. L-ribose
    b. D-ribose   d. D-deoxyribose
12. grape sugar, dextrose, or blood sugar
13. D-fructose

14. glycogen
15. a. glucose + H_2O → no reaction
    b. lactose + H_2O → glucose + galactose
    c. sucrose + H_2O → glucose + fructose
    d. starch + H_2O → many glucose molecules
    e. cellulose + H_2O → many glucose molecules
16. sucrose, a nonreducing sugar
17. carbohydrate(s)
18. Humans cannot make the enzyme cellulose.
19. lactose intolerance
20. D-glucose
21. alpha
22. Benedict's test
23. ribose or D-ribose
24. $C_5H_{10}O_5$
25. galactosemia
26. reducing sugars
27. disaccharides
28. liver and muscle tissues
29. Benedict's or Tollens'
30. Benedict's or Tollens'

Chapter 15

1. fatty acids
2. They are nonpolar molecules.
3. triglycerides (triacylglycerols)
4. They are long-chain monocarboxylic acids with an even number of carbon atoms.
5. glycerol + two fatty acids + a phosphate group
6. long-chain alcohol + fatty acid
7. chylomicrons, very low-density lipoproteins (VLDL), low-density lipoproteins (LDL), high-density lipoproteins (HDL)
8. glycerol + 3 fatty acids
9. HDL
10. cholesterol
11. hydrophobic (nonpolar) fatty acid tail
12. cholesterol
13. the degree of saturation and carbon chain length
14. $C_{11}H_{23}COOH + NaOH → C_{11}H_{23}COO^-Na^+ + H_2O$
15. the four fused rings known as the steroid nucleus
16. egg yolks, dairy products, liver, and other animal meats
17. progesterone
18. lipids
19. solid

20. ester + water
21. Ca^{2+} or Mg^{2+}
22. saturated solid fats
23. prostaglandins, leukotrienes, and thromboxanes
24. dilation
25. nonionic and nonpolar
26. polar and nonpolar
27. lecithin
28. high-density lipoproteins
29. cortisone
30. birth control

Chapter 16

1. glycine
2. phenylalanine
3. Enzymes are proteins that act as biological catalysts.
4. Antibodies, also called immunoglobulins, are specific protein molecules produced by special cells of the immune system in response to foreign antigens.
5. hemoglobin and myoglobin
6. keratin
7. a mixture of the amino acids
8.

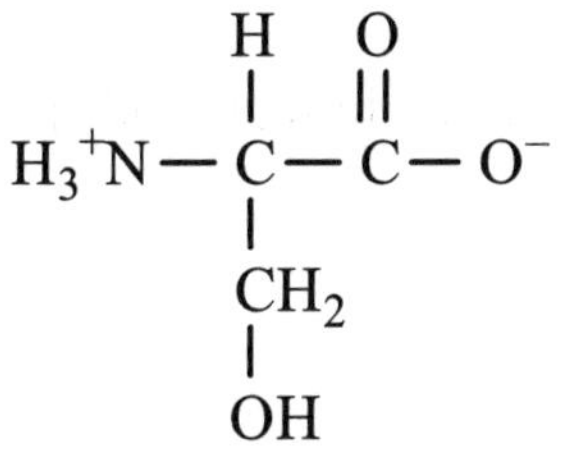

$$H_3{}^+N - \overset{\overset{\displaystyle H}{|}}{C} - \overset{\overset{\displaystyle O}{||}}{C} - O^-$$
$$\underset{\underset{\displaystyle OH}{|}}{\underset{\displaystyle CH_2}{|}}$$

9. aspartate and glutamate
10. lysine, arginine, and histidine
11. a. ala   b. cys   c. gly   d. his   e. lys
12. peptide bond or amide bond
13. primary structure
14. secondary structure
15. quaternary structure
16. α-helix or helices
17. proline
18. β-pleated sheet
19. disulfide bond
20. A hydrophobic molecule is nonpolar and thus is not water-soluble.
21. The enzyme lowers the activation energy of the reaction.
22. substrate
23. $2H_2O_2(l) \rightarrow 2H_2O(l) + O_2(g)$
24. a. succinate

b. sucrose

c. glycogen

25.

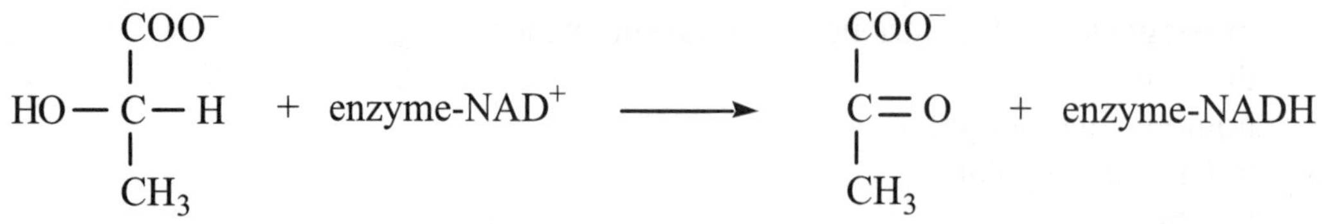

$$\begin{array}{c} COO^- \\ | \\ HO-C-H \\ | \\ CH_3 \end{array} + enzyme\text{-}NAD^+ \longrightarrow \begin{array}{c} COO^- \\ | \\ C=O \\ | \\ CH_3 \end{array} + enzyme\text{-}NADH$$

| lactate (reduced form) | lactate dehydrogenase (oxidized form) | pyruvate (oxidized form) | lactate dehydrogenase (reduced form) |

26. The enzyme lowers the energy of activation of a reaction.

27. formation of enzyme-substrate complex

28. For an enzyme that requires a nonprotein component in order to function in catalysis, the protein portion is called its apoenzyme and the nonprotein part is the cofactor.

29. Nicotinamide adenine dinucleotide, as either the oxidized form (NAD^+) or its reduced form ($NADH + H^+ = NADH$).

30. Most are used in some form as coenzymes.

31. pH optimum

32. denaturation

33. irreversible inhibitors and reversible, competitive inhibitors

34. The sulfa drugs are competitive inhibitors of a bacterial enzyme required for the synthesis of the required vitamin folic acid.

35. in the enzyme active site

36. the optimum temperature (37°C for most enzymes found in the human body)

Chapter 17

1. DNA

2. DNA

3. a. dATP

   b. CTP

   c. TMP

4. DNA → RNA → protein

5. Watson and Crick

6. a prime

7. hydrogen bonding between A and T and G and C

8. right angle

9. thymine

10. 3′ → 5′ direction

11. The two strands are complementary. One strand specifies the sequence of bases on the other.

12. pyrimidine dimers

13. semiconservative replication

14. If an error is made in copying the DNA, a daughter cell would inherit a mutant gene that could result in cell death. If the mutation were in an egg or sperm cell, it could lead to a genetic disorder in the offspring.
15. uracil
16. D-ribose
17. cancer
18. a. Messenger RNA—carries the genetic information for protein from DNA to the ribosomes
    b. Ribosomal RNA—a structural and functional component of the ribosomes
    c. Transfer RNA—responsible for translating the genetic code of the mRNA into the primary structure of a protein
19. transfer RNA
20. RNA polymerase
21. promoter
22. intervening sequences, or introns
23. exons
24. codons
25. Francis Crick
26. methionine and tryptophan
27. glutamate
28. xeroderma pigmentosum
29. a small and a large ribosomal subunit
30. polysomes, or polyribosomes
31. silent mutation

Chapter 18

1. glycolysis
2. ATP
3. 7.3 kcal/mole
4. anaerobic
5. two
6. NADH
7. ethanol
8. lactate
9. carbohydrates
10. amino acids
11. adenosine triphosphate, or ATP
12. glucose is phosphorylated
13. fructose-6-phosphate
14. cytoplasm
15. substrate-level phosphorylation
16. glucose-6-phosphate
17. a. glucose
18. e. glyceraldehyde-3-phosphate
19. c. pyruvate

20. F
21. T
22. mitochondria
23. energy need
24. oxaloacetate
25. two
26. three
27. one
28. two
29. mitochondria
30. electron transport system
31. cytochromes
32. hydrogen ions or protons
33. protons
34. F_1 catalyzes the phosphorylation of ADP to produce ATP.
35. NAD^+
36. It returns to be used again in the citric acid cycle.
37. oxygen, or O_2
38. ATP synthase
39. three
40. two

Chapter 19

1. acetyl coenzyme A
2. acetyl coenzyme A
3. fat globules
4. bile
5. micelle
6. cholate and chenodeoxycholate
7. colipase
8. chylomicrons
9. adipose tissue
10. 1 $FADH_2$ and 1 NADH
11. adipocyte
12. 129 ATP molecules
13. The abnormal rise in concentration of blood ketone bodies.
14. Starvation, a diet that is extremely low in carbohydrates, and the disease diabetes mellitus.
15. ketoacidosis
16. liver
17. heart muscle
18. fatty acid synthase
19. in the cytoplasm
20. VLDL

21. the phosphopantetheine group
22. in the matrix space of the mitochondria
23. 2 acetyl CoA, 1 FADH$_2$, and 1NADH
24. acyl CoA
25. acetyl CoA
26. liver, adipose, and muscle cells
27. glutamate, or glutamic acid
28. citric acid cycle, or Krebs cycle
29. oxalocetate and glutamate
30. liver

APPENDIX C

Answers to Vocabulary Quizzes

Chapter 1

1. chemistry
2. data
3. hypothesis
4. gaseous
5. potential energy
6. result
7. extensive
8. specific gravity
9. theory
10. weight

Chapter 2

1. mass number
2. cation
3. quantum number
4. electron
5. element
6. orbit
7. negative; anion
8. isotopes
9. neutron
10. proton
11. alkali metals
12. periodic table
13. period
14. families
15. valence
16. noble gases
17. electron affinity
18. ionization energy

Chapter 3

1. van der Waals
2. crystal
3. electrolyte
4. intermolecular
5. electron configuration
6. lone pair
7. polar
8. tetrahedral
9. triple bond

Chapter 4

1. Avogadro's number
2. chemical equation
3. chemical formula
4. formula unit
5. atomic mass
6. conservation of energy
7. mole
8. molecular weight
9. products
10. reactants

Chapter 5

1. thermodynamics
2. release
3. absorb
4. entropy
5. nutritional calorie
6. calorimeter
7. kinetics
8. activation energy
9. catalyst
10. LeChatelier's principle

Chapter 6

1. kinetic-molecular
2. Boyle's
3. Charles's
4. surfactants
5. amorphous
6. barometer
7. pressure
8. one
9. metallic
10. ideal

Chapter 7

1. crenation
2. electrolyte
3. hypertonic solution
4. hypotonic solution

5. molarity
6. osmosis
7. osmotic pressure
8. saturated
9. solute
10. supersaturated solution

Chapter 8

1. Brønsted-Lowry
2. autoionization
3. buffer solution
4. Brønsted-Lowry
5. polyprotic
6. pH
7. standard
8. titration

Chapter 9

1. alpha particle
2. beta particle
3. binding energy
4. Einstein's equation
5. fusion
6. gamma radiation
7. half-life
8. ionizing radiation
9. lethal dose
10. tracer

Chapter 10

1. alkane
2. isomers
3. alkyl group
4. parent chain
5. unsaturated hydrocarbon
6. halogenation
7. hydrocarbon
8. functional group
9. substitution reaction

Chapter 11

1. alkene
2. addition polymer
3. geometric isomers
4. phenyl group
5. parent compound
6. alkyne
7. addition reaction
8. heterocyclic aromatic compound
9. substitution reaction

Chapter 12

1. alcohols
2. thiols
3. carbinol carbon
4. ether
5. phenol
6. Zaitsev's rule
7. oxidation
8. carbonyl group
9. oxidation
10. Tollens' test

Chapter 13

1. ester
2. saponification
3. oxidation
4. carboxyl group
5. primary amine
6. amide bond
7. tertiary amine
8. alkylammonium ion
9. quaternary ammonium salt
10. amine

Chapter 14

1. galactosemia
2. Fischer Projection
3. chiral
4. enantiomers
5. glycogen
6. cellulose
7. glycosidic bond
8. reducing sugar
9. stereochemistry
10. lactose

Chapter 15

1. cholesterol
2. fluid mosaic model
3. phosphoglycerides
4. triglycerides
5. fatty acids
6. sphingomyelin
7. chylomicron
8. essential fatty acid
9. waxes
10. emulsifying agent

Chapter 16

1. hydrophobic
2. prosthetic group
3. glycoproteins
4. denaturation
5. tertiary structure
6. peptide bond
7. β-pleated sheet
8. active site
9. transition state
10. competitive inhibitor

Chapter 17

1. nucleotide
2. double helix
3. central dogma of molecular biology
4. messenger RNA
5. translation
6. ribosome
7. mutation
8. RNA polymerase
9. silent mutation

Chapter 18

1. adenosine triphosphate
2. glycolysis
3. gluconeogenesis
4. substrate level phosphorylation
5. catabolism
6. glycogen
7. oxidative phosphorylation
8. electron transport system
9. mitochondria

10. ATP synthase

Chapter 19

1. adipose tissue
2. triglyceride
3. bile
4. chylomicron
5. β-oxidation
6. coenzyme A
7. glucagon
8. insulin
9. urea cyle
10. ketone bodies